INTRODUCTION TO SPECTRAL THEORY IN HILBERT SPACE

GILBERT HELMBERG

Emeritus Professor of Mathematics
University of Innsbruck

DOVER PUBLICATIONS, INC.
Mineola, New York

Bibliographical Note

This Dover edition, first published in 2008, is an unabridged republication of the work originally published as Volume 6 in the North-Holland Series in Applied Mathematics and Mechanics by North-Holland Publishing Company, Amsterdam and London, and John Wiley & Sons, Inc., New York, in 1969.

Library of Congress Cataloging-in-Publication Data

Helmberg, Gilbert.
    Introduction to spectral theory in Hilbert space / Gilbert Helmberg. — Dover ed.
        p. cm.
    Originally published: Amsterdam : North-Holland Pub. ; New York : Wiley, 1969.
    Includes bibliographical references and index.
    ISBN-13: 978-0-486-46622-4
    ISBN-10: 0-486-46622-1
    1. Spectral theory (Mathematics). 2. Hilbert space. 3. Linear operators. I. Title.

QA322.H4 2008
515'.7222—dc22

2007053040

Manufactured in the United States of America
Dover Publications, Inc., 31 East 2nd Street, Mineola, N.Y. 11501

To Thea, Arno, and Wolfgang

To my parents

PREFACE

An invitation to write a book on Hilbert space and spectral theory is both tempting and dangerous. It is tempting since the topic is an attractive one on which to write a book. It is dangerous since a good number of books on this topic have already been written, and it is likely that a subsequent one will be superfluous.

In the present case temptation has prevailed over caution and the author now faces the difficulty of finding some justification for what he has done. Yielding to another temptation I am inclined to take the easy way out of this situation by expressing the hope that among the many different people interested in this subject there might be some who find this presentation particularly suited to their personal taste.

As far as the topic is concerned the aim of this book is to make the reader familiar with everything needed in order to understand, believe, and apply the spectral theorem for selfadjoint operators (not necessarily bounded) in Hilbert space. This implies that in due course answers will have to be supplied to such questions as:

What is a Hilbert space?

What is a bounded operator in Hilbert space?

What is a (not necessarily bounded) selfadjoint operator in Hilbert space?

What is the spectrum of such an operator?

What is meant by the spectral decomposition of such an operator?

Roughly speaking, the first question is answered in chapters I and II, the second in chapter III, while the answer to the third connects the chapters III and IV. The answer to the fourth question is provided in chapter IV, and the answer to the last (together with a rudimentary answer to the question: what use is the whole subject?) is contained in chapters V, VI and VII.

As far as the prospective reader is concerned I have tried to make the book intelligible also to someone who has only done some analysis and analytic geometry but who may possibly have forgotten part of it. (In fact in some rare and inessential places there also appear some cardinal numbers and references to properties of analytic functions.) It will greatly help the reader if he brings along some knowledge of Lebesgue integration, but it will also suffice if he is willing to rely on the summary given in appendix B whenever this becomes necessary. If neither of these two conditions applies the reader may very well skip the sections § 5 ($\mathfrak{L}_2$ Hilbert spaces), § 9 (Polynomial bases in $\mathfrak{L}_2$ spaces), § 16 (The Fourier-Plancherel operator), § 18 (Differentiation operators in $\mathfrak{L}_2$ spaces), § 19 (Multiplication operators in $\mathfrak{L}_2$ spaces), and § 29 (Fredholm integral equations). He will probably still be able to understand the rest of the book except for some of the examples, but the chances are that he will not see why one should worry about Hilbert space and spectral theory at all. Incidentally, anyone looking for the shortest possible route in this book to the spectral theorem for (not necessarily bounded) selfadjoint operators may also omit the entire chapter V (Spectral analysis of compact linear operators) and § 37 (The spectral decomposition of a bounded normal operator).

In any case, the book is intended to be intelligible also to someone who is interested in the topic but who lacks either the time or the desire to fill in gaps, to furnish proofs left as an exercise for the reader, or to work his way through an inspiring set of exercises considered to form an integral part of the text. I have to admit that in some places, in particular in section § 35 (Functions of a unitary operator), some of the proofs are not furnished in detail, but I have done so only where I felt it more desirable to refer to a reasoning already familiar to the reader than to bore him by repeating it. I also have to admit that every section is followed by an exercise consisting of some assertions to be proved and related to the foregoing material, but the reader may cheerfully omit any or all of them. If he wants to take notice of one exercise or other I hope he will find some of the assertions interesting and that they will add to his knowledge of the subject. Should he even feel challenged to do one or other of the exercises he might probably not find them difficult at all.

The purpose of this book is to be an introduction to the subject and no part of it is claimed to be original. In fact, it owes much (as I do myself) to the books of ACHIESER-GLASMANN [1], HALMOS [2], HEWITT-STROMBERG [6] and RIESZ-NAGY [9] cited in the bibliography. It also owes much to the active interest of students at the universities of Mainz, Innsbruck and Amsterdam, whose response in courses on the same subject has contributed both to the choice of the topics and to the mode of presentation.

My thanks are due to the North-Holland Publishing Company for the kind invitation to write this book and for the friendly compliance with my special wishes, and to my colleague Prof. Dr. H. A. LAUWERIER of the University of Amsterdam for having suggested this invitation and for encouraging me to accept it. I also have to thank my collaborators Dr. K. A. Post and Mr. F. H. Simons for reading the manuscript and the proofs, for chasing mistakes, and for offering valuable suggestions. The Department of Mathematics of the Technological University of Eindhoven has kindly provided me with all the facilities for the preparation of the manuscript any author could possibly want. In particular I have to thank Miss E. E. F. M. Weijers and Miss A. M. A. van Leuken for the skill with which they typed the manuscript and for the patience with which they modified it according to my various changes of mind.

There are also some persons who, without having anything to do with mathematics themselves, readily sacrificed some part of their personal life in order that these lines could be written. To them I therefore want to dedicate what might be worth dedicating in this book: to Thea, Arno, and Wolfgang, and to my parents.

Eindhoven–Innsbruck–Cavalese,
summer 1967.

Gilbert Helmberg

CONTENTS

REMARKS ON NOTATION

$a \in A$ is to be read: a is an element of the set A.

$\{a = (\alpha_1, \alpha_2): \alpha_1 \in \mathbf{R}, \alpha_2 \in \mathbf{R}\}$ denotes the set of all objects a which are ordered pairs (α_1, α_2) where α_1 and α_2 are real numbers. Similarly, the set of all objects a which have a certain property $\mathfrak{P}$ is denoted by $\{a: a$ has property $\mathfrak{P}\}$.

☐ signals the end of a proof.

Theorem 5 of § 17 (say) is referred to within § 17 by (cf. theorem 5) and outside of § 17 by (cf. § 17 theorem 5).

A complete index of symbols is contained at the end of this book.

CHAPTER I

THE CONCEPT OF HILBERT SPACE

§ 1. Finite dimensional Euclidean space

From elementary analytic geometry we are familiar with the notion of the two-dimensional Euclidean plane and, more generally, the *n-dimensional Euclidean space* (also called the *Euclidean n-space*). Roughly speaking, Hilbert space may be described as a possibly infinite-dimensional Euclidean space. In fact it turns out that to every cardinal number $\mathfrak{n}$ there corresponds a standard model of Hilbert space of dimension $\mathfrak{n}$ with the real numbers as scalars, and if $\mathfrak{n} = n$ is finite, then this Hilbert space coincides with the Euclidean space of the same dimension n.

While the phenomena of particular interest in the theory of Hilbert space are connected with infinite dimension, many features of finite-dimensional Euclidean space are shared by Hilbert space of arbitrary dimension. We shall therefore in this section recall some facts about Euclidean space which we shall then use to introduce in an axiomatic way the notion of Hilbert space as a generalization of Euclidean space.

In order to keep the notation compact we shall consider, as a special example, the two-dimensional Euclidean plane. Everything stated for this case will also be true for *n*-dimensional Euclidean space if pairs of coordinates are replaced by *n*-tuples of coordinates.

We shall denote by $\mathbf{R}$ the field of real numbers, at the same time considered as the one-dimensional Euclidean space. Correspondingly we shall denote by $\mathbf{R}^2$ the two-dimensional Euclidean plane and by $\mathbf{R}^n$ the *n*-dimensional Euclidean space. The elements of $\mathbf{R}^2$ may, according to taste or need, be visualised for example as points in the plane (in which we have singled out a point *o* called *origin* or *zero*) or as vectors pointing from *o* to arbitrary points in the plane. After having fixed a rectangular pair of coordinate axes through *o*, the positive direction and unit length on each axis, we may associate with any vector $a \in \mathbf{R}^2$ the coordinates α_1, α_2 of its endpoint with respect to these axes, and these coordinates uniquely determine the vector *a* (fig. 1). In a less picturesque way we may therefore regard $\mathbf{R}^2$ as the collection of all pairs of real numbers (α_1, α_2):

$$\mathbf{R}^2 = \{a = (\alpha_1, \alpha_2) : \alpha_1 \in \mathbf{R}, \alpha_2 \in \mathbf{R}\}.$$

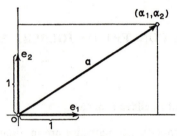

Fig. 1

If the vectors $a = (\alpha_1, \alpha_2)$, $b = (\beta_1, \beta_2)$ and the real number λ is given, then the sum $a + b$ and the *scalar multiple* λa are defined by

$$a + b = (\alpha_1 + \beta_1, \alpha_2 + \beta_2),$$
$$\lambda a = (\lambda\alpha_1, \lambda\alpha_2)$$

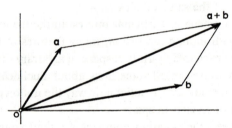

Fig. 2

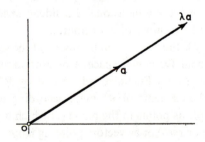

Fig. 3

(occasionally we shall for λa also use the notation $\lambda \cdot a$). The operations of addition and scalar multiplication correspond to the geometric operations of adding the vector b to a (the vector b is transferred with its initial point to the endpoint of a; fig. 2) and of expanding the vector to a λ-fold of its original length (fig. 3; for $|\lambda| < 1$ this actually amounts to shrinking the vector a, negative λ at the same time reverses the direction of a). If we denote by $e_1 = (1, 0)$ and $e_2 = (0, 1)$ the unit vectors in direction of the coordinate axes, then we have

$$a = (\alpha_1, \alpha_2) = \alpha_1 e_1 + \alpha_2 e_2,$$
$$\mathbf{R}^2 = \{\alpha_1 e_1 + \alpha_2 e_2 : \alpha_1 \in \mathbf{R}, \alpha_2 \in \mathbf{R}\}.$$

The following properties of addition and scalar multiplication are easily verified and summed up in the statement that $\mathbf{R}^2$ is a *linear space* (the stated equations hold for all real numbers α, β and for all vectors a, b, c):

(1)
$$
\begin{aligned}
(a + b) + c &= a + (b + c), & \alpha(\beta a) &= (\alpha\beta) a, \\
a + o &= o + a = a, & 1 \cdot a &= a, \\
a + (-a) &= (-a) + a = o, & \alpha(a + b) &= \alpha a + \alpha b, \\
a + b &= b + a, & (\alpha + \beta) a &= \alpha a + \beta a
\end{aligned}
$$

(if $a = (\alpha_1, \alpha_2)$, then $-a$ is defined by $-a = (-\alpha_1, -\alpha_2)$).

If $a = (\alpha_1, \alpha_2)$ and $b = (\beta_1, \beta_2)$, then the *"length"* of a is given by

$$(2) \qquad \|a\| = \sqrt{\alpha_1^2 + \alpha_2^2}$$

while the *"inner product"*

$$(3) \quad \langle a, b \rangle = \alpha_1 \beta_1 + \alpha_2 \beta_2 = \|a\| \, \|b\| \, \frac{\alpha_1 \beta_1 + \alpha_2 \beta_2}{\sqrt{\alpha_1^2 + \alpha_2^2} \sqrt{\beta_1^2 + \beta_2^2}} = \|a\| \, \|b\| \cos \widehat{ab}$$

determines the cosine of the angle between a and b. From (2) and (3) we deduce

(4)
$$\langle a, a \rangle = \|a\|^2,$$
$$|\langle a, b \rangle| \leq \|a\| \, \|b\| \qquad \text{(\emph{Cauchy's inequality}).}$$

The following statements again hold for all real numbers α and for all vectors a, b as is easily verified.

$$\|a\| \geq 0,$$
$$\|a\| = 0 \Leftrightarrow a = o,$$
$$\|\alpha a\| = |\alpha| \, \|a\|,$$
$$\|a + b\| \leq \|a\| + \|b\| \qquad \text{(\emph{triangle inequality})}$$

(the last inequality corresponds to the fact that in any triangle the length of every side is less or equal to the sum of the lengths of the two other sides; fig. 4),

$$\langle a + b, c \rangle = \langle a, c \rangle + \langle b, c \rangle,$$
$$\langle \alpha a, b \rangle = \alpha \langle a, b \rangle,$$
$$\langle a, b \rangle = \langle b, a \rangle,$$
$$\langle a, a \rangle \geqslant 0,$$
$$\langle a, a \rangle = 0 \Leftrightarrow a = o.$$

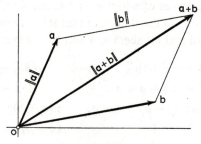

Fig. 4

Having recalled several algebraic properties of $\mathbf{R}^2$ we now turn our attention to some topological properties of $\mathbf{R}^2$ connected with the existence of a "*distance*" in $\mathbf{R}^2$.

The distance of two vectors a, b in $\mathbf{R}^2$ is given by the quantity $\|b-a\| = \|a-b\|$ (fig. 5).

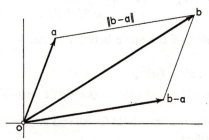

Fig. 5

A sequence $\{a_n\}_{n=1}^{\infty}$ in $\mathbf{R}^2$ *converges* to a vector $a \in \mathbf{R}^2$ (notation: $\lim_{n \to \infty} a_n = a$) if the distance $\|a - a_n\|$ converges to 0 as $n \to \infty$. Thus, the following statements are equivalent:

a) $\lim_{n \to \infty} a_n = a$;

b) $\lim_{n \to \infty} \|a - a_n\| = 0$;

    c) for every real number $\varepsilon > 0$ there exists an index $n(\varepsilon)$ such that $\|a - a_n\| < \varepsilon$ for all $n \geqslant n(\varepsilon)$.

    A sequence $\{a_n\}_{n=1}^{\infty}$ is called *fundamental* (or *Cauchy*) if the distances $\|a_m - a_n\|$ ultimately become small for large m and n, more precisely if for every real number $\varepsilon > 0$ there exists an index $n(\varepsilon)$ such that

$$\|a_m - a_n\| < \varepsilon \quad \text{for all } m \geqslant n(\varepsilon) \text{ and all } n \geqslant n(\varepsilon).$$

In short we shall write for this

$$\lim_{m,\,n \to \infty} \|a_m - a_n\| = 0.$$

    It is an important fact that every fundamental sequence in $\mathbf{R}^2$ converges to some vector in $\mathbf{R}^2$. This fact is also expressed by saying that $\mathbf{R}^2$ is (metrically) *complete*. Indeed, let $\{a_n\}_{n=1}^{\infty}$ be a fundamental sequence in $\mathbf{R}^2$ and let $a_n = (\alpha_{n,\,1}, \alpha_{n,\,2})$ for $n \geqslant 1$. Then we have

$$\|a_m - a_n\| = \sqrt{\{(\alpha_{m,\,1} - \alpha_{n,\,1})^2 + (\alpha_{m,\,2} - \alpha_{n,\,2})^2\}} \geqslant |\alpha_{m,\,k} - \alpha_{n,\,k}| \text{ for } k = 1, 2,$$
$$\lim_{m,\,n \to \infty} |\alpha_{m,\,k} - \alpha_{n,\,k}| = 0 \quad \text{for } k = 1, 2.$$

It is well known that $\mathbf{R}$ is complete. Therefore there exist two real numbers α_1, α_2 such that

$$\lim_{n \to \infty} \alpha_{n,\,k} = \alpha_k \quad \text{for } k = 1, 2.$$

For the vector $a = (\alpha_1, \alpha_2)$ we obtain

$$\lim_{n \to \infty} \|a - a_n\| = \lim_{n \to \infty} \sqrt{\{(\alpha_1 - \alpha_{n,\,1})^2 + (\alpha_2 - \alpha_{n,\,2})^2\}} = 0.$$

    Finally, let us consider the set $\mathfrak{A}$ of all vectors $a = (\alpha_1, \alpha_2)$ with rational coordinates α_1, α_2. From elementary set theory it is known that $\mathfrak{A}$ is a countable set. On the other hand, given any vector $b \in \mathbf{R}^2$ and any real number $\varepsilon > 0$ we can find a vector $a \in \mathfrak{A}$ such that $\|b - a\| < \varepsilon$, in other words the set $\mathfrak{A}$ is *everywhere dense* in $\mathbf{R}^2$. The fact that $\mathbf{R}^2$ contains a countable everywhere dense set is also expressed by saying that $\mathbf{R}^2$ is *separable*.

    Using these algebraic and topological properties of $\mathbf{R}^2$ as axioms we shall later define Hilbert space as a linear space with an inner product which is complete with respect to the "length" (then called *norm*) induced by the inner product. There is essentially only one separable infinite-dimensional Hilbert space appearing, however, in several disguises. This Hilbert space will be of special interest for us.

In order to arrive at the precise definition of Hilbert space with some insight into the immediate implications of this definition for the structure of Hilbert space we shall first introduce in § 2 the concept of an *inner product space* and investigate to some extent its algebraic properties. In § 3 we shall introduce the concept of a *normed linear space* and study some of its topological properties. Both concepts will then be combined in the definition of *Hilbert space*. The interplay of algebraic and topological properties will then essentially be the topic of our further study of Hilbert space in the second chapter.

§ 2. Inner product spaces

While in § 1 we have considered $\mathbf{R}^2$ as a linear space over the field of real numbers (as scalar multipliers) we shall henceforth also admit complex numbers as scalars. We shall denote by $\mathbf{C}$ the field of complex numbers and by $\mathbf{F}$ either of the fields $\mathbf{R}$ or $\mathbf{C}$ if we do not want to restrict our discussion to a particular one of these. If α is a complex scalar, then $\bar{\alpha}$ will denote the complex conjugate of α. Thus we have $\alpha = \bar{\alpha}$ iff $\alpha \in \mathbf{R}$.

We first formalize the linear operations and their properties mentioned in § 1 (1).

DEFINITION 1. A set $\mathfrak{L}$ is called a *commutative group under addition* with *zero element o* if with every ordered pair (a, b) of elements of $\mathfrak{L}$ there is associated a unique element $a+b \in \mathfrak{L}$, called the *sum* of a and b, in such a way that the following conditions are satisfied (a, b, c denote arbitrary elements of $\mathfrak{L}$):

a) $(a+b)+c = a+(b+c)$ (*associative law*);
b) $a+o = o+a = a$ (*existence of zero*);
c) for every $a \in \mathfrak{L}$ there exists an element $a' \in \mathfrak{L}$ such that
$a+a' = a'+a = o$ (*existence of inverse elements*);
d) $a+b = b+a$ (*commutative law*).

Remark 1. The inverse element a' of a is also written $-a$ and $b+(-a)$ is simply written $b-a$. It is well known that both the zero element and, for every $a \in \mathfrak{L}$, the inverse element $-a$ are determined uniquely by the conditions a), b), c) in definition 1. The reader who is familiar with algebra will also notice that a group may equivalently be defined by even somewhat weaker conditions than the ones just mentioned.

DEFINITION 2. A set $\mathfrak{L}$ is called a *linear space* (or *vector space*) *over* $\mathbf{F}$ if $\mathfrak{L}$ is a commutative group under addition and if a multiplication of the elements of $\mathfrak{L}$ with the elements of $\mathbf{F}$, called *scalar multiplication*, is defined in such a way that the following conditions are satisfied (a, b denote arbitrary elements of $\mathfrak{L}$, and α, β denote arbitrary numbers in $\mathbf{F}$):

a) $\alpha(\beta a) = (\alpha\beta) a$;
b) $1 \cdot a = a$;
c) $\alpha(a+b) = \alpha a + \alpha b$ $\Big\}$ (*distributive laws*).
d) $(\alpha+\beta) a = \alpha a + \beta a$

A linear space over $\mathbf{C}$ (or $\mathbf{R}$) is also called a *complex* (or, respectively, *real*) *linear space*. From now on, if nothing else is said, small greek letters $(\alpha, \beta, \gamma, ..., \lambda, \mu, ...)$ will denote arbitrary scalars in $\mathbf{F}$. Small italics $(a, b, c, ..., f, g, ...)$ will denote arbitrary elements of the linear space $\mathfrak{L}$, with exception of j, k, l, m, n which will be used for integers, in particular for indices.

THEOREM 1. a) $0 \cdot a = o$;
                b) $(-1) \cdot a = -a$.
PROOF.   a) $0 \cdot a = 0 \cdot a + (a-a) = (0+1) \cdot a - a = a - a = o$;
                b) $(-1) \cdot a = (-1) \cdot a + (1 \cdot a - a) = (-1+1) \cdot a - a = 0 \cdot a - a =$
                         $= o - a = -a$. $\square$

DEFINITION 3. The elements $f_k \in \mathfrak{L} (1 \leqslant k \leqslant n)$ are called *linearly dependent* if there exist scalars $\alpha_k \in \mathbf{F} (1 \leqslant k \leqslant n)$, not all equal to zero, such that $\sum_{k=1}^{n} \alpha_k f_k = o$. Otherwise (i.e. if $\sum_{k=1}^{n} \alpha_k f_k = o$ only for $\alpha_k = 0$, $1 \leqslant k \leqslant n$) the elements f_k are called *linearly independent*.

Remark 2. We shall call an *infinite* family of elements of $\mathfrak{L}$ *linearly independent* if every *finite* subfamily is linearly independent. Every family which includes the element o is obviously linearly dependent. Every finite sum $\sum_{k=1}^{n} \alpha_k f_k$ is called a (finite) *linear combination* of $f_1, ..., f_n$.

THEOREM 2. *If* $f_1, ..., f_n$ *are linearly independent and* $f_1, ..., f_n, f$ *are linearly dependent, then* f *is a linear combination of* $f_1, ..., f_n$.
PROOF. Since $f_1, ..., f_n, f$ are linearly dependent we have

$$\alpha f + \sum_{k=1}^{n} \alpha_k f_k = o$$

where not all scalars $\alpha_1, ..., \alpha_n, \alpha$ are equal to zero. Since $f_1, ..., f_n$ are linearly independent, α cannot be equal to zero. We conclude

$$f = \sum_{k=1}^{n} \left(-\frac{\alpha_k}{\alpha}\right) f_k . \square$$

DEFINITION 4. A linear space $\mathfrak{L}$ over $\mathbf{F}$ is called an *inner product space* over $\mathbf{F}$ if with every ordered pair (f, g) of elements of $\mathfrak{L}$ there is associated a unique scalar $\langle f, g \rangle \in \mathbf{F}$, called the *inner product* of f and g, in such a way that the following conditions are satisfied:

a) $\langle f_1 + f_2, g \rangle = \langle f_1, g \rangle + \langle f_2, g \rangle$;

b) $\langle \alpha f, g \rangle = \alpha \langle f, g \rangle$;

c) $\langle g, f \rangle = \overline{\langle f, g \rangle}$;

d) $\langle f, f \rangle \geqslant 0$;

e) $\langle f, f \rangle = 0 \Leftrightarrow f = o$.

Note that it follows from c) and it is implied by the formulation of d) that $\langle f, f \rangle \in \mathbf{R}$.

In the rest of this section $\mathfrak{L}$ will denote a fixed inner product space over $\mathbf{F}$ unless explicitly stated otherwise.

THEOREM 3. a) $\langle f, g_1 + g_2 \rangle = \langle f, g_1 \rangle + \langle f, g_2 \rangle$;

b) $\langle f, \alpha g \rangle = \bar{\alpha} \langle f, g \rangle$.

PROOF.   a) $\langle f, g_1 + g_2 \rangle = \overline{\langle g_1 + g_2, f \rangle} = \overline{\langle g_1, f \rangle} + \overline{\langle g_2, f \rangle}$
$$= \langle f, g_1 \rangle + \langle f, g_2 \rangle;$$

b) $\langle f, \alpha g \rangle = \overline{\langle \alpha g, f \rangle} = \bar{\alpha} \overline{\langle g, f \rangle} = \bar{\alpha} \langle f, g \rangle. \square$

COROLLARY 3.1. $\langle \sum_{k=1}^{n} \alpha_k f_k, \sum_{l=1}^{m} \beta_l g_l \rangle = \sum_{k=1}^{n} \sum_{l=1}^{m} \alpha_k \bar{\beta}_l \langle f_k, g_l \rangle$.

Corollary 3.1 states that the inner product of two sums may also be computed by pairwise "multiplying" the individual summands and adding up the results. This resembles the familiar rule for ordinary multiplication of sums and to some extent motivates the terminology "inner product".

Let us consider some examples of inner product spaces over $\mathbf{C}$.

Example 1. Let $n \geqslant 1$ be fixed and define $\mathfrak{L} = \{a = (\alpha_1, ..., \alpha_n): \alpha_k \in \mathbf{C}$ for $1 \leqslant k \leqslant n\}$. For $a = (\alpha_1, ..., \alpha_n)$ and $b = (\beta_1, ..., \beta_n)$ we define

$$a + b = (\alpha_1 + \beta_1, ..., \alpha_n + \beta_n),$$
$$\lambda a = (\lambda \alpha_1, ..., \lambda \alpha_n),$$
$$\langle a, b \rangle = \sum_{k=1}^{n} \alpha_k \bar{\beta}_k.$$

It is easy to check that all conditions mentioned in the definition of an inner product space over $\mathbf{C}$ are satisfied. Evidently the space $\mathfrak{L}$ is a complex version of the n-dimensional Euclidean space, also called the *n-dimensional unitary space* and denoted by $\mathbf{C}^n$. For $n = 1$ it coincides with the field of complex numbers $\mathbf{C}$.

Example 2. Let $\mathfrak{L}$ be the set of all "finite" sequences of complex numbers, i.e. of all sequences of complex numbers, the terms of which are ultimately zero:

$$\mathfrak{L} = \{a = \{\alpha_k\}_{k=1}^{\infty} : \alpha_k \in \mathbf{C} \quad \text{for } 1 \leqslant k < \infty, \alpha_k = 0 \quad \text{for } k > n(a)\}.$$

For $a = \{\alpha_k\}_{k=1}^{\infty}$ and $b = \{\beta_k\}_{k=1}^{\infty}$ in $\mathfrak{L}$ define sum, scalar multiple, and inner product analogously as in example 1:

$$a + b = \{\alpha_k + \beta_k\}_{k=1}^{\infty},$$
$$\lambda a = \{\lambda \alpha_k\}_{k=1}^{\infty},$$
$$\langle a, b \rangle = \sum_{k=1}^{\infty} \alpha_k \overline{\beta_k}$$

(the last series in fact reduces to a finite sum). Again, all conditions for an inner product space are satisfied.

Example 3. Let α and β be real numbers such that $\alpha < \beta$ and let $\mathfrak{L}$ be the set of all continuous complex-valued functions on the closed interval $[\alpha, \beta]$. For two functions f and g in $\mathfrak{L}$ the functions $f + g$ and λf are defined by

$$\left.\begin{array}{l} (f + g)(\xi) = f(\xi) + g(\xi) \\ (\lambda f)(\xi) = \lambda f(\xi) \end{array}\right\} \text{ for all } \xi \in [\alpha, \beta]$$

(i.e. addition and scalar multiplication are defined pointwise on $[\alpha, \beta]$). Furthermore we define

$$\langle f, g \rangle = \int_{\alpha}^{\beta} f(\xi) \overline{g(\xi)} \, d\xi.$$

Then again $\mathfrak{L}$ is an inner product space over $\mathbf{C}$ which we shall denote by $\mathfrak{C}[\alpha, \beta]$.

We proceed with the discussion of two concepts which in a natural way come up in connection with an inner product: the norm of a vector (cf. definition 5) and orthogonality of vectors (cf. definition 6).

DEFINITION 5. For every vector $f \in \mathfrak{L}$, the non-negative real number $\|f\| = \sqrt{\langle f, f \rangle}$ is called the *norm* of f (induced by the inner product $\langle \, , \, \rangle$ in $\mathfrak{L}$). A vector of norm 1 is called a *unit vector*.

THEOREM 4 (INEQUALITY OF CAUCHY–SCHWARZ–BUNJAKOWSKY).

$$|\langle f, g \rangle| \leqslant \|f\| \cdot \|g\|.$$

PROOF. For $\langle f, g \rangle = 0$ there is nothing to prove. Assume $\langle f, g \rangle \neq 0$ (which implies $g \neq o$ and $\langle g, g \rangle = \|g\|^2 \neq 0$) and let $\alpha = \langle f, g \rangle / \langle g, g \rangle$. Then we have

$$\frac{|\langle f, g \rangle|^2}{\langle g, g \rangle} = \alpha \langle g, f \rangle = \bar{\alpha} \langle f, g \rangle = |\alpha|^2 \langle g, g \rangle,$$

$$0 \leqslant \langle f - \alpha g, f - \alpha g \rangle = \langle f, f \rangle - \alpha \langle g, f \rangle - \bar{\alpha} \langle f, g \rangle + |\alpha|^2 \langle g, g \rangle$$

$$= \langle f, f \rangle - \frac{|\langle f, g \rangle|^2}{\langle g, g \rangle},$$

$$0 \leqslant \|f\|^2 - \frac{|\langle f, g \rangle|^2}{\|g\|^2},$$

$$|\langle f, g \rangle|^2 \leqslant \|f\|^2 \cdot \|g\|^2. \quad \square$$

Remark 3. If we would not have known from the very beginning that $\langle f, g \rangle \neq 0$ implies $\|g\| \neq 0$, we still could have deduced it from the inequality

$$0 \leqslant \langle f - \beta g, f - \beta g \rangle = \langle f, f \rangle - \beta \langle g, f \rangle - \bar{\beta} \langle f, g \rangle + |\beta|^2 \langle g, g \rangle,$$

valid for all $\beta \in \mathbf{F}$. Indeed, for $\langle g, g \rangle = \|g\|^2 = 0$ we could choose $\beta = (\langle f, f \rangle + 1)/2\langle g, f \rangle$, thereby obtaining $0 \leqslant -1$, a contradiction. This reasoning seems unnecessarily complicated but has an important consequence: if we ever should want to omit condition e) in the definition of an "inner product" (definition 4), the proof and therefore the statement of theorem 4 would still remain valid (cf. § 30 theorem 2).

COROLLARY 4.1. $|\langle f, g \rangle| = \|f\| \cdot \|g\|$ *iff f and g are linearly dependent.*
PROOF. If $|\langle f, g \rangle| = \|f\| \cdot \|g\|$ and if both f and g are different from zero (otherwise f and g are certainly linearly dependent), then we have $\langle f, g \rangle \neq 0$ and $\langle g, g \rangle \neq 0$. Choosing α as in the proof of theorem 4 we obtain $\alpha \neq 0$ and $\langle f - \alpha g, f - \alpha g \rangle = 0$, thus $f - \alpha g = o$.

Conversely, if f and g are linearly dependent, then we may assume without loss of generality $f = \lambda g$ (cf. theorem 2). We conclude

$$|\langle f, g \rangle| = |\lambda| \langle g, g \rangle = |\lambda| \|g\|^2 = \|\lambda g\| \cdot \|g\| = \|f\| \cdot \|g\|. \quad \square$$

COROLLARY 4.2. *If* $\mathfrak{L} \neq \{o\}$, *then* $\|f\| = \sup_{\|g\|=1} |\langle f, g \rangle|$.
PROOF. The assertion is trivial for $f = o$. Otherwise we have

$$\|f\| = \left\langle f, \frac{f}{\|f\|} \right\rangle \leqslant \sup_{\|g\|=1} |\langle f, g \rangle| \leqslant \sup_{\|g\|=1} \|f\| \cdot \|g\| = \|f\|. \quad \square$$

THEOREM 5. a) $\|f\| \geqslant 0$;

b) $\|f\| = 0 \Leftrightarrow f = o$;

c) $\|\lambda f\| = |\lambda|\, \|f\|$;

d) $\|f + g\| \leqslant \|f\| + \|g\|$ (*triangle inequality*).

PROOF.

a), b), c) follow directly from the definition of $\|f\|$;

d) $\begin{aligned}
\|f + g\|^2 &= \langle f + g, f + g \rangle = \|f\|^2 + \langle f, g \rangle + \langle g, f \rangle + \|g\|^2 \\
&= \|f\|^2 + 2\operatorname{Re}\langle f, g \rangle + \|g\|^2 \\
&\leqslant \|f\|^2 + 2|\langle f, g \rangle| + \|g\|^2 \\
&\leqslant \|f\|^2 + 2\|f\| \cdot \|g\| + \|g\|^2 \\
&= (\|f\| + \|g\|)^2. \quad \square
\end{aligned}$

COROLLARY 5.1. $\|f + g\| = \|f\| + \|g\|$ *iff* $g = o$ or $f = \lambda g$, $\lambda \geqslant 0$.

PROOF. If $\|f + g\| = \|f\| + \|g\|$ and $g \neq o$, then by the proof of theorem 5 d) we have $\operatorname{Re}\langle f, g \rangle = |\langle f, g \rangle| = \|f\| \cdot \|g\|$ which implies $\langle f, g \rangle = |\langle f, g \rangle| = \|f\| \cdot \|g\|$. By corollary 4.1 we further have $\alpha f + \beta g = o$ for some scalars α, β not both equal to zero. As $g \neq o$ we must necessarily have $\alpha \neq 0$ and $f = \lambda g \, (\lambda = -\beta/\alpha)$. From $|\langle f, g \rangle| = \langle f, g \rangle = \lambda \langle g, g \rangle$ we conclude $\lambda \geqslant 0$.

Conversely, for $f = \lambda g$ and $\lambda \geqslant 0$ we obtain

$$\|f + g\| = \|(1 + \lambda)g\| = (1 + \lambda)\|g\| = \|\lambda g\| + \|g\| = \|f\| + \|g\|. \quad \square$$

THEOREM 6 (PARALLELOGRAM LAW). $\|f + g\|^2 + \|f - g\|^2 = 2\|f\|^2 + 2\|g\|^2$.

PROOF. Straightforward computation. $\square$

In $\mathbf{R}^2$ theorem 6 admits an illustrative interpretation: the sum of the squares of the diagonals in any parallelogram equals the sum of the squares of the sides (fig. 6).

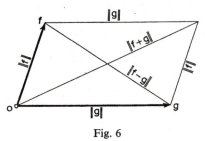

Fig. 6

DEFINITION 6. A vector f is called *orthogonal to* g (notation: $f \perp g$) if $\langle f, g \rangle = 0$. A family $\mathfrak{F} = \{f_\sigma\}_{\sigma \in \Sigma} \subset \mathfrak{L}$ is called *orthogonal* if $f_{\sigma_1} \perp f_{\sigma_2}$ for

$\sigma_1 \in \Sigma$, $\sigma_2 \in \Sigma$, $\sigma_1 \neq \sigma_2$; the family $\mathfrak{F}$ is called *orthonormal* if $\mathfrak{F}$ is orthogonal and $\| f_\sigma \| = 1$ for all $\sigma \in \Sigma$.

Example 4. Let $\mathfrak{L} = \mathfrak{C}[\alpha, \beta]$ as in example 3. For $k = 0, \pm 1, \pm 2, \ldots$ let $e_k \in \mathfrak{L}$ be defined by

$$e_k(\xi) = \frac{1}{\sqrt{(\beta - \alpha)}} \, e^{2\pi i k \, (\xi - \alpha)/(\beta - \alpha)} \quad \text{for all } \xi \in [\alpha, \beta].$$

Then $\{e_k\}_{k=-\infty}^{+\infty}$ is an orthonormal family. Indeed, we have

$$\langle e_k, e_l \rangle = \int_\alpha^\beta e_k(\xi) \, \overline{e_l(\xi)} \, d\xi$$

$$= \frac{1}{\beta - \alpha} \int_\alpha^\beta e^{2\pi i \, (k-l) \, (\xi - \alpha)/(\beta - \alpha)} \, d\xi$$

$$= \delta_{k,l} = \begin{cases} 1 \text{ for } k = l, \\ 0 \text{ for } k \neq l. \end{cases}$$

THEOREM 7. *An orthogonal family $\mathfrak{F}$ of non-zero vectors is linearly independent* (cf. remark 2).

PROOF. Suppose $f_k \in \mathfrak{F} (1 \leqslant k \leqslant n)$ and $\sum_{k=1}^n \alpha_k f_k = o$. For any l, $1 \leqslant l \leqslant n$, we have

$$0 = \langle o, f_l \rangle = \langle \sum_{k=1}^n \alpha_k f_k, f_l \rangle = \sum_{k=1}^n \alpha_k \langle f_k, f_l \rangle = \alpha_l \| f_l \|^2.$$

Because of $f_l \neq o$ this implies $\alpha_l = 0$. $\square$

THEOREM 8 (PYTHAGOREAN THEOREM). *If $f \perp g$, then $\| f + g \|^2 = \| f \|^2 + \| g \|^2$.*
PROOF. $\langle f + g, f + g \rangle = \langle f, f \rangle + \langle g, g \rangle$ since $\langle f, g \rangle = \langle g, f \rangle = 0$. $\square$

COROLLARY 8.1. *If $f \perp g$ and $\| f \| = \| g \| = 1$, then $\| f - g \| = \sqrt{2}$.*

COROLLARY 8.2. *If $\{ f_k \}_{k=1}^n$ is an orthogonal family of vectors, then*

$$\left\| \sum_{k=1}^n f_k \right\|^2 = \sum_{k=1}^n \| f_k \|^2.$$

PROOF. Induction on n. $\square$

In $\mathbf{R}^2$ also theorem 8 admits an illustrative interpretation: in every rectangular triangle the square of the hypothenuse equals the sum of the squares of the adjacent sides (fig. 7).

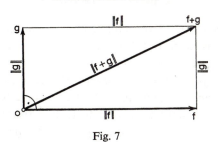

Fig. 7

THEOREM 9 (BESSEL'S INEQUALITY). *Let $\{e_k\}_{k=1}^n$ be an orthonormal family of vectors. Then for any $f \in \mathfrak{L}$*

$$\|f\|^2 \geqslant \sum_{k=1}^n |\langle f, e_k\rangle|^2 .$$

PROOF. Let $g = f - \sum_{k=1}^n \langle f, e_k\rangle e_k$. Then for any $l\,(1 \leqslant l \leqslant n)$ we have

$$\langle g, e_l\rangle = \langle f, e_l\rangle - \sum_{k=1}^n \langle f, e_k\rangle \langle e_k, e_l\rangle = \langle f, e_l\rangle - \langle f, e_l\rangle = 0 .$$

Thus $g \perp e_l$ for $1 \leqslant l \leqslant n$ and the vectors $g, \langle f, e_1\rangle e_1, \ldots, \langle f, e_n\rangle e_n$ form an orthogonal family. By corollary 8.2 we conclude

$$\|f\|^2 = \|g + \sum_{k=1}^n \langle f, e_k\rangle e_k\|^2$$

$$= \|g\|^2 + \sum_{k=1}^n |\langle f, e_k\rangle|^2 \geqslant \sum_{k=1}^n |\langle f, e_k\rangle|^2 . \;\square$$

Exercise. Let $\{e_k\}_{k=1}^n$ be an orthonormal family of vectors. Then

$$\|f\|^2 = \sum_{k=1}^n |\langle f, e_k\rangle|^2 \quad \text{iff } f = \sum_{k=1}^n \langle f, e_k\rangle e_k .$$

§ 3. Normed linear spaces

Having investigated to some extent the purely algebraic concepts entering in the definition of Hilbert space we now proceed to study the topological aspects of this definition. These aspects come into play as soon as we use the norm already introduced in § 2 definition 5 in order to define a "distance" of two vectors f, g by the quantity $\|f - g\|$. This concept of a distance allows us to define in our space neighbourhoods of a vector, converging sequences, and fundamental sequences of vectors analoguously as done by way of illustration in § 1.

In the process of carrying out this program it turns out that the concept on which these topological considerations hinge is only the norm with its properties as collected in § 2 theorem 5, but not the inner product which actually induced this norm in § 2 definition 5. This is an important point since we shall encounter linear spaces with a norm not (or not obviously) induced by an inner product. If we now separate the concept of a norm from the concept of an inner product, then we shall be able to apply everything that will follow in this section not only to Hilbert space but also to other linear spaces with a norm such as the algebra of bounded linear operators on a Hilbert space (cf. § 12 theorem 2).

Again, in what follows, **F** will alternatively denote the scalar field **R** or **C**.

DEFINITION 1. A linear space $\mathfrak{L}$ over **F** is called a *normed linear space* over **F** if with every vector $f \in \mathfrak{L}$ there is associated a unique number $\| f \| \in$ **R**, called the *norm* of f, in such a way that the following conditions are satisfied:

   a) $\| f \| \geqslant 0$;
   b) $\| f \| = 0 \Leftrightarrow f = o$;
   c) $\| \lambda f \| = |\lambda| \, \| f \|$;
   d) $\| f + g \| \leqslant \| f \| + \| g \|$.

By § 2 theorem 5 every inner product space is a normed linear space. We shall now exhibit an essentially different example of a normed linear space.

Example 1. Let $\mathfrak{L} = \mathbb{C}[0, 2\pi]$ with addition and scalar multiplication defined pointwise as in § 2 example 3. In contrast to what has been done there we define

$$\| f \| = \sup \{ |f(\xi)| : \xi \in [0, 2\pi] \}.$$

It is easy to check that this newly defined norm (also called the *uniform norm*) satisfies all conditions of definition 1. However, this norm is not induced by some inner product in $\mathfrak{L}$. Indeed, if this were the case, then the parallelogram law (§ 2 theorem 6) would have to hold. Choosing, however, $f = \max(\sin x, 0)$ and $g = \max(-\sin x, 0)$ (fig. 8) we have

$$\| f + g \| = \| f - g \| = \| f \| = \| g \| = 1,$$
$$2 = \| f + g \|^2 + \| f - g \|^2 \neq 2 \| f \|^2 + 2 \| g \|^2 = 4.$$

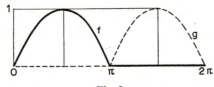

Fig. 8

Remark 1. We denote by x the function defined for all ξ in its domain by $x(\xi)=\xi$. Thus e.g. $\sin x$ is a *function* while its *value* at ξ is $\sin \xi$.

In the rest of this section $\mathfrak{L}$ will denote a fixed normed linear space over **F**, unless the contrary is stated.

THEOREM 1. $\big|\,\|f\| - \|g\|\,\big| \leqslant \|f-g\|$.

PROOF. The assertion follows from the two inequalities

$$\|f\| = \|(f-g)+g\| \leqslant \|f-g\| + \|g\|,$$
$$\|g\| = \|(g-f)+f\| \leqslant \|f-g\| + \|f\|. \;\square$$

DEFINITION 2. For any vector $f_0 \in \mathfrak{L}$ and for any $\varepsilon > 0$ the set $\{f: \|f-f_0\| < \varepsilon\}$ is called the *open sphere of radius* ε or shorter the *open* ε-*sphere about* f_0 (also ε-*neighbourhood of* f_0 or *open ball of radius* ε *centered at* f_0). A subset $\mathfrak{A} \subset \mathfrak{L}$ is called *open* if for every $f \in \mathfrak{A}$ there is some $\varepsilon > 0$ such that the open ε-sphere about f is contained in $\mathfrak{A}$.

Example 2. The open ε-sphere about a vector $f_0 \in \mathfrak{L}$ is an open subset of $\mathfrak{L}$. Indeed, let

$$\mathfrak{S} = \{f: \|f-f_0\| < \varepsilon\},$$

and let $f_1 \in \mathfrak{S}$ be given. Moreover, for $\varepsilon_1 = \varepsilon - \|f_1 - f_0\|$ let

$$\mathfrak{S}_1 = \{f: \|f-f_1\| < \varepsilon_1\}$$

be the open ε_1-sphere about f_1. Then for any $f \in \mathfrak{S}_1$ we have

$$\|f-f_0\| = \|(f-f_1)+(f_1-f_0)\|$$
$$\leqslant \|f-f_1\| + \|f_1-f_0\| < \varepsilon_1 + \|f_1-f_0\| = \varepsilon.$$

Therefore $\mathfrak{S}_1$ is contained in $\mathfrak{S}$.

In $\mathbf{R}^2$ the open ε-sphere about f is just the open disk with radius ε and center f. In the sequel, if we do not want to refer to a particular radius ε we shall simply speak of an *open sphere about* f. Furthermore $\emptyset$ will denote the empty set, $\bigcup$ will denote union and $\bigcap$ will denote intersection of sets. The complement of a subset $\mathfrak{A} \subset \mathfrak{L}$ will be denoted by $\mathfrak{A}^c$.

THEOREM 2. a) *The sets* $\mathfrak{L}$ *and* $\emptyset$ *are open.*

b) *If* $\{\mathfrak{A}_\sigma\}_{\sigma \in \Sigma}$ *is any family of open sets, then the set* $\bigcup_{\sigma \in \Sigma} \mathfrak{A}_\sigma$ *is open.*

c) *If* $\{\mathfrak{A}_k\}_{k=1}^n$ *is a finite family of open sets, then the set* $\bigcap_{k=1}^n \mathfrak{A}_k$ *is open.*

PROOF.   a) Trivial (there is no element in $\emptyset$).

        b) If $f \in \bigcup_{\sigma \in \Sigma} \mathfrak{A}_\sigma$ is given, then $f \in \mathfrak{A}_\sigma$ for some $\sigma \in \Sigma$ and there is an open ε-sphere about f which is contained in $\mathfrak{A}_\sigma$. This ε-sphere is then also contained in $\bigcup_{\sigma \in \Sigma} \mathfrak{A}_\sigma$.

        c) If $f \in \bigcap_{k=1}^{n} \mathfrak{A}_k$ is given, then for every $k\,(1 \leqslant k \leqslant n)$ there is a real number $\varepsilon_k > 0$ such that the open ε_k-sphere about f is contained in $\mathfrak{A}_k$. Choosing $\varepsilon = \min_{1 \leqslant k \leqslant n} \varepsilon_k$ we have $\varepsilon > 0$, and the open ε-sphere about f is contained in every set $\mathfrak{A}_k$ and therefore also in $\bigcap_{k=1}^{n} \mathfrak{A}_k$. $\square$

DEFINITION 3. A vector f_1 is called an *accumulation point of a subset* $\mathfrak{A} \subset \mathfrak{L}$ if every open sphere about f_1 contains at least one vector $f \in \mathfrak{A}$ different from f_1. A subset $\mathfrak{A} \subset \mathfrak{L}$ is called *closed* if it contains all its accumulation points.

Example 3. Let $\mathfrak{A} = \{ f : \| f - f_0 \| < \varrho \}\ (\varrho > 0)$ be an open sphere and consider the following positions of a vector f_1 with respect to $\mathfrak{A}$:

    a) $0 < \| f_1 - f_0 \| < \varrho$. Then $f_1 \in \mathfrak{A}$ and f_1 is an accumulation point of $\mathfrak{A}$. Indeed, given any $\varepsilon > 0$ let $\delta = \min(1, \varepsilon/2\varrho)$ and let $f = f_0 + (1 - \delta)\,(f_1 - f_0)$. Then we have

$$\| f - f_0 \| = (1 - \delta)\,\| f_1 - f_0 \| < \rho \,,$$

$$0 \neq \| f - f_1 \| = \delta\,\| f_1 - f_0 \| \leqslant \tfrac{1}{2}\varepsilon < \varepsilon \,.$$

We conclude that every open ε-sphere about f_1 contains a vector $f \in \mathfrak{A}$ different from f_1. Hence f_1 is an accumulation point of $\mathfrak{A}$.

    b) $0 < \| f_1 - f_0 \| = \varrho$. Then $f_1 \notin \mathfrak{A}$ but f_1 is still an accumulation point of $\mathfrak{A}$. The same reasoning applies as in a).

    c) $0 < \varrho < \| f_1 - f_0 \|$. Then $f_1 \notin \mathfrak{A}$ and f_1 is not an accumulation point of $\mathfrak{A}$. Indeed, let $\varepsilon = \| f_1 - f_0 \| - \varrho$ and suppose the open ε-sphere about f_1 contained a vector $f \in \mathfrak{A}$. Then we would get $\| f_1 - f_0 \| \leqslant \| f_1 - f \| + \| f - f_0 \| < \varepsilon + \varrho = \| f_1 - f_0 \|$, a contradiction.

THEOREM 3. *A subset* $\mathfrak{A} \subset \mathfrak{L}$ *is closed iff its complement* $\mathfrak{A}^c$ *in* $\mathfrak{L}$ *is open.*

PROOF. Let $\mathfrak{A}$ be closed and let $f \in \mathfrak{A}^c$ be given. Then f is not an accumulation point of $\mathfrak{A}$ and therefore there must be some open sphere about f entirely contained in $\mathfrak{A}^c$. Thus $\mathfrak{A}^c$ is open.

    Conversely if $\mathfrak{A}^c$ is open and if f is an accumulation point of $\mathfrak{A}$, then no open sphere about f can be entirely contained in $\mathfrak{A}^c$. Therefore f cannot belong to $\mathfrak{A}^c$ but must belong to $\mathfrak{A}$. $\square$

COROLLARY 3.1.   a) *The sets $\mathfrak{L}$ and $\emptyset$ are closed.*

          b) *If $\{\mathfrak{A}_\sigma\}_{\sigma \in \Sigma}$ is any family of closed sets, then the set $\bigcap_{\sigma \in \Sigma} \mathfrak{A}_\sigma$ is closed.*

          c) *If $\{\mathfrak{A}_k\}_{k=1}^n$ is a finite family of closed sets, then the set $\bigcup_{k=1}^n \mathfrak{A}_k$ is closed.*

PROOF.   a) Trivial.

          b) The set $(\bigcap_{\sigma \in \Sigma} \mathfrak{A}_\sigma)^c = \bigcup_{\sigma \in \Sigma} \mathfrak{A}_\sigma^c$ is a union of open sets by theorem 3 and therefore open by theorem 2 b). Now apply theorem 3 again.

          c) The set $(\bigcup_{k=1}^n \mathfrak{A}_k)^c = \bigcap_{k=1}^n \mathfrak{A}_k^c$ is a finite intersection of open sets by theorem 3 and therefore open by theorem 2 c). Now again apply theorem 3. □

THEOREM 4. *For any subset $\mathfrak{A} \subset \mathfrak{L}$ let $\overline{\mathfrak{A}}$ be the union of $\mathfrak{A}$ with the set of its accumulation points. Then $\overline{\mathfrak{A}}$ is closed.*

PROOF. Suppose $f_1 \notin \overline{\mathfrak{A}}$ is an accumulation point of $\overline{\mathfrak{A}}$. We shall show that in fact f_1 is already an accumulation point of $\mathfrak{A}$ and therefore $f_1 \in \overline{\mathfrak{A}}$. Let $\varepsilon > 0$ be given. Then there exists a vector $f \in \overline{\mathfrak{A}}$ such that $\| f - f_1 \| < \tfrac{1}{2}\varepsilon$. If $f \notin \mathfrak{A}$, then f is necessarily an accumulation point of $\mathfrak{A}$ and there exists a vector $g \in \mathfrak{A}$ such that $\| g - f \| < \tfrac{1}{2}\varepsilon$; if $f \in \mathfrak{A}$, then we simply put $g = f$. In any case we have

$$\| g - f_1 \| \leqslant \| g - f \| + \| f - f_1 \| < \varepsilon.$$

Thus every open sphere about f_1 contains a vector $g \in \mathfrak{A}$ (which is necessarily different from f_1) and f_1 is indeed an accumulation point of $\mathfrak{A}$. □

COROLLARY 4.1. *A subset $\mathfrak{A} \subset \mathfrak{L}$ is closed iff $\mathfrak{A} = \overline{\mathfrak{A}}$.*

Example 4. If $\mathfrak{A} = \{ f : \| f - f_0 \| < \varrho \}$ $(\varrho > 0)$, then from what has been shown in example 3 it follows that $\overline{\mathfrak{A}} = \{ f : \| f - f_0 \| \leqslant \varrho \}$. We shall call this set the *closed ϱ-sphere about f_0*, in particular for $\varrho = 1$ and $f_0 = o$ the *closed unit sphere* in $\mathfrak{L}$.

DEFINITION 4. If $\mathfrak{A}$ is any subset of $\mathfrak{L}$, then the union $\overline{\mathfrak{A}}$ of $\mathfrak{A}$ with the set of its accumulation points is called the *closure* of $\mathfrak{A}$. A set $\mathfrak{A}$ is called *everywhere dense* (in $\mathfrak{L}$) if $\overline{\mathfrak{A}} = \mathfrak{L}$. The space $\mathfrak{L}$ is called *separable* if $\mathfrak{L}$ contains a countable everywhere dense subset.

Recall that in § 1 we have exhibited a countable everywhere dense subset of $\mathbf{R}^2$. Similarly, the *n*-dimensional unitary space $\mathbf{C}^n$ is separable for every $n \geqslant 1$ (cf. § 2 example 1). In fact, the set of vectors with *complex rational* coordinates $\alpha_k = \beta_k + i\gamma_k$ (β_k, γ_k are rational numbers) is countable and everywhere dense.

DEFINITION 5. A sequence $\{f_n\}_{n=1}^\infty \subset \mathfrak{L}$ *converges to a vector* f, called its *limit* (notation: $\lim_{n \to \infty} f_n = f$), if for every $\varepsilon > 0$ there exists an index $n(\varepsilon)$ such that

$$\|f - f_n\| < \varepsilon \quad \text{for all } n \geqslant n(\varepsilon).$$

An infinite series $\sum_{k=1}^\infty g_k$ *converges to a vector* g, called its *sum* (notation: $\sum_{k=1}^\infty g_k = g$), if $\lim_{n \to \infty} \sum_{k=1}^n g_k = g$. A sequence or series that does not converge is said to *diverge*.

Remark 2. Equivalent formulations for $f = \lim_{n \to \infty} f_n$ are the following:

   a) For every $\varepsilon > 0$, the open ε-sphere about f contains all vectors f_n with sufficiently large index n.

   b) $\lim_{n \to \infty} \|f - f_n\| = 0$.

   Equivalent formulations for $g = \sum_{k=1}^\infty g_k$ are the following:

   c) The sequence of partial sums $\{\sum_{k=1}^n g_k\}_{n=1}^\infty$ converges to g.

   d) $\lim_{n \to \infty} \|g - \sum_{k=1}^n g_k\| = 0$.

   Note that a converging sequence uniquely determines its limit: if $f = \lim_{n \to \infty} f_n$ and $g = \lim_{n \to \infty} f_n$, then

$$\|f - g\| \leqslant \|f - f_n\| + \|g - f_n\| < 2\varepsilon \quad \text{for all } n \geqslant n(\varepsilon)$$

and therefore necessarily $f = g$.

THEOREM 5. *Let the vector* f *and the subset* $\mathfrak{A} \subset \mathfrak{L}$ *be given. Then* $f \in \overline{\mathfrak{A}}$ *iff there exists a sequence* $\{f_n\}_{n=1}^\infty \subset \mathfrak{A}$ *such that* $f = \lim_{n \to \infty} f_n$.
PROOF. If $f \in \overline{\mathfrak{A}}$, then for every $n \geqslant 1$ the open $1/n$-sphere about f contains some vector $f_n \in \mathfrak{A}$. From $\|f - f_n\| < 1/n$ we conclude $\lim_{n \to \infty} f_n = f$.

   Conversely, suppose there exists a sequence $\{f_n\}_{n=1}^\infty \subset \mathfrak{A}$ such that $f = \lim_{n \to \infty} f_n$. If $f \notin \mathfrak{A}$ and if $\varepsilon > 0$ is given, then from

$$\|f - f_n\| < \varepsilon \quad \text{for all } n \geqslant n(\varepsilon)$$

we conclude that $f_{n(\varepsilon)}$ is different from f and contained in the open ε-sphere about f. Thus f is an accumulation point of $\mathfrak{A}$ and therefore contained in $\overline{\mathfrak{A}}$. $\square$

It is a pleasant consequence of the various properties of the norm, already incorporated in its definition, that the linear operations (addition and scalar multiplication) as well as the norm itself and, in case the norm is induced by an inner product, also this inner product are continuous with respect to convergence as defined in definition 5. This is the content of the next theorem.

THEOREM 6. *Suppose*

$$\lim_{n\to\infty} f_n = f, \qquad \lim_{n\to\infty} \alpha_n = \alpha,$$

$$\lim_{n\to\infty} g_n = g, \qquad \lim_{n\to\infty} \beta_n = \beta.$$

Then the following statements hold:

a) $\lim_{n\to\infty} (\alpha_n f_n + \beta_n g_n) = \alpha f + \beta g$;

b) $\lim_{n\to\infty} \| f_n \| = \| f \|$;

c) *If $\mathfrak{L}$ is an inner product space and the norm is induced by the inner product $\langle\,,\rangle$, then*

$$\lim_{n\to\infty} \langle f_n, g_n \rangle = \langle f, g \rangle.$$

PROOF. a) $\|(\alpha f + \beta g) - (\alpha_n f_n + \beta_n g_n)\| \leqslant \|\alpha f - \alpha_n f_n\| + \|\beta g - \beta_n g_n\|$

$$\leqslant |\alpha - \alpha_n|\, \| f \| + |\alpha_n|\, \| f - f_n \| + |\beta - \beta_n|\, \|g\| + |\beta_n|\, \|g - g_n\|.$$

Since the converging sequences $\{\alpha_n\}_{n=1}^{\infty}$, $\{\beta_n\}_{n=1}^{\infty}$ are bounded, the last member converges to 0 as $n \to \infty$.

b) $\| \| f \| - \| f_n \| \| \leqslant \| f - f_n \|$ by theorem 1, and $\| f - f_n \|$ can be made arbitrarily small by taking n sufficiently large.

c) $|\langle f, g \rangle - \langle f_n, g_n \rangle| \leqslant |\langle f, g \rangle - \langle f, g_n \rangle| + |\langle f, g_n \rangle - \langle f_n, g_n \rangle|$

$$= |\langle f, g - g_n \rangle| + |\langle f - f_n, g_n \rangle|$$

$$\leqslant \| f \| \cdot \|g - g_n\| + \| f - f_n \| \cdot \|g_n\|.$$

From $\lim_{n\to\infty} \|g_n\| = \|g\|$ (cf. b)) we deduce that the sequence $\{\|g_n\|\}_{n=1}^{\infty}$ is bounded. Therefore the last member tends to 0 as $n \to \infty$. $\square$

DEFINITION 6. A sequence $\{ f_n \}_{n=1}^{\infty} \subset \mathfrak{L}$ is called *fundamental* if for every $\varepsilon > 0$ there exists an index $n(\varepsilon)$ such that

$$\| f_m - f_n \| < \varepsilon \quad \text{for all } m \geqslant n(\varepsilon) \text{ and all } n \geqslant n(\varepsilon)$$

(notation: $\lim_{m,n\to\infty} \| f_m - f_n \| = 0$).

Intuitively speaking, the terms of a fundamental sequence ultimately come very close to each other. Thus one might expect the sequence to converge. In fact, every converging sequence is fundamental, but the converse is not true in general.

THEOREM 7. *A converging sequence in $\mathfrak{L}$ is fundamental.*

PROOF. Suppose $\lim_{n\to\infty} f_n = f$. Then we have

$$\| f_m - f_n \| \leqslant \| f_m - f \| + \| f_n - f \|$$

and the right hand member converges to 0 as $m, n \to \infty$. $\square$

DEFINITION 7. A normed linear space $\mathfrak{L}$ is called *complete* if every fundamental sequence in $\mathfrak{L}$ converges to some vector in $\mathfrak{L}$. A complete normed linear space over $\mathbf{F}$ is called a *Banach space* over $\mathbf{F}$ (a *real Banach space* if $\mathbf{F} = \mathbf{R}$, a *complex Banach space* if $\mathbf{F} = \mathbf{C}$).

Remark 3. According to definition 7 a sequence $\{f_n\}_{n=1}^{\infty}$ in a Banach space converges iff it is fundamental; by the same token, a series $\sum_{k=1}^{\infty} g_k$ in a Banach space converges iff the sequence of partial sums $\{\sum_{k=1}^{n} g_k\}_{n=1}^{\infty}$ is fundamental, i.e. iff $\lim_{m,n \to \infty} \|\sum_{k=m}^{n} g_k\| = 0$.

Example 5. Let $\mathfrak{L} = \mathbf{C}^n$ be the n-dimensional unitary space (cf. § 2 example 1). Then $\mathfrak{L}$ is complete in the norm induced by the inner product. The proof is the same as given for $\mathbf{R}^2$ in § 1.

Example 6. Let $\mathfrak{L}$ be the inner product space of all "finite" sequences of complex numbers (cf. § 2 example 2). The space $\mathfrak{L}$ is not complete in the norm induced by the inner product. Indeed, let $a_n = \{1, 1/2, ..., 1/n, 0, 0, ...\}$ for $n \geq 1$. Then for $n \geq m$ we have

$$\|a_m - a_n\|^2 = \sum_{k=m+1}^{n} \frac{1}{k^2}.$$

Since the series $\sum_{k=1}^{\infty} 1/k^2$ converges we have $\lim_{m,n \to \infty} \|a_m - a_n\| = 0$. On the other hand, the sequence $\{a_n\}_{n=1}^{\infty}$ cannot converge to any element

$$a = \{\alpha_1, \alpha_2, ..., \alpha_j, 0, 0, ...\} \in \mathfrak{L}:$$

indeed, for all $n > j$ we would have $\|a - a_n\|^2 \geq 1/(j+1)^2 > 0$.

Example 7. Let $\mathfrak{L} = \mathfrak{C}[\alpha, \beta]$ be the inner product space as in § 2 example 3. Then $\mathfrak{L}$ is not complete. In order to simplify the notation, suppose $\alpha = -1$, $\beta = +1$, and let f_n be defined for $n \geq 1$ by

$$f_n(\xi) = \begin{cases} 0 & \text{for } -1 \leq \xi \leq 0 \\ n\xi & \text{for } 0 \leq \xi \leq 1/n \\ 1 & \text{for } 1/n \leq \xi \leq 1 \end{cases}$$

(fig. 9). Then we have

$$f_m(\xi) - f_n(\xi) = 0 \quad \text{for } -1 \leq \xi \leq 0,$$
$$|f_m(\xi) - f_n(\xi)| \leq 1 \quad \text{for } 0 \leq \xi \leq \max(1/m, 1/n),$$
$$f_m(\xi) - f_n(\xi) = 0 \quad \text{for } \max(1/m, 1/n) \leq \xi \leq 1.$$

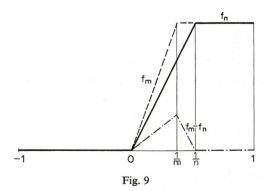

Fig. 9

We conclude

$$\|f_m - f_n\|^2 = \int\limits_{-1}^{+1} |f_m(\xi) - f_n(\xi)|^2 \, d\xi \leqslant \max\left(\frac{1}{m}, \frac{1}{n}\right),$$

$$\lim_{m,\,n \to \infty} \|f_m - f_n\| = 0.$$

If f were a continuous function on $[-1, +1]$ such that

$$\lim_{n \to \infty} \|f - f_n\|^2 = \lim_{n \to \infty} \int\limits_{-1}^{+1} |f(\xi) - f_n(\xi)|^2 \, d\xi = 0,$$

then necessarily

$$f(\xi) = \begin{cases} 0 & \text{for } -1 \leqslant \xi \leqslant 0 \\ 1 & \text{for } 0 \leqslant \xi \leqslant 1 \end{cases}$$

which is impossible by the continuity of f.

Example 8. Let $\mathfrak{L} = \mathfrak{C}[\alpha, \beta]$ be the normed linear space as in example 1 (there α, β have been chosen 0, 2π respectively). It is well known (and follows from the fact that a uniformly fundamental sequence of continuous functions converges to a continuous function) that $\mathfrak{L}$ is complete.

If a normed linear space is not complete, then it is possible to embed it in a larger normed linear space which is complete.

THEOREM 8. *Let $\mathfrak{L}$ be a (possibly not complete) normed linear space over* **F**. *Then there exists a normed linear space $\mathfrak{L}_1$ over* **F** *(with norm $\|\ \|_1$), called the completion of $\mathfrak{L}$, with the following properties:*

a) *$\mathfrak{L} \subset \mathfrak{L}_1$;*
b) *$\|f\|_1 = \|f\|$ for all $f \in \mathfrak{L}$;*

c) $\mathfrak{L}$ *is everywhere dense in* $\mathfrak{L}_1$;

d) $\mathfrak{L}_1$ *is complete.*

If the norm $\| \ \|$ *is induced by an inner product* $\langle \ , \ \rangle$ *in* $\mathfrak{L}$, *then the norm* $\| \ \|_1$ *is induced by an inner product* $\langle \ , \ \rangle_1$ *in* $\mathfrak{L}_1$ *and* $\langle f, g \rangle = \langle f, g \rangle_1$ *for all* $f \in \mathfrak{L}$ *and all* $g \in \mathfrak{L}$. *If* $\mathfrak{L}_1$ *and* $\mathfrak{L}_1'$ *are two normed linear spaces enjoying the properties* a), b), c), d), *then there is a norm-preserving one-to-one mapping of* $\mathfrak{L}_1'$ *onto* $\mathfrak{L}_1$ *which even preserves inner products if the norm* $\| \ \|$ *is induced by an inner product in* $\mathfrak{L}$.

According to the last statement, the space $\mathfrak{L}_1$ with the properties a), b), c), d) is essentially uniquely determined by $\mathfrak{L}$. It is therefore justified to call it "*the*" *completion* of $\mathfrak{L}$. Since we shall refer to this theorem only casually when dealing with special examples of Hilbert spaces in § 4 and § 5 we shall not enter here into its proof which is somewhat involved and technical. The basic idea is the same as the one used in the construction of the field of real numbers starting with the field of rational numbers. Intuitively speaking, with every fundamental sequence in $\mathfrak{L}$ which does not already converge to a point in $\mathfrak{L}$ (as one would like it to) one associates a new point (then becoming its limit point) which one adjoins to $\mathfrak{L}$ and the norm of which is determined uniquely by the requirement of continuity (theorem 6 b)); the same holds for the linear operations (theorem 6 a)). One only has to be careful to associate the same new point with all fundamental sequences in $\mathfrak{L}$ the terms of which ultimately come arbitrary close to each other. For details the reader is referred e.g. to [7] § 2.13, § 4.1.II.

Exercise. Let the vectors $g_1, ..., g_m$ be linearly independent (cf. § 2 definition 3) and define the subset $\mathfrak{M} \subset \mathfrak{L}$ by

$$\mathfrak{M} = \{f = \sum_{k=1}^{m} \alpha_k g_k : \alpha_k \in \mathbf{C} \quad \text{for } 1 \leqslant k \leqslant m\}.$$

Then the following statements hold:

a) $\alpha = \sup \{\| f \| : f = \sum_{k=1}^{m} \alpha_k g_k \in \mathfrak{M}, \sum_{k=1}^{m} |\alpha_k| = 1\} < \infty$.

b) $\beta = \inf \{\| f \| : f = \sum_{k=1}^{m} \alpha_k g_k \in \mathfrak{M}, \sum_{k=1}^{m} |\alpha_k| = 1\} > 0$.

(Hint: consider a sequence $\{f_n\}_{n=1}^{\infty} \subset \mathfrak{M}, f_n = \sum_{k=1}^{m} \alpha_{n,k} g_k$, such that $\lim_{n \to \infty} \| f_n \| = \beta$ and extract a subsequence for which every sequence $\{\alpha_{n,k}\}_{n=1}^{\infty} (1 \leqslant k \leqslant m)$ converges.)

c) If $f_n = \sum_{k=1}^{m} \alpha_{n,k} g_k$ for $1 \leqslant n < \infty$ and if the sequence $\{\| f_n \|\}_{n=1}^{\infty}$ is bounded, then every sequence $\{\alpha_{n,k}\}_{n=1}^{\infty} (1 \leqslant k \leqslant m)$ is bounded.

d) The set $\mathfrak{M}$ is closed.

§ 4. The Hilbert space ℓ_2

DEFINITION 1. An inner product space over $\mathbf{C}$ or $\mathbf{R}$ which is complete (in other words a Banach space) with respect to the norm induced by the inner product is called a *complex* (or, respectively, *real*) *Hilbert space*.

According to what has been said in § 3 example 5, the *n*-dimensional unitary space $\mathbf{C}^n$ is a complex Hilbert space; by the same token, the *n*-dimensional Euclidean space $\mathbf{R}^n$ is a real Hilbert space. In the present section we shall exhibit an important example of a separable complex Hilbert space which is infinite-dimensional in the sense that it contains an infinite orthonormal (and therefore linearly independent) family of vectors. In fact, in a sense explained in § 10 there is essentially only *one* infinite-dimensional separable complex Hilbert space of which this one will be the standard model. Henceforth, when speaking about Hilbert space without the explicit attribute "real" we shall always mean complex Hilbert space.

Consider the set ℓ_2 of all (absolutely) *square summable sequences* of complex numbers:

$$\ell_2 = \{a = \{\alpha_k\}_{k=1}^{\infty} : \alpha_k \in \mathbf{C} \quad \text{for } 1 \leqslant k \leqslant \infty, \ \sum_{k=1}^{\infty} |\alpha_k|^2 < \infty\}.$$

THEOREM 1. *For $a = \{\alpha_k\}_{k=1}^{\infty} \in \ell_2$ and $b = \{\beta_k\}_{k=1}^{\infty} \in \ell_2$ define*

$$a + b = \{\alpha_k + \beta_k\}_{k=1}^{\infty},$$
$$\lambda a = \{\lambda \alpha_k\}_{k=1}^{\infty}.$$

Then ℓ_2 is a complex linear space.

PROOF. Using CAUCHY's inequality (§ 2 theorem 4) in two-dimensional unitary space (as stated in § 1 (4)) we obtain

$$|\alpha_k \cdot 1 + \beta_k \cdot 1|^2 \leqslant (|\alpha_k|^2 + |\beta_k|^2) \cdot (1 + 1) = 2(|\alpha_k|^2 + |\beta_k|^2),$$

$$\sum_{k=1}^{\infty} |\alpha_k + \beta_k|^2 \leqslant 2(\sum_{k=1}^{\infty} |\alpha_k|^2 + \sum_{k=1}^{\infty} |\beta_k|^2) < \infty.$$

We conclude $a + b \in \ell_2$. Similarly from $\sum_{k=1}^{\infty} |\lambda \alpha_k|^2 = |\lambda|^2 \sum_{k=1}^{\infty} |\alpha_k|^2 < \infty$ we conclude $\lambda a \in \ell_2$. All conditions required in the definition of a linear space over $\mathbf{C}$ (§ 2 definition 2) are obviously satisfied. $\square$

THEOREM 2. *For $a = \{\alpha_k\}_{k=1}^{\infty} \in \ell_2$ and $b = \{\beta_k\}_{k=1}^{\infty} \in \ell_2$ the series $\sum_{k=1}^{\infty} |\alpha_k \beta_k|$ converges. With the inner product defined by*

$$\langle a, b \rangle = \sum_{k=1}^{\infty} \alpha_k \overline{\beta_k}$$

ℓ_2 is a complex inner product space.

PROOF. From $0 \leqslant (|\alpha_k| - |\beta_k|)^2$ we conclude

$$2|\alpha_k \bar{\beta_k}| \leqslant |\alpha_k|^2 + |\beta_k|^2$$

$$\sum_{k=1}^{\infty} |\alpha_k \bar{\beta_k}| \leqslant \tfrac{1}{2} \left(\sum_{k=1}^{\infty} |\alpha_k|^2 + \sum_{k=1}^{\infty} |\beta_k|^2 \right) < \infty .$$

Therefore the series $\sum_{k=1}^{\infty} \alpha_k \bar{\beta_k}$ converges. All properties required of an inner product (§ 2 definition 4) are checked easily. $\square$

THEOREM 3. ℓ_2 *is complete with respect to the norm induced by the inner product.*

PROOF. Let $a_n = \{\alpha_k^{(n)}\}_{k=1}^{\infty}$ and let $\{a_n\}_{n=1}^{\infty}$ be a fundamental sequence in ℓ_2. From

$$|\alpha_k^{(m)} - \alpha_k^{(n)}|^2 \leqslant \sum_{k=1}^{\infty} |\alpha_k^{(m)} - \alpha_k^{(n)}|^2 = \|a_m - a_n\|^2$$

we deduce that for every fixed $k \geqslant 1$ the sequence $\{\alpha_k^{(n)}\}_{n=1}^{\infty}$ is fundamental in $\mathbf{C}$ and therefore converges to a number $\alpha_k \in \mathbf{C}$. Let $a = \{\alpha_k\}_{k=1}^{\infty}$ (we have not yet shown that a belongs to ℓ_2). For every $\varepsilon > 0$ we have

$$\|a_m - a_n\|^2 = \sum_{k=1}^{\infty} |\alpha_k^{(m)} - \alpha_k^{(n)}|^2 < \varepsilon^2 \quad \text{for all } m \geqslant n(\varepsilon) \text{ and all } n \geqslant n(\varepsilon)$$

and therefore all the more for every fixed index $l \geqslant 1$

$$\sum_{k=1}^{l} |\alpha_k^{(m)} - \alpha_k^{(n)}|^2 < \varepsilon^2 \quad \text{for all } m \geqslant n(\varepsilon) \text{ and all } n \geqslant n(\varepsilon) .$$

Letting m tend to ∞ we obtain

$$\sum_{k=1}^{l} |\alpha_k - \alpha_k^{(n)}|^2 \leqslant \varepsilon^2 \quad \text{for all } n \geqslant n(\varepsilon) \text{ and for all } l \geqslant 1 .$$

Now we let l tend to ∞ and get

(1) $$\sum_{k=1}^{\infty} |\alpha_k - \alpha_k^{(n)}|^2 \leqslant \varepsilon^2 \quad \text{for all } n \geqslant n(\varepsilon) .$$

The sequence $a - a_n = \{\alpha_k - \alpha_k^{(n)}\}_{k=1}^{\infty}$ therefore belongs to ℓ_2 and, since ℓ_2 is a linear space, so does the sequence $a = (a - a_n) + a_n$. Furthermore from (1) we conclude

$$\|a - a_n\| \leqslant \varepsilon \quad \text{for all } n \geqslant n(\varepsilon),$$

$$\lim_{n \to \infty} a_n = a .$$

Thus every fundamental sequence in ℓ_2 converges to some element in ℓ_2. $\square$

COROLLARY 3.1. ℓ_2 *is a complex Hilbert space.*

THEOREM 4. ℓ_2 *contains a countably infinite orthonormal family.*
PROOF. Let $e_n = \{\delta_{n,k}\}_{k=1}^{\infty}$ for $1 \leqslant n < \infty$ ($\delta_{n,k} = 0$ for $n \neq k$, $\delta_{n,k} = 1$ for $n = k$). Then $\langle e_m, e_n \rangle = \delta_{m,n}$ for $1 \leqslant m \leqslant n < \infty$. $\square$

THEOREM 5. ℓ_2 *is separable.*
PROOF. Let ℓ' be the set of all "finite" sequences of complex rational numbers:

$$\ell' = \{a' = \{\alpha_k'\}_{k=1}^{\infty} : \alpha_k' \in \mathbf{C}, \text{ Re } \alpha_k' \text{ and Im } \alpha_k' \text{ are rational for } 1 \leqslant k < \infty,$$
$$\alpha_k' = 0 \text{ for } k \geqslant n(a')\}.$$

This subset ℓ' of ℓ_2 is countable. Indeed, the set of all complex rational numbers has the same cardinality as the set of all pairs of rational numbers and therefore is countable. For every $n \geqslant 1$, the subset of ℓ' consisting of these elements $a' = \{\alpha_k'\}_{k=1}^{\infty} \in \ell'$ having the property that $\alpha_k' = 0$ for all $k \geqslant n$ is countable. The set ℓ' is the countable union of all these countable subsets (for $1 \leqslant n < \infty$) and therefore is countable.

The set ℓ' is everywhere dense in ℓ_2. Indeed, let $a = \{\alpha_k\}_{k=1}^{\infty} \in \ell_2$ and $\varepsilon > 0$ be given. Choose $n \geqslant 1$ such that $\sum_{k=n+1}^{\infty} |\alpha_k|^2 < \frac{1}{2}\varepsilon^2$ and choose $a' = \{\alpha_k'\}_{k=1}^{\infty} \in \ell'$ such that $\alpha_k' = 0$ for $k \geqslant n+1$ and $|\alpha_k - \alpha_k'|^2 \leqslant \varepsilon^2/2n$ for $1 \leqslant k \leqslant n$ (this is possible since the complex rational numbers are everywhere dense in $\mathbf{C}$). Then we have

$$\|a - a'\|^2 = \sum_{k=1}^{n} |\alpha_k - \alpha_k'|^2 + \sum_{k=n+1}^{\infty} |\alpha_k|^2 < n \cdot \frac{\varepsilon^2}{2n} + \frac{1}{2}\varepsilon^2 = \varepsilon^2. \square$$

Remark 1. The proof of theorem 5 at the same time shows that the complex inner product space of all finite sequences of complex numbers exhibited in § 2 example 2 is everywhere dense in ℓ_2. In view of § 3 theorem 8 it follows that ℓ_2 is the completion of this inner product space.

By a somewhat peculiar modification of this model Hilbert space ℓ_2 it is possible to produce an example of a non-separable Hilbert space. In place of sequences $\{\alpha_k\}_{k=1}^{\infty}$ of complex numbers (which provide the elements of ℓ_2) we consider families $\{\alpha_\xi\}_{\xi \in \mathbf{R}}$ of complex numbers, where the index ξ runs through the real numbers. Such a family $a = \{\alpha_\xi\}_{\xi \in \mathbf{R}}$ may also be visualized as a function on $\mathbf{R}$ if we define $a(\xi) = \alpha_\xi$. Let $\mathfrak{L}$ be the set of all such functions a on $\mathbf{R}$ which have the following properties:

a) the function a is zero on $\mathbf{R}$ except on a set of points ($=$ indices) which is countable ($=$ finite or countable infinite) and which may depend on a;

b) the sum of the squares of the absolute function values at these points is finite.

Both properties a), b) together may in short be expressed by the requirement

$$\sum_{\xi \in \mathbf{R}} |a(\xi)|^2 = \sum_{\xi \in \mathbf{R}} |\alpha_\xi|^2 < \infty.$$

If addition and scalar multiplication are defined pointwise $((a+b)(\xi)=$ $=a(\xi)+b(\xi)=\alpha_\xi+\beta_\xi, (\lambda a)(\xi)=\lambda a(\xi)=\lambda\alpha_\xi)$ then the same reasoning as applied in the proofs of theorems 1, 2, 3 shows that $\mathfrak{L}$ is a complex linear space, that

$$\langle a, b \rangle = \sum_{\xi \in \mathbf{R}} a(\xi) \overline{b(\xi)} = \sum_{\xi \in \mathbf{R}} \alpha_\xi \overline{\beta_\xi}$$

properly defines an inner product in $\mathfrak{L}$, and that $\mathfrak{L}$ is a Hilbert space with respect to this inner product.

Denoting by $e_\nu (\nu \in \mathbf{R})$ the function defined on $\mathbf{R}$ by $e_\nu(\xi)=\delta_{\nu, \xi}$ we see that the family $\{e_\nu : \nu \in \mathbf{R}\} \subset \mathfrak{L}$ which has the cardinality of the continuum is orthonormal. For $\nu \neq \mu$ we therefore have $\|e_\nu - e_\mu\| = \sqrt{2}$ (cf. § 2 corollary 8.1). Let $\mathfrak{S}_\nu$ be the open sphere about e_ν with radius $\frac{1}{2}\sqrt{2}$. We assert that for $\nu \neq \mu$ the spheres $\mathfrak{S}_\nu$ and $\mathfrak{S}_\mu$ are disjoint. Indeed, if a were an element common to $\mathfrak{S}_\nu$ and $\mathfrak{S}_\mu$, then we would get

$$\|e_\nu - e_\mu\| \leqslant \|e_\nu - a\| + \|a - e_\nu\| < \tfrac{1}{2}\sqrt{2} + \tfrac{1}{2}\sqrt{2} = \sqrt{2}$$

which we know to be false.

Now let $\mathfrak{L}'$ be any subset which is everywhere dense in $\mathfrak{L}$. Then every sphere $\mathfrak{S}_\nu (\nu \in \mathbf{R})$ must at least contain one element of $\mathfrak{L}'$. Since the spheres $\mathfrak{S}_\nu (\nu \in \mathbf{R})$ are pairwise disjoint, $\mathfrak{L}'$ must contain at least as many elements as there are points ν in $\mathbf{R}$. Therefore any everywhere dense subset of $\mathfrak{L}$ has at least the cardinality of the continuum and $\mathfrak{L}$ is certainly not separable.

If we replace the index set $\mathbf{R}$ by any set of given cardinal number $\mathfrak{n}$, then an analogous reasoning shows that there exists a Hilbert space containing an orthonormal family of cardinal number $\mathfrak{n}$ and having the property that any everywhere dense subset has at least the cardinal number $\mathfrak{n}$.

Exercise. Define the set $\mathfrak{A}$ of sequences of complex numbers by

$$\mathfrak{A} = \{a = \{\alpha_k\}_{k=1}^\infty : \alpha_k \in \mathbf{C}, |\alpha_k| \leqslant \frac{1}{k} \text{ for } 1 \leqslant k \leqslant \infty\}.$$

a) $\mathfrak{A}$ is a subset of ℓ_2.

b) Every sequence $\{a_n\}_{n=1}^\infty \subset \mathfrak{A}$ contains a subsequence which converges in ℓ_2.

c) Let e_n be defined for $1 \leqslant n < \infty$ by $e_n = \{\delta_{n,k}\}_{k=1}^\infty$ (cf. the proof of theorem 4). Every subsequence of the sequence $\{e_n\}_{n=1}^\infty$ diverges.

§ 5. $\mathfrak{L}_2$ Hilbert spaces

In applications of Hilbert space theory e.g. to integral equations there turn up Hilbert spaces the elements of which are functions (more precisely classes of functions) on certain subsets of the real line, of the plane, or more general of the n-dimensional Euclidean space. The study of this special mode of appearance of Hilbert space leans heavily on the theory of Lebesgue integration. For the reader who is not familiar with this theory the pertinent definitions and theorems are collected in appendix B. Still, he may also choose to simply accept the theorems and corollaries in this section on belief and to skip the proofs. This will not seriously affect his understanding of the rest of the book. The same holds with respect to later sections such as § 9, § 16, § 18, § 19 and § 29.

From now on an integral sign will always signify Lebesgue integration, unless explicitly stated otherwise. This will not do any harm in those sections where we are only concerned with Riemann integrable (e.g. continuous) functions, since for those functions the Lebesgue integral and the Riemann integral coincide (cf. appendix B6). Lebesgue measurable functions need only be defined *"almost everywhere"* (a.e.), i.e. everywhere except possibly on a set of Lebesgue measure zero, and their real and imaginary parts may assume the "values" $+\infty$ and $-\infty$. This will not cause any difficulty for the algebraic operations performed with these functions, since for the functions considered in the sequel these "values" also can be assumed only on a set of Lebesgue measure zero. Two functions which coincide a.e. will always be identified, such that on a set of Lebesgue measure zero we may always redefine a function according to our needs (cf. appendix B5, B6).

Let the real numbers α, β ($\alpha < \beta$) be given. We shall denote the set of all Lebesgue measurable complex-valued functions on $[\alpha, \beta]$ by $\mathfrak{F}(\alpha, \beta)$ (cf. appendix B5) and the subset of all Lebesgue integrable functions on $[\alpha, \beta]$ by $\mathfrak{L}_1(\alpha, \beta)$ (cf. appendix B6). Let $\mathfrak{L}_2(\alpha, \beta)$ be the subset of all (absolutely) *square integrable complex-valued functions* on $[\alpha, \beta]$:

$$\mathfrak{L}_2(\alpha, \beta) = \{f \in \mathfrak{F}(\alpha, \beta): \int_\alpha^\beta |f(\xi)|^2 \, d\xi < \infty\}$$

(note that the functions in $\mathfrak{F}(\alpha, \beta)$ and $\mathfrak{L}_2(\alpha, \beta)$ need not be defined in α and β; as far as these two sets are concerned it therefore does not make any difference whether we consider the closed interval $[\alpha, \beta]$ or the open interval $]\alpha, \beta[$). Analogously as done in § 4 we shall show that $\mathfrak{L}_2(\alpha, \beta)$ is a separable infinite-dimensional complex Hilbert space, in fact the completion of the inner product space $\mathfrak{C}[\alpha, \beta]$ (§ 2 example 3, cf. § 3 example 7).

THEOREM 1. *The set $\mathfrak{L}_2(\alpha, \beta)$ is a complex linear space under pointwise addition and scalar multiplication.*

PROOF. For $f \in \mathfrak{L}_2(\alpha, \beta)$, $g \in \mathfrak{L}_2(\alpha, \beta)$ and $\lambda \in \mathbf{C}$ the functions $f + g$ and λf belong to $\mathfrak{F}(\alpha, \beta)$. As in the proof of § 4 theorem 1 we obtain

$$|f(\xi) + g(\xi)|^2 \leqslant 2(|f(\xi)|^2 + |g(\xi)|^2),$$

$$\int_\alpha^\beta |f(\xi) + g(\xi)|^2 \, d\xi \leqslant 2(\int_\alpha^\beta |f(\xi)|^2 \, d\xi + \int_\alpha^\beta |g(\xi)|^2 \, d\xi) < \infty,$$

$$\int_\alpha^\beta |\lambda f(\xi)|^2 \, d\xi = |\lambda| \int_\alpha^\beta |f(\xi)|^2 \, d\xi < \infty.$$

We conclude $f + g \in \mathfrak{L}_2(\alpha, \beta)$ and $\lambda f \in \mathfrak{L}_2(\alpha, \beta)$. Again, all conditions mentioned in the definition of a complex linear space (§ 2 definition 2) are obviously satisfied. $\square$

THEOREM 2. *For $f \in \mathfrak{L}_2(\alpha, \beta)$ and $g \in \mathfrak{L}_2(\alpha, \beta)$ the function $f\bar{g}$ is integrable on $[\alpha, \beta]$. With the inner product defined by*

$$\langle f, g \rangle = \int_\alpha^\beta f(\xi) \, \overline{g(\xi)} \, d\xi$$

$\mathfrak{L}_2(\alpha, \beta)$ *is a complex inner product space.*

PROOF. The function $f\bar{g}$ is Lebesgue measurable on $[\alpha, \beta]$ (cf. appendix B5). Similarly as in the proof of § 4 theorem 2 we find

$$2|f(\xi) g(\xi)| \leqslant |f(\xi)|^2 + |g(\xi)|^2$$

$$\int_\alpha^\beta |f(\xi) \overline{g(\xi)}| \, d\xi \leqslant \tfrac{1}{2}(\int_\alpha^\beta |f(\xi)|^2 \, d\xi + \int_\alpha^\beta |g(\xi)|^2 \, d\xi) < \infty.$$

Thus $f\bar{g}$ is integrable on $[\alpha, \beta]$ and the Lebesgue integral $\int_\alpha^\beta f(\xi) \overline{g(\xi)} \, d\xi$ exists (cf. appendix B6). Again all properties required of an inner product are checked easily. $\square$

COROLLARY 2.1. *Every function $f \in \mathfrak{L}_2(\alpha, \beta)$ $(-\infty < \alpha < \beta < +\infty)$ is integrable on $[\alpha, \beta]$, in short: $\mathfrak{L}_2(\alpha, \beta) \subset \mathfrak{L}_1(\alpha, \beta)$.*

PROOF. Choose $g \equiv 1$ in theorem 2. $\square$

Remark 1. While the statement of corollary 2.1 remains true only for a finite interval $[\alpha, \beta]$, the proof of theorem 2 actually remains valid for $-\infty \leqslant \alpha < \beta \leqslant +\infty$. Thus for any (possibly infinite) interval $]\alpha, \beta[$, if f and g are Lebesgue measurable and (absolutely) square integrable on $]\alpha, \beta[$, then fg is integrable on $]\alpha, \beta[$.

THEOREM 3. $\mathfrak{L}_2(\alpha, \beta)$ *is complete with respect to the norm induced by the inner product.*

PROOF. First note that for any $f \in \mathfrak{L}_2(\alpha, \beta)$ we have

$$(1) \qquad \int_\alpha^\beta |f(\xi)| \, d\xi = \langle |f|, 1 \rangle \leqslant \|f\| \cdot \|1\| = \sqrt{\beta - \alpha} \cdot \|f\|$$

by CAUCHY's inequality (§ 2 theorem 4).

Suppose now that $\{f_n\}_{n=1}^\infty$ is a fundamental sequence in $\mathfrak{L}_2(\alpha, \beta)$. Then there exists an index n_1 such that

$$\|f_m - f_n\| < \tfrac{1}{2} \quad \text{for all } m \geqslant n_1 \text{ and all } n \geqslant n_1.$$

By induction we construct an increasing sequence of natural numbers $n_1 < n_2 < \cdots < n_k < \cdots$ such that

$$\|f_m - f_n\| < 1/2^k \quad \text{for all } m \geqslant n_k \text{ and all } n \geqslant n_k.$$

Consider the sequence $\{f_{n_{k+1}} - f_{n_k}\}_{k=1}^\infty \subset \mathfrak{L}_2(\alpha, \beta)$. By LEBESGUE's theorem on monotone convergence (cf. appendix B7) we have

$$(2) \qquad \int_\alpha^\beta \sum_{k=1}^\infty |f_{n_{k+1}}(\xi) - f_{n_k}(\xi)| \, d\xi = \sum_{k=1}^\infty \int_\alpha^\beta |f_{n_{k+1}}(\xi) - f_{n_k}(\xi)| \, d\xi$$

$$\leqslant \sqrt{\beta - \alpha} \cdot \sum_{k=1}^\infty \|f_{n_{k+1}} - f_{n_k}\|$$

$$\leqslant \sqrt{\beta - \alpha} \cdot \sum_{k=1}^\infty \frac{1}{2^k} = \sqrt{\beta - \alpha} < \infty.$$

By B. LEVI's theorem (cf. appendix B7) the series $\sum_{k=1}^\infty |f_{n_{k+1}}(\xi) - f_{n_k}(\xi)|$ converges for almost every $\xi \in [\alpha, \beta]$ and so does the series

$$f_{n_1}(\xi) + \sum_{k=1}^\infty [f_{n_{k+1}}(\xi) - f_{n_k}(\xi)] = \lim_{k \to \infty} f_{n_k}(\xi).$$

As a consequence the function f defined a.e. on $[\alpha, \beta]$ by $f(\xi) = \lim_{k \to \infty} f_{n_k}(\xi)$ is finite a.e. and Lebesgue measurable on $[\alpha, \beta]$ (cf. appendix B5). Furthermore we have (using LEBESGUE's theorem on monotone convergence again)

$$\int_\alpha^\beta |f(\xi) - f_{n_l}(\xi)|^2 \, d\xi \le \int_\alpha^\beta \left[\sum_{k=l}^\infty |f_{n_{k+1}}(\xi) - f_{n_k}(\xi)| \right]^2 d\xi$$

$$= \lim_{m \to \infty} \int_\alpha^\beta \left[\sum_{k=l}^m |f_{n_{k+1}}(\xi) - f_{n_k}(\xi)| \right]^2 d\xi$$

$$= \lim_{m \to \infty} \| \sum_{k=l}^m |f_{n_{k+1}} - f_{n_k}| \|^2$$

$$\le \lim_{m \to \infty} \left(\sum_{k=l}^m \|f_{n_{k+1}} - f_{n_k}\| \right)^2$$

$$\le \lim_{m \to \infty} \left(\sum_{k=l}^m \frac{1}{2^k} \right)^2 = \left(\frac{1}{2^{l-1}} \right)^2.$$

We conclude $(f - f_{n_l}) \in \mathfrak{L}_2(\alpha, \beta)$ and $\|f - f_{n_l}\| \le 1/2^{l-1}$. Therefore we have $f = (f - f_{n_l}) + f_{n_l} \in \mathfrak{L}_2(\alpha, \beta)$. For $n \ge n_l$ we obtain

$$\|f - f_n\| \le \|f - f_{n_l}\| + \|f_n - f_{n_l}\| \le \frac{1}{2^{l-1}} + \frac{1}{2^l} < \frac{1}{2^{l-2}}.$$

This shows $f = \lim_{n \to \infty} f_n$. Thus every fundamental sequence converges to a limit and the inner product space $\mathfrak{L}_2(\alpha, \beta)$ is complete. $\square$

COROLLARY 3.1. $\mathfrak{L}_2(\alpha, \beta)$ is a complex Hilbert space.

Remark 2. If only real-valued functions and only real scalars are admitted, then we arrive at a real Hilbert space $\mathfrak{L}_2^r(\alpha, \beta)$.

Remark 3. The proof of theorem 3 at the same time shows that if the sequence $\{f_n\}_{n=1}^\infty \subset \mathfrak{L}_2(\alpha, \beta)$ converges to a function $g \in \mathfrak{L}_2(\alpha, \beta)$ (in the sense of the norm in $\mathfrak{L}_2(\alpha, \beta)$), then there exists a subsequence $\{f_{n_k}\}_{k=1}^\infty$ which converges a.e. on $[\alpha, \beta]$ to g. Indeed, such a sequence $\{f_n\}_{n=1}^\infty$ is fundamental (cf. § 3 theorem 7). If the subsequence $\{f_{n_k}\}_{k=1}^\infty$ and the function f are constructed as in the proof of theorem 3, then we have

$$f(\xi) = \lim_{k \to \infty} f_{n_k}(\xi) \quad \text{for almost all (a.a.) } \xi \in [\alpha, \beta],$$

$$f = \lim_{n \to \infty} f_n,$$

$$\|f - g\| \le \|f - f_n\| + \|g - f_n\| < \varepsilon \quad \text{for all } n \ge n(\varepsilon).$$

The last inequality implies $f = g$ (both functions considered as elements of

$\mathfrak{L}_2(\alpha, \beta))$, that is

$$\int\limits_{\alpha}^{\beta} |f(\xi) - g(\xi)|^2 \, \mathrm{d}\xi = \|f - g\|^2 = 0$$

$$f(\xi) = g(\xi) \quad \text{for a.a. } \xi \in [\alpha, \beta]$$

(cf. appendix B6). This proves the assertion.

THEOREM 4. $\mathfrak{L}_2(\alpha, \beta)$ *contains a countably infinite orthonormal family.*
PROOF. The family $\{e_k\}_{k=-\infty}^{+\infty} \subset \mathfrak{C}[\alpha, \beta]$ exhibited in § 2 example 4 has the required properties. $\square$

THEOREM 5. $\mathfrak{L}_2(\alpha, \beta)$ *is separable.*
PROOF. Let $\mathfrak{L}'$ be the set of all finite linear combinations $\sum_{k=-m}^{m} \alpha_k' e_k$ of the just mentioned functions e_k with complex rational coefficients α_k'. The set $\mathfrak{L}'$ is countable (cf. the proof of § 4 theorem 5). We shall now show that $\mathfrak{L}'$ is everywhere dense in $\mathfrak{L}_2(\alpha, \beta)$.

Let $f \in \mathfrak{L}_2(\alpha, \beta)$ and $\varepsilon > 0$ be given. In order to show that $f = \operatorname{Re} f + \mathrm{i} \operatorname{Im} f$ can be approximated by elements of $\mathfrak{L}'$ within ε it suffices to do this separately for the real functions $\operatorname{Re} f = \frac{1}{2}(f + \bar{f}) \in \mathfrak{L}_2(\alpha, \beta)$ and $\operatorname{Im} f = -\frac{1}{2}\mathrm{i}(f - \bar{f}) \in \mathfrak{L}_2(\alpha, \beta)$. We may therefore assume without loss of generality that f is real-valued. For $n \geqslant 1$ define the function f_n on $[\alpha, \beta]$ by

$$f_n(\xi) = \begin{cases} -n & \text{whenever} & f(\xi) < -n, \\ f(\xi) & \text{whenever} & -n \leqslant f(\xi) \leqslant +n, \\ n & \text{whenever} & n < f(\xi). \end{cases}$$

Then we have $f_n \in \mathfrak{F}(\alpha, \beta)$ and

$$\int\limits_{\alpha}^{\beta} |f_n(\xi)|^2 \, \mathrm{d}\xi \leqslant \int\limits_{\alpha}^{\beta} |f(\xi)|^2 \, \mathrm{d}\xi < \infty,$$

therefore $f_n \in \mathfrak{L}_2(\alpha, \beta) \subset \mathfrak{L}_1(\alpha, \beta)$ (cf. corollary 2.1). Moreover by LEBESGUE'S theorem on monotone convergence we have

$$\lim_{n \to \infty} \|f - f_n\|^2 = \lim_{n \to \infty} \int\limits_{\alpha}^{\beta} |f(\xi) - f_n(\xi)|^2 \, \mathrm{d}\xi = 0.$$

We now choose an index n such that $\|f - f_n\| < \frac{1}{4}\varepsilon$. Next by LUZIN's theorem (cf. appendix B5) we choose a continuous function h on $[\alpha, \beta]$ which coincides with f_n on $[\alpha, \beta]$ except on a set of Lebesgue measure smaller than $\varepsilon^2/64n$. Without loss of generality we may assume that h is real-valued and only assumes values lying in the interval $[-n, n]$ (replacing, if necessary, the function h by $h^r = \operatorname{Re} h$ and then h^r by h_n^r, defined analogously as f_n, we do not change the original function h where it already coincides with f_n). "Cutting off linearly" the function h in the interval $[\beta - \varepsilon^2/64n, \beta]$ (fig. 10)

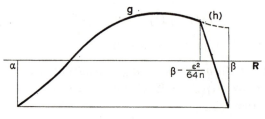

Fig. 10

we obtain a real valued function g on $[\alpha, \beta]$ satisfying $g(\alpha) = g(\beta)$, assuming values lying in the interval $[-n, n]$, and coinciding with f_n on $[\alpha, \beta]$ except on a set Y of Lebesgue measure smaller than $\varepsilon^2/32n$. We find

$$\|f_n - g\|^2 = \int_\alpha^\beta |f_n(\xi) - g(\xi)|^2 \, d\xi \leqslant \int_Y 2n \, d\xi \leqslant 2n \cdot \frac{\varepsilon^2}{32n} = \frac{\varepsilon^2}{16}.$$

We then choose a finite complex linear combination $\sum_{k=-m}^{+m} \alpha_k e_k$ such that

$$|g(\xi) - \sum_{k=-m}^{m} \alpha_k e_k(\xi)| < \frac{\varepsilon}{4\sqrt{(\beta - \alpha)}} \quad \text{for all } \xi \in [\alpha, \beta]$$

by WEIERSTRASS' approximation theorem (cf. appendix B1). Finally we choose complex rational numbers $\alpha_k'(-m \leqslant k \leqslant m)$ in such a way that

$$|\alpha_k - \alpha_k'| < \frac{\varepsilon}{4(2m + 1)\sqrt{(\beta - \alpha)}} \quad \text{for } -m \leqslant k \leqslant m.$$

We obtain

$$|g(\xi) - \sum_{k=-m}^{m} \alpha_k' e_k(\xi)| \leqslant |g(\xi) - \sum_{k=-m}^{m} \alpha_k e_k(\xi)| + \sum_{k=-m}^{m} |\alpha_k - \alpha_k'| \, |e_k(\xi)|$$

$$\leqslant \frac{\varepsilon}{4\sqrt{(\beta - \alpha)}} + \frac{\varepsilon}{4\sqrt{(\beta - \alpha)}} = \frac{\varepsilon}{2\sqrt{(\beta - \alpha)}} \quad \text{for all } \xi \in [\alpha, \beta],$$

$$\|g - \sum_{k=-m}^{m} \alpha_k' e_k\|^2 \leqslant \int_\alpha^\beta \left(\frac{\varepsilon}{2\sqrt{(\beta - \alpha)}}\right)^2 d\xi = \tfrac{1}{4}\varepsilon^2,$$

$$\|f - \sum_{k=-m}^{m} \alpha_k' e_k\| \leqslant \|f - f_n\| + \|f_n - g\| + \|g - \sum_{k=-m}^{m} \alpha_k' e_k\|$$

$$\leqslant \tfrac{1}{4}\varepsilon + \tfrac{1}{4}\varepsilon + \tfrac{1}{2}\varepsilon = \varepsilon. \quad \square$$

COROLLARY 5.1. *The set $\mathfrak{C}[\alpha, \beta]$ is everywhere dense in $\mathfrak{L}_2(\alpha, \beta)$.*

Corollary 5.1 also shows that $\mathfrak{L}_2(\alpha, \beta)$ is the completion of $\mathfrak{C}[\alpha, \beta]$ (cf. § 3 theorem 8). Theorem 4 and theorem 5 also hold for the real Hilbert space $\mathfrak{L}_2^r(\alpha, \beta)$ in place of $\mathfrak{L}_2(\alpha, \beta)$. This may be proved by using in place of the exponential functions e_k their real and imaginary parts, i.e. cosine and sine functions (cf. § 8 example 4). At the same time one sees that $\mathfrak{L}_2^r(\alpha, \beta)$ is the completion of the real inner product space of real-valued continuous functions on $[\alpha, \beta]$.

Let us have a look what happens if we now admit $-\infty$ in place of α or $+\infty$ in place of β, thus considering the sets $\mathfrak{L}_2(\alpha, +\infty)$, $\mathfrak{L}_2(-\infty, \beta)$, and $\mathfrak{L}_2(-\infty, +\infty)$ of all (absolutely) square integrable complex-valued functions on $[\alpha, +\infty[,]-\infty, \beta]$ and $]-\infty, +\infty[$ respectively. The reasoning in the proofs of theorem 1 and theorem 2 applies without change. In the proof of theorem 3, (1) and therefore also (2) only make sense for a finite interval $[\alpha, \beta]$. Since, however, (1) and (2) hold for every finite interval $[\alpha, \beta]$ contained in $[\alpha, +\infty[,]-\infty, \beta]$, and $]-\infty, +\infty[$, respectively, the eventual result that the limit

$$f(\xi) = \lim_{k \to \infty} f_{n_k}(\xi)$$

exists and is finite almost everywhere still holds in all of these cases. The further reasoning remains unaltered, thus showing that $\mathfrak{L}_2(\alpha, +\infty)$, $\mathfrak{L}_2(-\infty, \beta)$, and $\mathfrak{L}_2(-\infty, +\infty)$ are Hilbert spaces.

For all of these spaces it is easy to see that they contain a countably infinite orthonormal family. Indeed, for $\mathfrak{L}_2(\alpha, +\infty)$, choose any $\beta > \alpha$, define the functions $e_k(-\infty < k < +\infty)$ on $[\alpha, \beta]$ as in § 2 example 4, and define $e_k(\xi) = 0$ for $\beta < \xi < \infty$. Then $\{e_k\}_{k=-\infty}^{+\infty}$ is a countably infinite orthonormal family in $\mathfrak{L}_2(\alpha, +\infty)$. A similar reasoning applies for $\mathfrak{L}_2(-\infty, \beta)$ and $\mathfrak{L}_2(-\infty, +\infty)$.

Next we show that all of these spaces are separable. In fact it suffices to do this for $\mathfrak{L}_2(-\infty, +\infty)$. Restricting all functions considered to the intervals $[\alpha, +\infty[$ or $]-\infty, \beta]$, the corresponding statement for $\mathfrak{L}_2(\alpha, +\infty)$ or $\mathfrak{L}_2(-\infty, \beta)$ is established.

Observe that for any $f \in \mathfrak{L}_2(-\infty, +\infty)$ we have

$$\|f\|^2 = \int_{-\infty}^{+\infty} |f(\xi)|^2 \, d\xi = \lim_{n \to \infty} \int_{-n}^{+n} |f(\xi)|^2 \, d\xi$$

by LEBESGUE's theorem on monotone convergence. For every Hilbert space $\mathfrak{L}_2(-n, n)$ (n a natural number) we choose a countable everywhere dense subset $\mathfrak{L}'_n$ of functions in $\mathfrak{L}_2(-n, n)$ which we then extend to the functions in $\mathfrak{L}_2(-\infty, +\infty)$ by defining them to vanish outside $[-n, n]$. The set $\mathfrak{L}' = \bigcup_{n=1}^{\infty} \mathfrak{L}'_n$ is countable. Given any $f \in \mathfrak{L}_2(-\infty, +\infty)$ and any $\varepsilon > 0$ we first choose n in such a way that

$$\int_{-\infty}^{-n} |f(\xi)|^2 \, d\xi + \int_{n}^{+\infty} |f(\xi)|^2 \, d\xi < \tfrac{1}{2}\varepsilon^2 \, .$$

We then choose $g \in \mathfrak{L}'_n \subset \mathfrak{L}'$ in such a way that

$$\int_{-n}^{+n} |f(\xi) - g(\xi)|^2 \, d\xi < \tfrac{1}{2}\varepsilon^2 \, .$$

We obtain

$$\|f - g\|^2 = \int_{-n}^{+n} |f(\xi) - g(\xi)|^2 \, d\xi + \int_{-\infty}^{-n} |f(\xi)|^2 \, d\xi + \int_{n}^{+\infty} |f(\xi)|^2 \, d\xi < \varepsilon^2 \, .$$

Thus $\mathfrak{L}'$ is everywhere dense in $\mathfrak{L}_2(-\infty, +\infty)$.

By a similar reasoning one arrives at separable infinite-dimensional real Hilbert spaces $\mathfrak{L}_2^r(\alpha, +\infty)$, $\mathfrak{L}_2^r(-\infty, \beta)$, and $\mathfrak{L}_2^r(-\infty, +\infty)$.

Finally we consider the set $\mathfrak{L}_2(X)$ of all complex-valued functions that are Lebesgue measurable and square integrable on a "square"

$$X = \,]\alpha, \beta[\,\times\,]\alpha, \beta[\,= \{(\xi, \eta) \in \mathbf{R}^2 : \alpha < \xi < \beta, \alpha < \eta < \beta\}$$

in the Euclidean plane $\mathbf{R}^2$ (we also allow that α and β may be replaced by $-\infty$ and $+\infty$ respectively). Then the statements and the proofs of theorem 1 and theorem 2 apply for $\mathfrak{L}_2(X)$ if integration over $]\alpha, \beta[$ is replaced by integration over X. Thus, for any two elements f and g in $\mathfrak{L}_2(X)$ we have $f\bar{g} \in \mathfrak{L}_1(X)$ and the inner product is defined by

$$\langle f, g \rangle = \int_X f(\xi, \eta) \, \overline{g(\xi, \eta)} \, d(\xi, \eta)$$

$$= \int_{\eta=\alpha}^{\beta} \int_{\xi=\alpha}^{\beta} f(\xi, \eta) \, \overline{g(\xi, \eta)} \, d\xi \, d\eta = \int_{\xi=\alpha}^{\beta} \int_{\eta=\alpha}^{\beta} f(\xi, \eta) \, \overline{g(\xi, \eta)} \, d\eta \, d\xi$$

(the integral over X may be replaced by an iterated integral with respect to ξ and η in any order by FUBINI's theorem; cf. appendix B8). The proof of theorem 3 goes through for $\mathfrak{L}_2(X)$ in case of a "finite" square X if we furthermore take into account that in (1) and (2) we now have $\|1\| = \beta - \alpha$ in place of $\sqrt{(\beta - \alpha)}$. In case of an "infinite square" X the same reasoning still holds for every "finite" square $]\alpha, \beta[\times]\alpha, \beta[$ contained in X. As a consequence, the limit

$$f(\xi, \eta) = \lim_{k \to \infty} f_{n_k}(\xi, \eta)$$

again exists and is finite a.e. on X. The further reasoning showing that $\lim_{n \to \infty} \|f - f_n\| = 0$ exactly parallels the one demonstrated in the proof of theorem 3. Thus again we arrive at the conclusion that $\mathfrak{L}_2(X)$ is a complex Hilbert space.

If $\{e_k\}_{k=1}^\infty$ is a countably infinite orthonormal family in $\mathfrak{L}_2(\alpha, \beta)$, then define the functions $e_{k,l}$ on X for $1 \leqslant k < \infty$, $1 \leqslant l < \infty$ by

$$e_{k,l}(\xi, \eta) = e_k(\xi) \, \overline{e_l(\eta)}$$

(the reason for taking the complex conjugate will become clear in § 14 example 4). Then $e_{k,l}$ is Lebesgue measurable and square integrable on X. In fact, the family $\{e_{k,l}\}_{1 \leqslant k, l < \infty}$ is even orthonormal since

$$\int_X e_{k,l}(\xi, \eta) \, \overline{e_{m,n}(\xi, \eta)} \, \mathrm{d}(\xi, \eta) = \int_\alpha^\beta e_k(\xi) \, \overline{e_m(\xi)} \, \mathrm{d}\xi \cdot \int_\alpha^\beta \overline{e_l(\eta)} \, e_n(\eta) \, \mathrm{d}\eta = \delta_{k,m} \cdot \delta_{l,n}.$$

Theorem 4 therefore also holds for $\mathfrak{L}_2(X)$. The fact that $\mathfrak{L}_2(X)$ is separable will follow later from some more general theorems on orthonormal families in Hilbert space (cf. § 9 theorem 10, § 9 corollary 10.1).

Exercise. If the sequence $\{f_n\}_{n=1}^\infty$ is fundamental in $\mathfrak{L}_2(\alpha, \beta)$ (α and β may be replaced by $-\infty$ and $+\infty$, respectively) and if for some subsequence $\{f_{n_k}\}_{k=1}^\infty$ the limit

$$f(\xi) = \lim_{k \to \infty} f_{n_k}(\xi)$$

exists and is finite a.e. on $]\alpha, \beta[$, then $f \in \mathfrak{L}_2(\alpha, \beta)$ and

$$f = \lim_{n \to \infty} f_n.$$

CHAPTER II

SPECIFIC GEOMETRY OF HILBERT SPACE

§ 6. Subspaces

In the present section and in the following ones we consider an arbitrary but fixed Hilbert space $\mathfrak{H}$. We shall stick to our practice of denoting complex scalars by small greek letters and elements of $\mathfrak{H}$ by small italics.

As promised in § 1 we shall now begin a systematic study of the interaction of the algebraic and topological properties of a Hilbert space. The notion of a subspace already combines these two aspects:

DEFINITION 1. A non-empty subset $\mathfrak{M}$ of $\mathfrak{H}$ is called a *linear manifold* if $f+g\in\mathfrak{M}$ for all $f\in\mathfrak{M}$ and all $g\in\mathfrak{M}$, and $\lambda f\in\mathfrak{M}$ for all $f\in\mathfrak{M}$ and all $\lambda\in\mathbf{C}$ (or $\mathbf{R}$ in case of a real Hilbert space). A closed linear manifold is called a *subspace*. A subspace is called *proper* if it does not coincide with $\mathfrak{H}$.

It is well known that in n-dimensional unitary space $\mathbf{C}^n$ every linear manifold is closed (cf. also § 3 exercise). In this case the notions of linear manifold and subspace therefore coincide. This is not true for Hilbert space in general, as the following examples 1 and 3 will show.

Example 1. Let $\mathfrak{H}=\ell_2$ and let $\mathfrak{M}$ be the subset of all "finite" sequences in ℓ_2 (cf. § 2 example 2). Obviously $\mathfrak{M}$ is a linear manifold, but since $\mathfrak{M}$ is everywhere dense in ℓ_2 we have $\overline{\mathfrak{M}}=\ell_2\neq\mathfrak{M}$ (cf. § 4 remark 1). Thus $\mathfrak{M}$ is not closed.

Example 2. Let $\mathfrak{H}=\ell_2$ and let $\mathfrak{M}$ be the subset of all $a=\{\alpha_k\}_{k=1}^{\infty}\in\ell_2$ with $\alpha_1=0$. Obviously $\mathfrak{M}$ is a linear manifold. It is easy to show that $\mathfrak{M}$ is even a subspace: If $b=\{\beta_k\}_{k=1}^{\infty}\in\ell_2$ is an accumulation point of $\mathfrak{M}$, then for every $\varepsilon>0$ there is an element $a\in\mathfrak{M}$ such that $\|b-a\|<\varepsilon$. Since we have $|\beta_1|=|\beta_1-\alpha_1|\leqslant\|b-a\|$ we conclude $\beta_1=0$ and $b\in\mathfrak{M}$. Thus $\mathfrak{M}$ is closed.

Example 3. Let $\mathfrak{H}=\mathfrak{L}_2(\alpha,\beta)$ and let $\mathfrak{M}=\mathfrak{C}[\alpha,\beta]\subset\mathfrak{L}_2(\alpha,\beta)$ as in § 2 example 3. Again, $\mathfrak{M}$ is a linear manifold but not a subspace: we have $\overline{\mathfrak{M}}=\mathfrak{H}\neq\mathfrak{M}$ (cf. § 5 corollary 5.1).

Example 4. Let $\mathfrak{H}=\mathfrak{L}_2(\alpha,\beta)$ $(-\infty\leqslant\alpha<\beta\leqslant+\infty)$ and let Y be a Lebesgue measurable subset (e.g. a subinterval) of the interval $]\alpha, \beta[$. We define

$$\mathfrak{M} = \{f \in \mathfrak{L}_2(\alpha, \beta) : f(\xi) = 0 \quad \text{for almost all } \xi \in Y\}$$

(recall that two functions which are equal a.e. on $]\alpha, \beta[$ are identified). Obviously $\mathfrak{M}$ is a linear manifold. We proceed to show that $\mathfrak{M}$ is a subspace. If Y is a set of Lebesgue measure zero we have $\mathfrak{M} = \mathfrak{L}_2(\alpha, \beta)$ and there is nothing to show. Suppose therefore Y has positive Lebesgue measure. If $g \in \mathfrak{M}$ is given, then for arbitrary $\varepsilon > 0$ there exists an $f \in \mathfrak{M}$ such that $\|g - f\|^2 = \int_\alpha^\beta |g(\xi) - f(\xi)|^2 \, d\xi < \varepsilon$. This implies

$$\int_Y |g(\xi)|^2 \, d\xi = \int_Y |g(\xi) - f(\xi)|^2 \, d\xi \leqslant \int_\alpha^\beta |g(\xi) - f(\xi)|^2 \, d\xi < \varepsilon.$$

We conclude $\int_Y |g(\xi)|^2 \, d\xi = 0$ and therefore $g(\xi) = 0$ a.e. on Y (cf. appendix B6). As a consequence, $g \in \mathfrak{M}$ and $\mathfrak{M}$ is closed.

Remark 1. The entire space $\mathfrak{H}$ and the subset $\mathfrak{O}$ consisting of the single element o are obviously subspaces. These two subspaces are called *trivial*. All other subspaces are called *non-trivial* subspaces.

Remark 2. Definition 1, remark 1, and the following definitions and theorems in this section also apply if the Hilbert space is replaced by any Banach space $\mathfrak{L}$ (cf. § 3 definition 7).

The notion of a subspace is important for several reasons. One reason lies in the fact that a subspace $\mathfrak{M}$ is a Hilbert space in its own right: any fundamental sequence $\{f_n\}_{n=1}^\infty \subset \mathfrak{M}$ converges to some element f in $\mathfrak{H}$ which, since $\mathfrak{M}$ is closed, must belong to $\mathfrak{M}$; thus $\mathfrak{M}$ is complete.

The question as to existence of subspaces in any given Hilbert space admits a nice answer: if we start out with an arbitrary linear manifold $\mathfrak{M}$ in $\mathfrak{H}$, then its closure $\overline{\mathfrak{M}}$ is a subspace. In fact, while $\mathfrak{M}$ may be contained in many subspaces, $\overline{\mathfrak{M}}$ is in a natural sense the "smallest" subspace containing $\mathfrak{M}$. More generally, for every given subset $\mathfrak{A}$ of $\mathfrak{H}$ there exists a unique "smallest" subspace containing $\mathfrak{A}$. After having proved the existence of this subspace we shall identify it explicitly under various assumptions about the initial set $\mathfrak{A}$.

THEOREM 1. *Let $\mathfrak{M}$ be a linear manifold in $\mathfrak{H}$. Then $\overline{\mathfrak{M}}$ is a subspace.*
PROOF. We have to show that, for $f \in \overline{\mathfrak{M}}$, $g \in \overline{\mathfrak{M}}$ and $\lambda \in \mathbf{C}$, we have $f + g \in \overline{\mathfrak{M}}$ and $\lambda f \in \overline{\mathfrak{M}}$. Given $\varepsilon > 0$ we choose $f_1 \in \mathfrak{M}$ and $g_1 \in \mathfrak{M}$ in such a way that

$\|f - f_1\| < \frac{1}{2}\varepsilon$ and $\|g - g_1\| < \frac{1}{2}\varepsilon$. Then we have $f_1 + g_1 \in \mathfrak{M}$ and

$$\|(f + g) - (_1 + g_1)\| \leqslant \|f - f_1\| + \|g - g_1\| < \varepsilon.$$

Thus, if $f + g \notin \mathfrak{M}$, then $f + g$ is an accumulation point of $\mathfrak{M}$ and therefore $f + g \in \overline{\mathfrak{M}}$. A similar reasoning leading to the assertion $\lambda f \in \overline{\mathfrak{M}}$ is left to the reader. $\square$

THEOREM 2. *Let* $\{\mathfrak{M}_\sigma\}_{\sigma \in \Sigma}$ *be a non-empty family of linear manifolds. Then the set* $\mathfrak{M} = \bigcap_{\sigma \in \Sigma} \mathfrak{M}_\sigma$ *is a linear manifold. If* $\{\mathfrak{M}_\sigma\}_{\sigma \in \Sigma}$ *is a family of subspaces, then* $\mathfrak{M}$ *is a subspace.*

PROOF. The set $\mathfrak{M}$ is not empty since it contains o. For $f \in \mathfrak{M}$, $g \in \mathfrak{M}$ and $\alpha \in \mathbf{C}$, $\beta \in \mathbf{C}$ we have $\alpha f + \beta g \in \mathfrak{M}_\sigma$ for all $\sigma \in \Sigma$, therefore $\alpha f + \beta g \in \mathfrak{M}$. If all sets $\mathfrak{M}_\sigma$ are closed $(\sigma \in \Sigma)$ then $\mathfrak{M}$ is closed by § 3 corollary 3.1 b). $\square$

THEOREM 3. *Given any subset* $\mathfrak{A} \subset \mathfrak{H}$ *there exists a unique subspace* $\mathfrak{M}$ *with the following properties*:

    a) $\mathfrak{A} \subset \mathfrak{M}$;
    b) *if* $\mathfrak{N}$ *is any subspace containing* $\mathfrak{A}$, *then* $\mathfrak{N} \supset \mathfrak{M}$.

PROOF. Consider the family $\{\mathfrak{M}_\sigma\}_{\sigma \in \Sigma}$ of all subspaces containing $\mathfrak{A}$. This family is not empty since $\mathfrak{H}$ belongs to it. Let $\mathfrak{M} = \bigcap_{\sigma \in \Sigma} \mathfrak{M}_\sigma$. Then $\mathfrak{M}$ is a subspace by theorem 2 which contains $\mathfrak{A}$ (property a)), and $\mathfrak{M} \subset \mathfrak{M}_\sigma$ for all $\sigma \in \Sigma$ (property b)). Furthermore, if $\mathfrak{M}'$ is another subspace enjoying the properties a) and b), then $\mathfrak{M}' \subset \mathfrak{M}$ since $\mathfrak{M}$ contains $\mathfrak{A}$ and $\mathfrak{M}'$ has property b). Similarly we conclude $\mathfrak{M} \subset \mathfrak{M}'$. Thus $\mathfrak{M}' = \mathfrak{M}$ and $\mathfrak{M}$ is determined uniquely by a) and b). $\square$

DEFINITION 2. If $\mathfrak{A}$ is a subset of $\mathfrak{H}$, then the subspace $\mathfrak{M}$ associated with $\mathfrak{A}$ according to theorem 3 is called the subspace *spanned* by $\mathfrak{A}$ or the *span* of $\mathfrak{A}$ (notation: $\mathfrak{M} = \bigvee \mathfrak{A}$).

THEOREM 4. *If* $\mathfrak{A}$ *is a subset of* $\mathfrak{H}$, *then*

$$\bigvee \mathfrak{A} = \left\{ \sum_{k=1}^{n} \alpha_k f_k : f_k \in \mathfrak{A}, \alpha_k \in \mathbf{C} \text{ for } 1 \leqslant k \leqslant n, n \geqslant 1 \right\}^-.$$

PROOF (for technical reasons the upper bar marking closure is put behind the set in question). The set of all finite complex linear combinations of elements in $\mathfrak{A}$ is obviously a linear manifold contained in $\bigvee \mathfrak{A}$. The closure of this set is then a subspace contained in $\bigvee \mathfrak{A}$ (cf. theorem 1). By the minimality of the latter (property b)) this subspace must coincide with $\bigvee \mathfrak{A}$. $\square$

If $\mathfrak{M}_1, \mathfrak{M}_2, \ldots$ are subspaces, then we shall denote the subspaces spanned by $\mathfrak{M}_1 \cup \mathfrak{M}_2$ and $\bigcup_{k=1}^{\infty} \mathfrak{M}_k$ by $\mathfrak{M}_1 \vee \mathfrak{M}_2$ and $\vee_{k=1}^{\infty} \mathfrak{M}_k$ respectively. These "spans" of subspaces should well be distinguished from the "sums" of subspaces which we are now going to define.

DEFINITION 3. If $\mathfrak{M}_1$ and $\mathfrak{M}_2$ are linear manifolds, then the set

$$\mathfrak{M}_1 + \mathfrak{M}_2 = \{f_1 + f_2 : f_1 \in \mathfrak{M}_1, f_2 \in \mathfrak{M}_2\}$$

is called the *vector sum* of $\mathfrak{M}_1$ and $\mathfrak{M}_2$. If $\{\mathfrak{M}_k\}_{k=1}^{\infty}$ is a sequence of linear manifolds, then the set

$$\sum_{k=1}^{\infty} \mathfrak{M}_k = \{f : f \in \mathfrak{H}, f = \sum_{k=1}^{\infty} f_k, f_k \in \mathfrak{M}_k \text{ for } 1 \leqslant k < \infty\}$$

is called the *vector sum* of the linear manifolds $\mathfrak{M}_k (1 \leqslant k < \infty)$.

In other words, the set $\sum_{k=1}^{\infty} \mathfrak{M}_k$ is the set of sums of all converging series $\sum_{k=1}^{\infty} f_k$ with $f_k \in \mathfrak{M}_k$ for $1 \leqslant k < \infty$. Obviously, this set is a linear manifold, as well as the set $\mathfrak{M}_1 + \mathfrak{M}_2$. The vector sum of two linear manifolds may also be considered as a special case of a vector sum of a sequence of linear manifolds, taking $\mathfrak{M}_k = \mathfrak{O}$ for $k > 2$ (cf. remark 1). Similarly we shall write

$$\sum_{k=1}^{n} \mathfrak{M}_k = \mathfrak{M}_1 + \mathfrak{M}_2 + \cdots + \mathfrak{M}_n$$

if we take $\mathfrak{M}_k = \mathfrak{O}$ for $k > n$. There is a simple connection between vector sums and spans of subspaces.

THEOREM 5. *If* $\{\mathfrak{M}_k\}_{k=1}^{\infty}$ *is a sequence of subspaces, then*

$$\mathfrak{M}_1 \vee \mathfrak{M}_2 = \overline{\mathfrak{M}_1 + \mathfrak{M}_2},$$

$$\bigvee_{k=1}^{\infty} \mathfrak{M}_k = \overline{\sum_{k=1}^{\infty} \mathfrak{M}_k}.$$

PROOF. The assertion follows immediately from theorem 1 since $\mathfrak{M}_1 + \mathfrak{M}_2$ and $\sum_{k=1}^{\infty} \mathfrak{M}_k$ are linear manifolds. $\square$

It is a surprising fact that not even the vector sum of only two subspaces need again be a subspace. An example of such a non-closed vector sum is given e.g. in [2] § 15 and reproduced in the exercise below.

Exercise. Let the subsets $\mathfrak{M}_1$ and $\mathfrak{M}_2$ of $\mathfrak{H} = \ell_2$ be defined by

$$\mathfrak{M}_1 = \{a = \{\alpha_k\}_{k=1}^{\infty} \in \ell_2 : \alpha_{2l} = 0 \text{ for } 1 \leqslant l < \infty\},$$

$$\mathfrak{M}_2 = \{b = \{\beta_k\}_{k=1}^{\infty} \in \ell_2 : \beta_{2l-1} = \delta_l \cos \frac{1}{l}, \beta_{2l} = \delta_l \sin \frac{1}{l} \text{ for } 1 \leqslant l < \infty\}.$$

Furthermore, let $c = \{\gamma_k\}_{k=1}$ where $\gamma_{2l-1} = 0$, $\gamma_{2l} = \sin 1/l$ for $1 \leqslant l < \infty$.
Prove the following assertions.

a) $\mathfrak{M}_1$ and $\mathfrak{M}_2$ are subspaces.
b) $\mathfrak{M}_1 \vee \mathfrak{M}_2 = \mathfrak{H}$.
c) $c \in \mathfrak{H} = \ell_2$.
d) $c \notin \mathfrak{M}_1 + \mathfrak{M}_2$.

§ 7. Orthogonal subspaces

One way to obtain a subspace, demonstrated in the preceding section, is to start with an arbitrary subset $\mathfrak{A} \subset \mathfrak{H}$ and to take its span $\vee \mathfrak{A}$ (cf. § 6 definition 2, § 6 theorem 4). Another way, as it will turn out presently, is to form the set of all vectors which are orthogonal to every vector in $\mathfrak{A}$ (cf. § 2 definition 6).

DEFINITION 1. A vector g is called *orthogonal to a subset* $\mathfrak{A} \subset \mathfrak{H}$ (notation: $g \perp \mathfrak{A}$) if $g \perp f$ for all $f \in \mathfrak{A}$. Two subsets $\mathfrak{A}$ and $\mathfrak{B}$ of $\mathfrak{H}$ are called *mutually orthogonal* (notation: $\mathfrak{A} \perp \mathfrak{B}$) if $f \perp g$ for all $f \in \mathfrak{A}$ and all $g \in \mathfrak{B}$. The set

$$\mathfrak{A}^\perp = \{g \in \mathfrak{H} : g \perp \mathfrak{A}\}$$

is called the *orthogonal complement* of $\mathfrak{A}$ (in $\mathfrak{H}$). If $\mathfrak{M}$ and $\mathfrak{N}$ are subspaces and $\mathfrak{M} \subset \mathfrak{N}$, then the set

$$\mathfrak{N} - \mathfrak{M} = \{g \in \mathfrak{N} : g \perp \mathfrak{M}\}$$
$$= \mathfrak{N} \cap \mathfrak{M}^\perp$$

is called the *orthogonal complement* of $\mathfrak{M}$ in $\mathfrak{N}$.

Remark 1. Note that the formal subtraction of one subspace from another one (which must contain the first one!) introduced in definition 1 is defined in a way essentially different from the way addition has been introduced in § 6 definition 3 (this is sometimes underlined by writing $\mathfrak{N} \ominus \mathfrak{M}$ in place of $\mathfrak{N} - \mathfrak{M}$). The connection between these two operations with subspaces will be cleared up in corollary 8.1 which will eventually justify the terminology.

Remark 2. If $\mathfrak{M}$ is a subspace, then combining the notations introduced in definition 1 we can also write

$$\mathfrak{M}^\perp = \mathfrak{H} - \mathfrak{M}.$$

Remark 3. If $\mathfrak{A}$ and $\mathfrak{B}$ are two subsets of $\mathfrak{H}$, then $\mathfrak{A} \subset \mathfrak{B}$ obviously implies $\mathfrak{A}^\perp \supset \mathfrak{B}^\perp$.

THEOREM 1. *If $\mathfrak{A}$ is a subset of $\mathfrak{H}$, then $\mathfrak{A}^\perp$ is a subspace and $\mathfrak{A} \cap \mathfrak{A}^\perp$ is either the trivial subspace $\mathfrak{O} = \{o\}$ or empty (if $o \notin \mathfrak{A}$).*

PROOF. Suppose $g_1 \in \mathfrak{A}^\perp$ and $g_2 \in \mathfrak{A}^\perp$. From $\langle f, g_1 \rangle = \langle f, g_2 \rangle = 0$ for all $f \in \mathfrak{A}$ we conclude $\langle f, \alpha_1 g_1 + \alpha_2 g_2 \rangle = \alpha_1 \langle f, g_1 \rangle + \alpha_2 \langle f, g_2 \rangle = 0$ for all $f \in \mathfrak{A}$ and therefore $\alpha_1 g_1 + \alpha_2 g_2 \in \mathfrak{A}^\perp$. Thus $\mathfrak{A}^\perp$ is a linear manifold. Let now $g \in \overline{\mathfrak{A}^\perp}$ be given and let $g = \lim_{n \to \infty} g_n$, $g_n \in \mathfrak{A}^\perp$ for all $n \geqslant 1$. Then for every $f \in \mathfrak{A}$ we have

$$\langle f, g \rangle = \lim_{n \to \infty} \langle f, g_n \rangle = \lim_{n \to \infty} 0 = 0$$

(cf. § 3 theorem 6 c)) and therefore $g \in \mathfrak{A}^\perp$. Thus $\mathfrak{A}^\perp$ is closed and a subspace. If $\mathfrak{A} \cap \mathfrak{A}^\perp$ is not empty and if $f \in \mathfrak{A} \cap \mathfrak{A}^\perp$, then we must have $f \perp f$, i.e. $\langle f, f \rangle = 0$ and therefore $f = o$. $\square$

THEOREM 2. *If $\mathfrak{M}$ and $\mathfrak{N}$ are orthogonal subspaces, then*

$$\mathfrak{M} \vee \mathfrak{N} = \mathfrak{M} + \mathfrak{N}.$$

PROOF. In view of § 6 theorem 5 we only have to show that, under our hypothesis, $\mathfrak{M} + \mathfrak{N}$ is closed. Suppose we have $f \in \overline{\mathfrak{M} + \mathfrak{N}}$, $f = \lim_{n \to \infty} f_n$, $f_n = g_n + h_n$, $g_n \in \mathfrak{M}$ and $h_n \in \mathfrak{N}$ for all $n \geqslant 1$ (cf. § 3 theorem 5). Because of $g_n \perp h_n$ we conclude from the Pythagorean theorem (§ 2 theorem 8)

$$\|f_m - f_n\|^2 = \|g_m - g_n\|^2 + \|h_m - h_n\|^2.$$

From the convergence of the sequence $\{f_n\}_{n=1}^\infty$ we conclude

$$\lim_{m, n \to \infty} \|f_m - f_n\|^2 = 0.$$

It follows that also

$$\lim_{m, n \to \infty} \|g_m - g_n\|^2 = 0,$$

$$\lim_{m, n \to \infty} \|h_m - h_n\|^2 = 0.$$

The sequences $\{g_n\}_{n=1}^\infty$ and $\{h_n\}_{n=1}^\infty$ are therefore fundamental and converge to limits g and h in $\mathfrak{H}$ respectively. Since $\mathfrak{M}$ and $\mathfrak{N}$ are closed we must have $g \in \mathfrak{M}$, $h \in \mathfrak{N}$ (cf. § 3 theorem 5) and

$$f = \lim_{n \to \infty} f_n = \lim_{n \to \infty} (g_n + h_n) = g + h \in \mathfrak{M} + \mathfrak{N}. \square$$

An analogous statement of course applies for a finite family of mutually orthogonal subspaces. The proof just given can also be extended to apply in the case of countably many summands. The basic idea of this extension has in fact already been used in the proof of § 4 theorem 3. We start with a generalization of the Pythagorean theorem just used in the proof of theorem 2.

THEOREM 3. *Let $\{g_k\}_{k=1}^{\infty}$ be an orthogonal family of vectors.*

a) *The series $\sum_{k=1}^{\infty} g_k$ converges iff $\sum_{k=1}^{\infty} \|g_k\|^2 < \infty$.*
b) *If the series $\sum_{k=1}^{\infty} g_k$ converges to the vector f, then $\|f\|^2 = \sum_{k=1}^{\infty} \|g_k\|^2$.*

PROOF. Suppose $\sum_{k=1}^{\infty} \|g_k\|^2 < \infty$. Then we have

$$\lim_{m, n \to \infty} \| \sum_{k=m}^{n} g_k \|^2 = \lim_{m, n \to \infty} \sum_{k=m}^{n} \|g_k\|^2 = 0$$

and by the completeness of $\mathfrak{H}$ the series $\sum_{k=1}^{\infty} g_k$ converges (cf. § 3 remark 3). On the other hand, if this series converges with sum f, then we get

$$\|f\|^2 = \langle f, f \rangle = \langle \lim_{n \to \infty} \sum_{k=1}^{n} g_k, \lim_{m \to \infty} \sum_{l=1}^{m} g_l \rangle$$

$$= \lim_{n \to \infty} \sum_{k=1}^{n} \langle g_k, g_k \rangle = \sum_{k=1}^{\infty} \|g_k\|^2. \quad \square$$

THEOREM 4. *If $\{\mathfrak{M}_k\}_{k=1}^{\infty}$ is a sequence of mutually orthogonal subspaces, then*

$$\bigvee_{k=1}^{\infty} \mathfrak{M}_k = \sum_{k=1}^{\infty} \mathfrak{M}_k.$$

PROOF. By § 6 theorem 5 it suffices to show that $\sum_{k=1}^{\infty} \mathfrak{M}_k$ is closed. Suppose we have

$$f \in \overline{\sum_{k=1}^{\infty} \mathfrak{M}_k},$$

$$f = \lim_{n \to \infty} f_n,$$

$$f_n = \sum_{k=1}^{\infty} g_{n, k}, \quad g_{n, k} \in \mathfrak{M}_k \quad \text{for all } n \geqslant 1 \text{ and all } k \geqslant 1.$$

The same reasoning as used in the proof of theorem 2 leads to

(1) $\|f_m - f_n\|^2 = \sum_{k=1}^{\infty} \|g_{m, k} - g_{n, k}\|^2 < \varepsilon^2 \quad \text{for all } m \geqslant n(\varepsilon) \text{ and all } n \geqslant n(\varepsilon)$

(we have applied theorem 3 b) in place of the Pythagorean theorem),

$$\lim_{m, n \to \infty} \|g_{m, k} - g_{n, k}\|^2 = 0 \quad \text{for all } k \geqslant 1.$$

Therefore there exist the limits

$$g_k = \lim_{n \to \infty} g_{n, k} \in \mathfrak{M}_k \quad \text{for all } k \geqslant 1.$$

From (1) we conclude

$$\sum_{k=1}^{l} \|g_{m,k} - g_{n,k}\|^2 < \varepsilon^2 \quad \text{for all } m \geqslant n(\varepsilon), \text{ all } n \geqslant n(\varepsilon) \text{ and all } l \geqslant 1,$$

$$\sum_{k=1}^{l} \|g_k - g_{n,k}\|^2 \leqslant \varepsilon^2 \quad \text{for all } n \geqslant n(\varepsilon) \text{ and all } l \geqslant 1,$$

$$(2) \quad \sum_{k=1}^{\infty} \|g_k - g_{n,k}\|^2 \leqslant \varepsilon^2 \quad \text{for all } n \geqslant n(\varepsilon).$$

By theorem 3 a) the series $\sum_{k=1}^{\infty}(g_k - g_{n,k})$ converges and so does the series

$$\sum_{k=1}^{\infty} g_k = \sum_{k=1}^{\infty} (g_k - g_{n,k}) + \sum_{k=1}^{\infty} g_{n,k}.$$

Furthermore (2) shows that the sum of the last series is the limit of the sequence $\{f_n\}_{n=1}^{\infty}$. We obtain

$$f = \lim_{n \to \infty} f_n = \sum_{k=1}^{\infty} g_k \in \sum_{k=1}^{\infty} \mathfrak{M}_k. \quad \square$$

Suppose $\{\mathfrak{M}_k\}_{k=1}^{\infty}$ is a family of mutually orthogonal subspaces. Apart from the fact that the subspace spanned by this family already coincides with its vector sum we note another remarkable fact: the representation of any vector in this vector sum as a sum of vectors in the "component spaces" $\mathfrak{M}_k$ is unique (cf. theorem 5 below). The special features of this particular situation are sometimes underlined by writing $\mathfrak{M}_1 \oplus \mathfrak{M}_2$ and $\sum_{k=1}^{\infty} \oplus \mathfrak{M}_k$ in place of $\mathfrak{M}_1 + \mathfrak{M}_2$ and $\sum_{k=1}^{\infty} \mathfrak{M}_k$.

THEOREM 5. *Let* $\{\mathfrak{M}_k\}_{k=1}^{\infty}$ *be a sequence of mutually orthogonal subspaces. Then for every vector* $f \in \bigvee_{k=1}^{\infty} \mathfrak{M}_k$ *there exists a unique vector* $f_k \in \mathfrak{M}_k$ *for all* $k \geqslant 1$ *such that*

$$f = \sum_{k=1}^{\infty} f_k.$$

PROOF. Suppose we have

$$f = \sum_{k=1}^{\infty} f'_k = \sum_{k=1}^{\infty} f''_k, \quad \left.\begin{array}{l} f'_k \in \mathfrak{M}_k \\ f''_k \in \mathfrak{M}_k \end{array}\right\} \text{ for all } k \geqslant 1.$$

It follows that

$$\sum_{k=1}^{\infty} (f'_k - f''_k) = o,$$

$$\sum_{k=1}^{\infty} \|f'_k - f''_k\|^2 = 0 \quad \text{(by theorem 3 b))},$$

$$f'_k = f''_k \quad \text{for all } k \geqslant 1. \quad \square$$

COROLLARY 5.1. *Suppose $\mathfrak{M}_1$ and $\mathfrak{M}_2$ are orthogonal subspaces. The decomposition of any vector $f \in \mathfrak{M}_1 \vee \mathfrak{M}_2 = \mathfrak{M}_1 + \mathfrak{M}_2$ into a sum*

$$f = f_1 + f_2, f_k \in \mathfrak{M}_k \quad \text{for } k = 1, 2$$

is unique.

While talking about orthogonal complements in definition 1 and theorem 1 we have not yet made sure whether there exists a non-zero vector orthogonal to a given subset $\mathfrak{A}$ at all. If for instance $\mathfrak{A} = \mathfrak{H}$, then this is obviously not the case. The theorems which follow lead up to the statement that if $\mathfrak{M}$ is a subspace, then $\mathfrak{M}^\perp$ is just big enough to guarantee $\mathfrak{M} + \mathfrak{M}^\perp = \mathfrak{H}$.

THEOREM 6. *Let $\mathfrak{M}$ be a subspace and let $f \in \mathfrak{H}$ be given. If $\delta = \inf \{\|f - g\| : g \in \mathfrak{M}\}$ then there exists a unique vector $P_\mathfrak{M} f \in \mathfrak{M}$, called the projection of f upon $\mathfrak{M}$, such that $\|f - P_\mathfrak{M} f\| = \delta$.*
PROOF. Let $\{g_n\}_{n=1}^\infty \subset \mathfrak{M}$ be a sequence of vectors in $\mathfrak{M}$ such that $\lim_{n \to \infty} \|f - g_n\| = \delta$. We apply the parallelogram law (§ 2 theorem 6) to the vectors $(f - g_m)$ and $(f - g_n)$:

$$\|2f - (g_m + g_n)\|^2 + \|g_m - g_n\|^2 = 2\|f - g_m\|^2 + 2\|f - g_n\|^2,$$

$$\|g_m - g_n\|^2 = 2\|f - g_m\|^2 + 2\|f - g_n\|^2 - 4\|f - \tfrac{1}{2}(g_m + g_n)\|^2.$$

Because of $\tfrac{1}{2}(g_m + g_n) \in \mathfrak{M}$ we necessarily have $\|f - \tfrac{1}{2}(g_m + g_n)\| \geq \delta$ and

$$\|g_m - g_n\|^2 \leq 2\|f - g_m\|^2 + 2\|f - g_n\|^2 - 4\delta^2.$$

The right member in this inequality tends to zero as m and n tend to ∞. Therefore there exists the limit

$$P_\mathfrak{M} f = \lim_{n \to \infty} g_n \in \mathfrak{M}$$

and we have

$$\|f - P_\mathfrak{M} f\| = \lim_{n \to \infty} \|f - g_n\| = \delta.$$

Finally, suppose f_1 and f_2 are two vectors in $\mathfrak{M}$ satisfying

$$\|f - f_1\| = \|f - f_2\| = \delta.$$

Applying the parallelogram law to $(f - f_1)$ and $(f - f_2)$ we obtain

$$\|2f - (f_1 + f_2)\|^2 + \|f_1 - f_2\|^2 = 2\|f - f_1\|^2 + 2\|f - f_2\|^2,$$

$$\|f_1 - f_2\|^2 = 4\delta^2 - 4\|f - \tfrac{1}{2}(f_1 + f_2)\|^2 \leq 0$$

since $\tfrac{1}{2}(f_1 + f_2) \in \mathfrak{M}$ and therefore $\|f - \tfrac{1}{2}(f_1 + f_2)\| \geq \delta$. We conclude $f_1 = f_2$. □

The situation described in theorem 6 and in theorem 7 below is illustrated in fig. 11.

THEOREM 7. *Let $\mathfrak{M}$ be a subspace and for any given vector $f \in \mathfrak{H}$ let $P_m f$ be the projection of f upon $\mathfrak{M}$. Then*

$$(f - P_{\mathfrak{M}}f) \perp \mathfrak{M}.$$

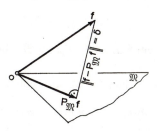

Fig. 11

PROOF. Let $f_0 = f - P_{\mathfrak{M}} f$ and $\|f_0\| = \delta$ (cf. theorem 6). For every vector $g \in \mathfrak{M}$ and for every $\alpha \in \mathbf{C}$ we have $P_{\mathfrak{M}} f + \alpha g \in \mathfrak{M}$ and therefore

$$\delta^2 = \|f_0\|^2 \leqslant \|f - (P_{\mathfrak{M}}f + \alpha g)\|^2 = \|f_0 - \alpha g\|^2$$
$$= \|f_0\|^2 - \alpha \langle g, f_0 \rangle - \bar{\alpha} \langle f_0, g \rangle + |\alpha|^2 \|g\|^2,$$
$$0 \leqslant - \alpha \langle g, f_0 \rangle - \bar{\alpha} \langle f_0, g \rangle + |\alpha|^2 \|g\|^2.$$

Suppose there were a vector $g \in \mathfrak{M}$ such that $\langle f_0, g \rangle \neq 0$ (which also implies $g \neq o$). Choosing $\alpha = \langle f_0, g \rangle / \|g\|^2$ we would obtain

$$0 \leqslant - 2 \frac{|\langle f_0, g \rangle|^2}{\|g\|^2} + \frac{|\langle f_0, g \rangle|^2}{\|g\|^2} = - \frac{|\langle f_0, g \rangle|^2}{\|g\|^2},$$

a contradiction. $\square$

COROLLARY 7.1. *Let $\mathfrak{M}$ be a linear manifold contained in a subspace $\mathfrak{N}$. Then $\overline{\mathfrak{M}} \neq \mathfrak{N}$ iff there exists a non-zero vector $f \in \mathfrak{N}$ orthogonal to $\mathfrak{M}$.*
PROOF. If $\overline{\mathfrak{M}} \neq \mathfrak{N}$, then take any non-zero vector f belonging to $\mathfrak{N}$ but not to $\overline{\mathfrak{M}}$. Then the vector $f_0 = f - P_{\overline{\mathfrak{M}}} f$ has all desired properties by theorem 7.

On the other hand, suppose $\overline{\mathfrak{M}} = \mathfrak{N}$ and let f be any vector in $\mathfrak{N}$ orthogonal to $\mathfrak{M}$. We shall show that then necessarily $f = o$. Indeed, we have $f = \lim_{n \to \infty} f_n$, $f_n \in \mathfrak{M}$ for all $n \geqslant 1$ (cf. § 3 theorem 5), and therefore

$$\langle f, f \rangle = \lim_{n \to \infty} \langle f, f_n \rangle = \lim_{n \to \infty} 0 = 0.$$

We conclude $f = o$. $\square$

COROLLARY 7.2. *Let $\mathfrak{M}$ be a linear manifold. Then $\overline{\mathfrak{M}} = \mathfrak{H}$ iff $\mathfrak{M}^\perp = \mathfrak{O}$.*

THEOREM 8 (PROJECTION THEOREM). *If $\mathfrak{M}$ is a subspace, then*

$$\mathfrak{H} = \mathfrak{M} + \mathfrak{M}^\perp .$$

PROOF. For every vector $f \in \mathfrak{H}$ we have

$$= P_{\mathfrak{M}} f + (f - P_{\mathfrak{M}} f)$$

where $P_{\mathfrak{M}} f \in \mathfrak{M}$ by theorem 6 and $(f - P_{\mathfrak{M}} f) \in \mathfrak{M}^\perp$ by theorem 7. $\square$

Taking into account corollary 5.1 we can rephrase theorem 8 as follows:

THEOREM 8′. *If $\mathfrak{M}$ is a subspace, then every vector $f \in \mathfrak{H}$ can in a unique way be decomposed into a sum*

$$f = f_1 + f_2$$

such that $f_1 (= P_{\mathfrak{M}} f) \in \mathfrak{M}$ and $f_2 (= P_{\mathfrak{M}^\perp} f) \in \mathfrak{M}^\perp$.

If in theorem 8 we replace $\mathfrak{H}$ by a subspace $\mathfrak{N}$ containing $\mathfrak{M}$, then the set $\mathfrak{M}^\perp = \{ f \in \mathfrak{H} : f \perp \mathfrak{M} \}$ has to be replaced by the set $\mathfrak{N} \cap \mathfrak{M}^\perp = \{ f \in \mathfrak{N} : f \perp \mathfrak{M} \}$. The resulting statement then justifies the notation $\mathfrak{N} \cap \mathfrak{M}^\perp = \mathfrak{N} - \mathfrak{M}$ (cf. remark 1, remark 2):

COROLLARY 8.1. *If $\mathfrak{M}$ and $\mathfrak{N}$ are subspaces and if $\mathfrak{M} \subset \mathfrak{N}$, then*

$$\mathfrak{N} = \mathfrak{M} + (\mathfrak{N} \cap \mathfrak{M}^\perp)$$
$$= \mathfrak{M} + (\mathfrak{N} - \mathfrak{M}) .$$

Example 1. Let $\mathfrak{H} = \mathfrak{L}_2 (\alpha, \beta)$ $(-\infty \leqslant \alpha < \beta \leqslant +\infty)$ and let Y be a Lebesgue measurable subset of the interval $]\alpha, \beta[$. For every $f \in \mathfrak{H}$ we define the Lebesgue measurable functions f_1, f_2 on $]\alpha, \beta[$ by

$$f_1(\xi) = \begin{cases} 0 & \text{for } \xi \in Y, \\ f(\xi) & \text{for } \xi \in]\alpha, \beta[\backslash Y, \end{cases}$$

$$f_2(\xi) = \begin{cases} f(\xi) & \text{for } \xi \in Y, \\ 0 & \text{for } \xi \in]\alpha, \beta[\backslash Y. \end{cases}$$

From

$$\int_\alpha^\beta |f_k(\xi)|^2 \, d\xi \leqslant \int_\alpha^\beta |f(\xi)|^2 \, d\xi < \infty \quad \text{for } k = 1, 2$$

we conclude $f_k \in \mathfrak{L}_2(\alpha, \beta)$ for $k = 1, 2$. Furthermore we have

$$f_1 \in \mathfrak{M}_1 = \{g \in \mathfrak{L}_2(\alpha, \beta): g(\xi) = 0 \quad \text{for almost all } \xi \in Y\},$$

$$f_2 \in \mathfrak{M}_2 = \{g \in \mathfrak{L}_2(\alpha, \beta): g(\xi) = 0 \quad \text{for almost all } \xi \in]\alpha, \beta[\backslash Y\},$$

$$f = f_1 + f_2,$$

$$\langle f_1, f_2 \rangle = \int_\alpha^\beta f_1(\xi) \overline{f_2(\xi)} \, d\xi = 0$$

(cf. § 6 example 4). Since the decomposition of any vector f into a sum of a vector in $\mathfrak{M}_k$ and a vector in $\mathfrak{M}_k^\perp$ is unique by theorem 8′ we conclude $f_k = P_{\mathfrak{M}_k} f$ for $k = 1, 2$ and $\mathfrak{M}_2 = \mathfrak{M}_1^\perp$.

In the sequel we shall write $\mathfrak{A}^{\perp\perp}$ for $(\mathfrak{A}^\perp)^\perp$.

THEOREM 9. *Let $\mathfrak{A}$ be a subset of $\mathfrak{H}$. Then $\bigvee \mathfrak{A} = \mathfrak{A}^{\perp\perp}$ and, as a consequence, $\mathfrak{A}^\perp = \mathfrak{A}^{\perp\perp\perp}$. In particular, $\mathfrak{A}$ is a subspace iff $\mathfrak{A} = \mathfrak{A}^{\perp\perp}$.*

PROOF. Let $f \in \bigvee \mathfrak{A}$ be the limit of a converging sequence $\{f_n\}_{n=1}^\infty$ where every vector f_n is a finite linear combination of vectors in $\mathfrak{A}$ (cf. § 6 theorem 4). Then for every $g \in \mathfrak{A}^\perp$ we have $\langle f, g \rangle = \lim_{n \to \infty} \langle f_n, g \rangle = 0$ and therefore $f \perp g$. This implies $\bigvee \mathfrak{A} \perp \mathfrak{A}^\perp$ and $\bigvee \mathfrak{A} \subset \mathfrak{A}^{\perp\perp}$. If $\bigvee \mathfrak{A} \neq \mathfrak{A}^{\perp\perp}$ then by corollary 7.1 there would exist a non-zero vector $f \in \mathfrak{A}^{\perp\perp} \cap \mathfrak{A}^\perp$. This, however, is impossible by theorem 1. The second statement in theorem 9 is obtained by replacing $\mathfrak{A}$ in the first statement by $\mathfrak{A}^\perp$. □

Exercise. Let $\mathfrak{H} = \ell_2$, define for $n \geq 1$ the element $e_n \in \ell_2$ by $e_n = \{\delta_{n,k}\}_{k=1}^\infty$ (cf. the proof of § 4 theorem 4) and let $\mathfrak{A} = \{e_{2l-1} + e_{2l}\}_{l=1}^\infty$.

a) Identify the subspaces $\bigvee \mathfrak{A}$ and $\mathfrak{A}^\perp$ in ℓ_2.

b) If $a = \{\alpha_k\}_{k=1}^\infty \in \ell_2$, then

$$P_{\bigvee \mathfrak{A}} a = \{\beta_k\}_{k=1}^\infty, \quad \text{where } \beta_{2l-1} = \beta_{2l} = \tfrac{1}{2}(\alpha_{2l-1} + \alpha_l) \quad \text{for } 1 \leq l < \infty,$$

$$P_{\mathfrak{A}^\perp} a = \{\gamma_k\}_{k=1}^\infty, \quad \text{where } \gamma_{2l-1} = -\gamma_{2l} = \tfrac{1}{2}(\alpha_{2l-1} - \alpha_{2l}) \quad \text{for } 1 \leq l < \infty.$$

§ 8. Bases

DEFINITION 1. Let $\mathfrak{M}$ be a subspace of a Hilbert space $\mathfrak{H}$. An orthonormal family $\{e_\sigma\}_{\sigma \in \Sigma} \subset \mathfrak{M}$ is called *maximal* in $\mathfrak{M}$ if the only vector in $\mathfrak{M}$ orthogonal to all $e_\sigma (\sigma \in \Sigma)$ is the zero vector o. A maximal orthonormal family in $\mathfrak{M}$ is also called a *basis* of $\mathfrak{M}$.

If $\mathfrak{M} = \mathfrak{H}$ and if there is no danger of confusion we shall omit the attributes (maximal) "in $\mathfrak{H}$" and (basis) "of $\mathfrak{H}$" and simply talk about a basis. Recall that any orthonormal family is linearly independent (cf. § 2 theorem 7).

Example 1. Let $\mathfrak{H}$ be the n-dimensional unitary space $\mathbf{C}^n$. Every n-tuple of orthonormal vectors is a basis.

Example 2. Let $\mathfrak{H} = \ell_2$ and for every natural number k let e_k be the unit vector all the components of which are zero except the k-th one which is 1 (cf. the proof of § 4 theorem 4). Then $\{e_k\}_{k=1}^\infty$ is a basis (we shall refer to it as the *standard basis* of ℓ_2). Indeed, suppose $\langle f, e_k \rangle = 0$ for all $k \geqslant 1$. Then all components of f must vanish.

Example 3. Let $\mathfrak{H} = \mathfrak{L}_2(0, 1)$ and for every integer k let $e_k = e^{2\pi i k x}$ (cf. § 3 remark 1). Then $\{e_k\}_{k=-\infty}^{+\infty}$ is an orthonormal family (cf. § 2 example 4). Suppose for some $f \in \mathfrak{H}$ we have $\langle f, e_k \rangle = 0$ for every integer k. Recall that the set $\mathfrak{L}'$ of all finite linear combinations $\sum_{k=-m}^m \alpha_k' e_k$ with complex rational coefficients α_k' is everywhere dense (cf. the proof of § 5 theorem 5). Obviously we have $\langle f, g \rangle = 0$ for every vector $g \in \mathfrak{L}'$. If f were different from o we could choose a vector $g \in \mathfrak{L}'$ such that $\|f - g\| < \|f\|$. This would imply

$$\|f\|^2 = \langle f, f \rangle = \langle f, f \rangle - \langle f, g \rangle = \langle f, f - g \rangle \leqslant \|f\| \, \|f - g\| < \|f\|^2,$$

a contradiction. We conclude that $\{e_k\}_{k=-\infty}^{+\infty}$ is a basis.

Example 4. Let $\mathfrak{H} = \mathfrak{L}_2(0, 1)$ as in example 3 and let

$$\begin{aligned} e_0 &\equiv 1, \\ f_k &= \sqrt{2} \cos 2\pi k x \\ g_k &= \sqrt{2} \sin 2\pi k x \end{aligned} \right\} \quad \text{for } k = 1, 2, \dots .$$

A straightforward computation shows that the family $\mathfrak{F} = \{e_0\} \cup \{f_k\}_{k=1}^\infty \cup \{g_k\}_{k=1}^\infty$ is orthonormal. This also follows from the fact that, in the notation of example 3, we have

$$f_k = \frac{e_k + e_{-k}}{\sqrt{2}}, \qquad g_k = \frac{e_k - e_{-k}}{i\sqrt{2}} \qquad \text{for all } k \geqslant 1.$$

If f is any vector orthogonal to e_0 and to f_k and g_k for all $k \geqslant 1$, then we conclude $f \perp e_k$ for every integer k. Since the family $\{e_k\}_{k=-\infty}^{+\infty}$ is a basis we have $f = o$. As a consequence the family $\mathfrak{F}$ also is a basis. Restricting our attention to real-valued functions only we see at the same time that the family $\mathfrak{F}$ also is a basis of the real Hilbert space $\mathfrak{L}_2^r(0, 1)$ (cf. § 5 remark 2).

Does every Hilbert space admit a basis? We shall show that for a separable Hilbert space the answer is "yes". In fact, if one assumes the axiom of choice or, equivalently, ZORN's lemma, the answer is "yes" in general, but we shall not dwell on this statement any further (cf. e.g. [2] § 14). For a separable Hilbert space there is even a method to construct explicitly a basis from any given countable everywhere dense subset. This method, called the *Gram-Schmidt orthonormalization process*, is set out in the following two theorems. In order to treat the two cases of a finite and of a countably infinite basis both at the same time we shall temporarily use the letter κ to denote either a natural number or the symbol ∞.

THEOREM 1 (GRAM-SCHMIDT ORTHOGONALIZATION). *Let* $\mathfrak{F} = \{f_k\}_{k=1}^{\kappa}$ *be a countable linearly independent family of vectors (finite for* $\kappa < \infty$, *infinite for* $\kappa = \infty$). *Then there exists an orthogonal family* $\mathfrak{G} = \{g_k\}_{k=1}^{\kappa}$ *(i.e. of same cardinal number as* $\mathfrak{F}$) *such that* $g_k \neq o$ *and* g_k *is a linear combination of* $f_1, ..., f_k$ *for every* k *(*$1 \leqslant k \leqslant \kappa < \infty$ *or* $1 \leqslant k < \kappa = \infty$ *respectively).*

PROOF. We construct the family $\mathfrak{G}$ by induction. Let $g_1 = f_1$ (by hypothesis we have $f_1 \neq o$). Suppose $g_1, ..., g_{k-1}$ are mutually orthogonal non-zero vectors already chosen so as to comply with the mentioned conditions $(k-1 < \kappa)$ and define

(1)
$$g_k = f_k - \sum_{l=1}^{k-1} \frac{\langle f_k, g_l \rangle}{\|g_l\|^2} g_l.$$

Then we have

$$\langle g_k, g_l \rangle = \langle f_k, g_l \rangle - \frac{\langle f_k, g_l \rangle}{\|g_l\|^2} \langle g_l, g_l \rangle = 0 \quad \text{for } 1 \leqslant l \leqslant k - 1.$$

Moreover, g_k is a linear combination of $f_1, ..., f_k$. Since $f_1, ..., f_k$ are linearly independent, (1) implies $g_k \neq o$. □

COROLLARY 1.1. *Let* $\mathfrak{F} = \{f_k\}_{k=1}^{\kappa}$ *and* $\mathfrak{G} = \{g_k\}_{k=1}^{\kappa}$ *be as in theorem 1. Then the following statements hold:*

a) f_k *is a linear combination of* $g_1, ..., g_k$ *for* $1 \leqslant k_{(\leqslant)} \kappa$.

b) $\bigvee \{f_k\}_{k=1}^{\kappa} = \bigvee \{g_k\}_{k=1}^{\kappa}$.

c) *The family* $\{e_k = g_k / \|g_k\|\}_{k=1}^{\kappa}$ *is an orthonormal family meeting the requirements mentioned in theorem 1 (GRAM-SCHMIDT ORTHONORMALIZATION).*

d) *If* $\{h_k\}_{k=1}^{\kappa}$ *is any other orthogonal family of non-zero vectors that meets the requirements mentioned in theorem 1, then* $h_k = \alpha_k g_k$ *and* $\alpha_k \neq 0$ *for* $1 \leqslant k_{(\leqslant)} \kappa$.

PROOF. a) The assertion is obviously true for $k=1$. Assuming it to be true for $k-1$ in place of k we have, by the requirement mentioned in theorem 1,

$$g_k = \sum_{l=1}^{k} \alpha_{k,l} f_l = \alpha_{k,k} f_k + \sum_{l=1}^{k-1} \beta_{k,l} g_l.$$

Since the mutually orthogonal non-zero vectors $g_1, \ldots, g_k$ are linearly independent (cf. § 2 theorem 7) we must have $\alpha_{k,k} \neq 0$. We conclude

$$f_k = \frac{1}{\alpha_{k,k}} \left(g_k - \sum_{l=1}^{k-1} \beta_{k,l} g_l \right).$$

b) By a) and theorem 1, every finite linear combination of vectors in $\mathfrak{F}$ is also a finite linear combination of vectors in $\mathfrak{G}$ and conversely. The assertion follows from § 6 theorem 4.

c) Clear.

d) Since h_k is a linear combination of $f_1, \ldots, f_k$ and since each of these vectors is a linear combination of $g_1, \ldots, g_k$, we have

$$h_k = \sum_{l=1}^{k} \alpha_{k,l} g_l \quad \text{for } 1 \leqslant k \, (\leqq) \, \kappa.$$

On the other hand, by the same token we have

$$g_l = \sum_{j=1}^{l} \beta_{l,j} h_j \quad \text{for } 1 \leqslant l \leqslant k.$$

We conclude

$$\alpha_{k,l} = \langle h_k, g_l \rangle / \|g_l\|^2 = 0 \quad \text{for } 1 \leqslant l < k,$$
$$h_k = \alpha_{k,k} g_k.$$

Because of $h_k \neq o$ we have $\alpha_{k,k} \neq 0$. $\square$

THEOREM 2. *A Hilbert space $\mathfrak{H}$ is separable iff it has a countable (finite or infinite) basis.*

PROOF. Suppose $\mathfrak{H}$ is separable and let $\{f_n\}_{n=1}^{\infty}$ be an everywhere dense sequence in $\mathfrak{H}$. We first select a subsequence $\mathfrak{F} = \{f_{n_k}\}_{k=1}^{\kappa} (\kappa < \infty \text{ or } \kappa = \infty)$ of linearly independent vectors as follows. Let n_1 be the smallest index n such that $f_n \neq o$. If $n_1 < \cdots < n_k$ have already been chosen in such a way that $f_{n_1}, \ldots, f_{n_k}$ are linearly independent and f_n is a linear combination of $f_1, \ldots, f_{n_k}$ for $1 \leqslant n \leqslant n_k$, then let n_{k+1} be the smallest index $n > n_k$ such that $f_{n_1}, \ldots, f_{n_k}, f_n$ are linearly independent (if such an index n exists at all; otherwise the process stops with $\kappa = k$). Obviously every vector $f_n (1 \leqslant n < \infty)$ is a finite linear combination of some of the vectors in $\mathfrak{F}$. Let $\mathfrak{E} = \{e_k\}_{k=1}^{\kappa}$ be an orthonormal family obtained by orthogonalizing $\mathfrak{F}$ (cf. corollary 1.1 c)). Then every vector $f_n (1 \leqslant n < \infty)$ is a finite linear combination of some of the vectors in $\mathfrak{E}$. Suppose for some $f \in \mathfrak{H}$ we have $f \perp e_k$ for all $e_k \in \mathfrak{E}$. We then also have $f \perp f_n$ for all $n \geqslant 1$. Choose a subsequence $\{f_n'\}_{n=1}^\infty$ of the everywhere dense sequence $\{f_n\}_{n=1}^\infty$ such that $f = \lim_{n \to \infty} f_n'$. Then we have $\langle f, f \rangle = = \lim_{n \to \infty} \langle f, f_n' \rangle = 0$ and therefore $f = o$. Thus $\mathfrak{E}$ is a basis.

Conversely, suppose $\mathfrak{E} = \{e_k\}_{k=1}^\kappa$ is a countable basis. Let $\mathfrak{M}$ be the linear manifold of all finite linear combinations of vectors in $\mathfrak{E}$. If $f \perp \mathfrak{M}$ for some vector $f \in \mathfrak{H}$, then $f \perp e_k$ for all $e_k \in \mathfrak{E}$ and therefore $f = o$. By § 7 corollary 7.2 we conclude that $\mathfrak{M}$ is everywhere dense in $\mathfrak{H}$. Let $\mathfrak{M}'$ be the subset of $\mathfrak{M}$ consisting of all finite linear combinations of vectors in $\mathfrak{E}$ with complex rational coefficients. The set $\mathfrak{M}'$ is countable (cf. the proof of § 4 theorem 5). We shall show that even $\mathfrak{M}'$ is everywhere dense in $\mathfrak{H}$. Indeed, for any given $f \in \mathfrak{H}$ and $\varepsilon > 0$, we may first choose an element $g = \sum_{k=1}^n \alpha_k e_k \in \mathfrak{M}$ such that $\|f - g\| < \frac{1}{2}\varepsilon$. Next for $1 \leqslant k \leqslant n$ we choose the complex rational number α_k' such that $|\alpha_k - \alpha_k'| < \varepsilon/2n$. For $g' = \sum_{k=1}^n \alpha_k' e_k \in \mathfrak{M}'$ we get

$$\|f - g'\| \leqslant \|f - g\| + \|g - g'\|$$
$$\leqslant \tfrac{1}{2}\varepsilon + \sum_{k=1}^n |\alpha_k - \alpha_k'| < \tfrac{1}{2}\varepsilon + n \cdot \frac{\varepsilon}{2n} = \varepsilon. \quad \square$$

Why should we be interested in a basis? One reason lies in the fact that, as soon as the vectors of $\mathfrak{H}$ have been expanded into "Fourier series" with respect to this basis (as we shall do presently), the at first sight somewhat chaotic bundle of vectors called separable infinite-dimensional Hilbert space develops a surprisingly clear structure. Its vectors turn out to be disguised square summable sequences and the whole Hilbert space simply turns out to be a disguised copy of ℓ_2 (cf. § 10 corollary 1.1).

THEOREM 3. *Let $\{e_k\}_{k=1}^\infty$ be an orthonormal family in $\mathfrak{H}$. The following statements are equivalent:*

a) $\{e_k\}_{k=1}^\infty$ *is a basis.*

b) $f \perp e_k$ *for all $k \geqslant 1$ implies $f = o$.*

c) $\mathfrak{H} = \vee \{e_k\}_{k=1}^\infty.$

d) $f = \sum_{k=1}^\infty \langle f, e_k \rangle e_k$ *for all $f \in \mathfrak{H}$* (FOURIER EXPANSION).

e) $\langle f, g \rangle = \sum_{k=1}^\infty \langle f, e_k \rangle \overline{\langle g, e_k \rangle}$ *for all $f \in \mathfrak{H}$ and all $g \in \mathfrak{H}$* (PARSEVAL'S IDENTITY).

f) $\|f\|^2 = \sum_{k=1}^\infty |\langle f, e_k \rangle|^2$ *for all $f \in \mathfrak{H}$* (PARSEVAL'S IDENTITY).

PROOF. a)$\Rightarrow$b) Clear by definition 1.

b) $\Rightarrow$c) If $\bigvee \{e_k\}_{k=1}^{\infty} \neq \mathfrak{H}$, then $\mathfrak{H}$ would contain a non-zero vector $f \perp \bigvee \{e_k\}_{k=1}^{\infty}$ (cf. § 7 corollary 7.2), in contradiction to b).

c) $\Rightarrow$d) The sets $\mathfrak{M}_k = \{\alpha e_k : \alpha \in \mathbf{C}\}$ $(1 \leqslant k < \infty)$ are mutually orthogonal subspaces (cf. also § 25 remark 1). By § 7 theorem 4 we have $\mathfrak{H} = \bigvee_{k=1}^{\infty} \mathfrak{M}_k = \sum_{k=1}^{\infty} \mathfrak{M}_k$ and every vector f admits a unique expansion in a convergent series

$$f = \sum_{k=1}^{\infty} \alpha_k(f) e_k$$

(cf. § 7 theorem 5). Taking the inner product with e_k on both sides we obtain

$$\langle f, e_k \rangle = \alpha_k(f).$$

d) $\Rightarrow$e) $\langle f, g \rangle = \langle \sum_{k=1}^{\infty} \langle f, e_k \rangle e_k, g \rangle = \sum_{k=1}^{\infty} \langle f, e_k \rangle \overline{\langle g, e_k \rangle}$ (cf. § 3 theorem 6 c)).

e) $\Rightarrow$f) $\langle f, f \rangle = \sum_{k=1}^{\infty} \langle f, e_k \rangle \overline{\langle f, e_k \rangle}$.

f) $\Rightarrow$a) If $f \perp e_k$ for all $k \geqslant 1$, then $\|f\|^2 = \sum_{k=1}^{\infty} |\langle f, e_k \rangle|^2 = 0$, therefore $f = o$. $\square$

Remark 1. In d) the vector f is expanded into a *Fourier series* with respect to the basis $\{e_k\}_{k=1}^{\infty}$. In this expansion $\langle f, e_k \rangle$ is called the *Fourier coefficient* corresponding to e_k. Although usually statement f) is called PARSEVAL's identity, this name is sometimes also assigned to statement e). In fact, this statement only appears to be more general than statement f), since both statements are equivalent by theorem 3.

Remark 2. All statements of theorem 3 remain true if everywhere the index k is allowed only to run from 1 to a natural number $\kappa < \infty$.

Example 5. Let $\mathfrak{H} = \mathfrak{L}_2(0, 1)$ and consider the basis exhibited in example 4. For every real-valued function $f \in \mathfrak{L}_2(0, 1)$ we define

$$\alpha_0 = \int_0^1 f(\xi) \, d\xi,$$

$$\alpha_k = \int_0^1 f(\xi) \cos 2\pi k \xi \, d\xi \quad \text{for } k = 1, 2, \ldots,$$

$$\beta_k = \int_0^1 f(\xi) \sin 2\pi k \xi \, d\xi \quad \text{for } k = 1, 2, \ldots.$$

In the notation of example 4 we have

$$\alpha_0 = \langle f, e_0 \rangle,$$

$$\alpha_k = \frac{1}{\sqrt{2}} \langle f, f_k \rangle \quad \text{for } k = 1, 2, \dots,$$

$$\beta_k = \frac{1}{\sqrt{2}} \langle f, g_k \rangle \quad \text{for } k = 1, 2, \dots.$$

PARSEVAL's identity then becomes

$$\int_0^1 f^2(\xi) \, d\xi = \alpha_0^2 + 2 \sum_{k=1}^{\infty} (\alpha_k^2 + \beta_k^2).$$

The following two corollaries of theorem 3 are two of many versions of what is sometimes called the RIESZ-FISCHER THEOREM.

COROLLARY 3.1. *If $\mathfrak{H}$ is a separable complex (or real) Hilbert space and if $\{e_k\}_{k=1}^{\infty}$ is a basis, then*

$$\mathfrak{H} = \{ \sum_{k=1}^{\infty} \alpha_k e_k : \alpha_k \in \mathbf{C} \ (or \ \mathbf{R}) \ for \ 1 \leqslant k < \infty, \ \sum_{k=1}^{\infty} |\alpha_k|^2 < \infty \}.$$

PROOF. If $\sum_{k=1}^{\infty} |\alpha_k|^2 < \infty$, then the series $\sum_{k=1}^{\infty} \alpha_k e_k$ converges by § 7 theorem 3 a). On the other hand, every vector in $\mathfrak{H}$ admits a Fourier expansion having the stated properties by theorem 3 d), f). □

COROLLARY 3.2. *Let $\{e_k\}_{k=1}^{\infty}$ be a basis of a separable complex (or real) Hilbert space $\mathfrak{H}$, and let $\{\alpha_k\}_{k=1}^{\infty}$ be a sequence in $\mathbf{C}$ (or $\mathbf{R}$) such that $\sum_{k=1}^{\infty} |\alpha_k|^2 < \infty$. Then there exists a unique vector $f \in \mathfrak{H}$ such that $\langle f, e_k \rangle = \alpha_k$ for all $k \geqslant 1$.*
PROOF. The vector $f = \sum_{k=1}^{\infty} \alpha_k e_k$ meets the requirements, and by theorem 3 d) it is the only one to do so. □

The cardinal number of a basis seems to be a reasonable candidate for what we would like to call the dimension of $\mathfrak{H}$. First, however, we have to make sure that this cardinal number does not depend on the choice of the basis.

THEOREM 4. *Any two bases of a separable Hilbert space $\mathfrak{H}$ have the same cardinal number.*
PROOF. By theorem 2 $\mathfrak{H}$ contains a countable basis $\mathfrak{E} = \{e_k\}_{k=1}^{\kappa}$ ($\kappa < \infty$ or $\kappa = \infty$).

a) $\kappa < \infty$. For $1 \leqslant k \leqslant \kappa$ let $\mathfrak{M}_k = \{\alpha e_k : \alpha \in \mathbf{C}\}$. Then $\mathfrak{H} = \sum_{k=1}^{\kappa} \mathfrak{M}_k$ and every subset of $\mathfrak{H}$ contains at most κ linearly independent vectors. For any other basis $\mathfrak{E}' = \{e'_k\}_{k=1}^{\kappa'}$ we must therefore have $\kappa' \leqslant \kappa$. By the same token we get $\kappa \leqslant \kappa'$ and therefore $\kappa = \kappa'$.

b) $\kappa = \infty$. By a) every other basis $\mathfrak{E}'$ must also be infinite. If $\mathfrak{E}'$ were uncountable, then (since the distance of any two orthonormal vectors is $\sqrt{2}$) $\mathfrak{H}$ could not be separable by the same reasoning as applied in § 4 when constructing a non-separable Hilbert space. Therefore $\mathfrak{E}'$ must be countably infinite. □

Without proof we mention that the statement of theorem 4 remains true even if the word "separable" is omitted (cf. [2] § 16 theorem 1).

DEFINITION 2. The cardinal number of a basis of a Hilbert space $\mathfrak{H}$ is called the *dimension* of $\mathfrak{H}$.

By the proof of theorem 4, part a), for the unitary space $\mathbf{C}^n$ the dimension in the sense of definition 2 is n, as one would certainly like it to be.

It follows from § 4 theorem 4, § 4 theorem 5, § 5 theorem 4, and § 5 theorem 5 that the spaces ℓ_2 and $\mathfrak{L}_2(\alpha, \beta)\,(-\infty \leqslant \alpha < \beta \leqslant +\infty)$ have dimension $\aleph_0$ (= the cardinal number of countably infinite sets). In fact, every separable infinite-dimensional Hilbert space has dimension $\aleph_0$. The fact that all of these spaces are essentially only different copies of one and the same will be established in § 10.

In § 4 there has been given an example of a non-separable Hilbert space. Applying definition 1 one sees easily that the family of functions $\{e_\nu\}_{\nu \in \mathbf{R}}$ exhibited there is a basis. As a consequence, the dimension of this particular Hilbert space is equal to the cardinal number of the continuum. Similarly, for the Hilbert space obtained in § 4 by replacing the index set $\mathbf{R}$ by any set of given cardinal number $\mathfrak{n}$ the dimension turns out to be exactly $\mathfrak{n}$.

Exercise. Let $\mathfrak{H}$ be a separable Hilbert space.

a) Every subspace $\mathfrak{M}$ is again separable as a Hilbert space in its own right. (Hint: if the sequence $\{f_n\}_{n=1}^{\infty}$ is everywhere dense in $\mathfrak{H}$, then consider the projection of this sequence upon $\mathfrak{M}$.)

b) Let $\{\mathfrak{M}_k\}_{k=1}^{\infty}$ be a sequence of mutually orthogonal subspaces such that $\mathfrak{H} = \sum_{k=1}^{\infty} \mathfrak{M}_k$ and let $\mathfrak{E}_k$ be a basis of $\mathfrak{M}_k$ for $1 \leqslant k < \infty$. Then $\bigcup_{k=1}^{\infty} \mathfrak{E}_k$ is a basis of $\mathfrak{H}$.

§ 9. Polynomial bases in $\mathfrak{L}_2$ spaces

Let α, β be real numbers such that $\alpha < \beta$. We already know that each of the spaces $\mathfrak{L}_2(\alpha, \beta)$, $\mathfrak{L}_2(\alpha, +\infty)$, $\mathfrak{L}_2(-\infty, \beta)$, and $\mathfrak{L}_2(-\infty, +\infty)$ is a Hilbert space of dimension $\aleph_0$ (cf. § 8). Therefore, for each of these spaces, every basis is countably infinite. For $\mathfrak{L}_2(0, 1)$ we already have exhibited a basis that is particularly nice in § 8 example 3, and a "real brother" of it in § 8 example 4. Applying the substitution $x = (y - \alpha)/(\beta - \alpha)$ and re-normalizing the functions we obtain analogous bases for $\mathfrak{L}_2(\alpha, \beta)$ (cf. § 2 example 4).

One reason why these bases seem "nice" is that they consist of functions which are well-known and analytically comparatively easy to handle. With this idea in mind it seems reasonable to look for bases that essentially consist of polynomials. To be more specific let us ask the following question: does there exist a basis $\{e_n\}_{n=1}^{\infty}$ of $\mathfrak{L}_2(\alpha, \beta)$ such that e_n is a polynomial of degree less than or equal to n?

The answer is yes and the idea underlying the construction of such a basis is simple: First, show that the functions $x^n (0 \leqslant n < \infty)$ are linearly independent in $\mathfrak{L}_2(\alpha, \beta)$. Secondly, show that they span $\mathfrak{L}_2(\alpha, \beta)$. Finally, apply Gram-Schmidt orthonormalization (cf. § 8 theorem 1, § 8 corollary 1.1 c)).

THEOREM 1. *The sequence $\{x^n\}_{n=0}^{\infty}$ is a linear independent family in $\mathfrak{L}_2(\alpha, \beta)$.*
PROOF. From $\|\sum_{k=0}^{n} \alpha_k x^k\| = 0$ we deduce

$$\sum_{k=0}^{n} \alpha_k \xi^k = 0 \quad \text{for almost all } \xi \in [\alpha, \beta]$$

(cf. appendix B6). Since a polynomial of degree at most n which does not vanish identically can have at most n roots we conclude $\alpha_k = 0$ for $0 \leqslant k \leqslant n$. $\square$

THEOREM 2. $\bigvee \{x^n\}_{n=0}^{\infty} = \mathfrak{L}_2(\alpha, \beta)$.
PROOF. By § 6 theorem 4 we have to show that finite complex linear combinations of the functions x^n, in other words complex polynomials in x, are everywhere dense in $\mathfrak{L}_2(\alpha, \beta)$. Given any $f \in \mathfrak{L}_2(\alpha, \beta)$ and any $\varepsilon > 0$ there is a function $g \in \mathfrak{C}[\alpha, \beta]$ such that $\|f - g\| < \frac{1}{2}\varepsilon$ (cf. § 5 corollary 5.1). By WEIERSTRASS' approximation theorem (cf. appendix B1) there is a polynomial h in x such that

$$|g(\xi) - h(\xi)| < \frac{\varepsilon}{2\sqrt{(\beta - \alpha)}} \quad \text{for all } \xi \in [\alpha, \beta].$$

We conclude

$$\|g - h\|^2 = \int_{\alpha}^{\beta} |g(\xi) - h(\xi)|^2 \, d\xi \leqslant \tfrac{1}{4}\varepsilon^2,$$

$$\|f - h\| \leqslant \|f - g\| + \|g - h\| < \tfrac{1}{2}\varepsilon + \tfrac{1}{2}\varepsilon = \varepsilon. \square$$

In place of now orthogonalizing step by step the sequence $\{x^n\}_{n=0}^\infty$ we simply present the result and then verify that it has been obtained in this way.

THEOREM 3. *For $0 \leqslant n < \infty$ let*

$$p_n = \frac{1}{\gamma_n} \frac{d^n \left[(x - \alpha)(x - \beta)\right]^n}{dx^n},$$

where $\gamma_n \in \mathbf{R}$ is chosen in such a way that $\|p_n\| = 1$. Then p_n is a polynomial of degree n for $0 \leqslant n < \infty$ and $\{p_n\}_{n=0}^\infty$ is a basis of $\mathfrak{L}_2(\alpha, \beta)$ obtained by applying Gram-Schmidt orthonormalization to the sequence $\{x^n\}_{n=0}^\infty$.

PROOF. Since $[(x-\alpha)(x-\beta)]^n$ is a polynomial of degree $2n$, p_n is a polynomial of degree n, in other words a linear combination of the linearly independent functions $x^0 \equiv 1, x, x^2, \ldots, x^n$. If we now show that the sequence $\{p_n\}_{n=0}^\infty$ is an orthogonal family, then by § 8 corollary 1.1 d) it coincides with the family obtained by orthogonalizing the sequence $\{x^n\}_{n=0}^\infty$. By § 8 corollary 1.1 b) it spans the same subspace as the sequence $\{x^n\}_{n=0}^\infty$, that is the entire Hilbert space $\mathfrak{L}_2(\alpha, \beta)$ by theorem 2. By § 8 theorem 3 a), c) the family $\{p_n\}_{n=0}^\infty$ is a basis of $\mathfrak{L}_2(\alpha, \beta)$.

It therefore remains to verify the equation

$$\langle x^m, \gamma_n p_n \rangle = \int_\alpha^\beta \xi^m \frac{d^n \left[(\xi - \alpha)(\xi - \beta)\right]^n}{d\xi^n} \, d\xi = 0 \quad \text{for } 0 \leqslant m < n < \infty.$$

Indeed, by repeated integration by parts we obtain (with trivial modifications for $m=0$)

$$\int_\alpha^\beta \xi^m \frac{d^n \left[(\xi - \alpha)(\xi - \beta)\right]^n}{d\xi^n} \, d\xi$$

$$= \xi^m \frac{d^{n-1} \left[(\xi - \alpha)(\xi - \beta)\right]^n}{d\xi^{n-1}} \bigg|_\alpha^\beta - m \int_\alpha^\beta \xi^{m-1} \frac{d^{n-1} \left[(\xi - \alpha)(\xi - \beta)\right]^n}{d\xi^{n-1}} \, d\xi$$

$$= - m \int_\alpha^\beta \xi^{m-1} \frac{d^{n-1} \left[(\xi - \alpha)(\xi - \beta)\right]^n}{d\xi^{n-1}} \, d\xi = \cdots$$

$$= (-1)^m m! \int_\alpha^\beta \frac{d^{n-m} \left[(\xi - \alpha)(\xi - \beta)\right]^n}{d\xi^{n-m}} \, d\xi$$

$$= (-1)^m m! \frac{d^{n-m-1} \left[(\xi - \alpha)(\xi - \beta)\right]^n}{d\xi^{n-m-1}} \bigg|_\alpha^\beta = 0. \quad \square$$

The polynomial p_n (or a convenient scalar multiple of it) is called the *Legendre polynomial* of degree n (of course it still depends on α and β; usually one considers the case $\alpha = -1$ and $\beta = +1$, cf. [12] V § 10).

If we want to apply an analogous reasoning to $\mathfrak{L}_2(-\infty, +\infty)$, then we have to replace the functions $x^n (0 \leqslant n < \infty)$ which do not belong to $\mathfrak{L}_2(-\infty, +\infty)$ by suitable functions which do belong to $\mathfrak{L}_2(-\infty, +\infty)$. A natural choice seems to be the family of functions $x^n e^{-\frac{1}{2}x^2} (0 \leqslant n < \infty)$ (the factor $\frac{1}{2}$ in the exponent is chosen for purely formal reasons).

THEOREM 4. *The sequence $\{x^n e^{-\frac{1}{2}x^2}\}_{n=0}^{\infty}$ is a linearly independent family in $\mathfrak{L}_2(-\infty, +\infty)$.*

PROOF. Because of $\lim_{|\xi| \to \infty} \xi^n e^{-\frac{1}{2}\xi^2} = 0$ the continuous function $x^n e^{-\frac{1}{2}x^2}$ is bounded on $]-\infty, +\infty[$ by some constant $\beta_n > 0$. We conclude

$$\int_{-\infty}^{+\infty} \xi^{2n} e^{-\xi^2} d\xi \leqslant \beta_{2n} \int_{-\infty}^{+\infty} e^{-\frac{1}{2}\xi^2} d\xi = \beta_{2n} \sqrt{2\pi} < \infty$$

and therefore $x^n e^{-\frac{1}{2}x^2} \in \mathfrak{L}_2(-\infty, +\infty)$ for $0 \leqslant n < \infty$. As in the proof of theorem 1, from $\|\sum_{k=0}^{n} \alpha_k x^k e^{-\frac{1}{2}x^2}\| = 0$ we deduce

$$\sum_{k=0}^{n} \alpha_k \xi^k e^{-\frac{1}{2}\xi^2} = 0 \quad \text{for almost all } \xi \in]-\infty, +\infty[$$

and therefore $\alpha_k = 0$ for $0 \leqslant k \leqslant n$. $\square$

THEOREM 5. $\bigvee \{x^n e^{-\frac{1}{2}x^2}\}_{n=0}^{\infty} = \mathfrak{L}_2(-\infty, +\infty)$.

PROOF. It suffices to show that for any function $f \in \mathfrak{L}_2(-\infty, +\infty)$ the relation $f \perp \bigvee \{x^n e^{-\frac{1}{2}x^2}\}_{n=0}^{\infty}$ implies $f = o$ (cf. § 7 corollary 7.2). Let therefore $f \in \mathfrak{L}_2(-\infty, +\infty)$ be given and suppose $\langle f, x^n e^{-\frac{1}{2}x^2}\rangle = 0$ for all $n \geqslant 0$.

For any given $\eta \in \mathbf{R}$ the function $e^{-i\eta x} e^{-\frac{1}{2}x^2}$ also belongs to $\mathfrak{L}_2(-\infty, +\infty)$ since

$$\int_{-\infty}^{+\infty} |e^{-i\xi\eta}|^2 e^{-\xi^2} d\xi = \int_{-\infty}^{+\infty} e^{-\xi^2} d\xi < \infty.$$

Consider the inner product

$$(1) \qquad \langle f, e^{-i\eta x} e^{-\frac{1}{2}x^2}\rangle = \int_{-\infty}^{+\infty} f(\xi) e^{i\xi\eta} e^{-\frac{1}{2}\xi^2} d\xi$$

$$= \int_{-\infty}^{+\infty} \left[\sum_{n=0}^{\infty} f(\xi) \frac{(i\xi\eta)^n}{n!} e^{-\frac{1}{2}\xi^2} \right] d\xi.$$

In order to check whether we may interchange integration and summation we compute (using LEBESGUE's theorem on monotone convergence, cf. appendix B7)

$$\sum_{n=0}^{\infty} \int_{-\infty}^{+\infty} \left| f(\xi) \frac{(i\xi\eta)^n}{n!} e^{-\frac{1}{2}\xi^2} \right| d\xi = \int_{-\infty}^{+\infty} |f(\xi)| e^{-\frac{1}{2}\xi^2} \sum_{n=0}^{\infty} \frac{|\xi\eta|^n}{n!} d\xi$$

$$= \int_{-\infty}^{+\infty} |f(\xi)| e^{|\xi\eta| - \frac{1}{2}\xi^2} d\xi .$$

Since the function $|f|$ as well as the function $e^{|\eta x| - \frac{1}{2}x^2} = e^{-\frac{1}{2}(|x| - |\eta|)^2} \cdot e^{\frac{1}{2}\eta^2}$ belongs to $\mathfrak{L}_2(-\infty, +\infty)$ the last integral is finite and coincides with $\langle |f|, e^{-\frac{1}{2}((|x| - |\eta|)^2 - \eta^2)} \rangle$ (cf. § 5 remark 1). Applying now B. LEVI's theorem to the integral (1) we get

$$(2) \qquad \langle f, e^{-i\eta x} e^{-\frac{1}{2}x^2} \rangle = \sum_{n=0}^{\infty} \frac{(i\eta)^n}{n!} \int_{-\infty}^{+\infty} f(\xi) \xi^n e^{-\frac{1}{2}\xi^2} d\xi$$

$$= \sum_{n=0}^{\infty} \frac{(i\eta)^n}{n!} \langle f, x^n e^{-\frac{1}{2}x^2} \rangle = 0 \quad \text{for all } \eta \in \mathbf{R}.$$

Next, for every natural number n we define a periodic Lebesgue measurable function g_n with period $2n$ on $\mathbf{R}$ by

$$g_n(\xi) = \operatorname{sgn} f(\xi) = \begin{cases} +1 & \text{if } f(\xi) > 0 \\ 0 & \text{if } f(\xi) = 0 \\ -1 & \text{if } f(\xi) < 0 \end{cases} \text{ for } \xi \in [-n, +n[.$$

Note that

$$\lim_{n \to \infty} g_n(\xi) = \operatorname{sgn} f(\xi) \quad \text{for all } \xi \in \mathbf{R},$$

$$|f(\xi) g_n(\xi) e^{-\frac{1}{2}\xi^2}| \leqslant |f(\xi)| e^{-\frac{1}{2}\xi^2} \quad \text{for all } \xi \in \mathbf{R} \text{ and all } n \geqslant 1.$$

From $|f| \in \mathfrak{L}_2(-\infty, +\infty)$ and $e^{-\frac{1}{2}x^2} \in \mathfrak{L}_2(-\infty, +\infty)$ we conclude $|f| e^{-\frac{1}{2}x^2} \in \mathfrak{L}_1(-\infty, +\infty)$ (cf. § 5 remark 1). Applying LEBESGUE's theorem on dominated convergence we get

$$\lim_{n \to \infty} \int_{-\infty}^{+\infty} f(\xi) g_n(\xi) e^{-\frac{1}{2}\xi^2} d\xi = \int_{-\infty}^{+\infty} |f(\xi)| e^{-\frac{1}{2}\xi^2} d\xi .$$

For every given $\varepsilon > 0$ it is therefore possible to find a natural number n such that

$$(3) \qquad |\int_{-\infty}^{+\infty} |f(\xi)| \, e^{-\frac{1}{2}\xi^2} \, d\xi - \int_{-\infty}^{+\infty} f(\xi) g_n(\xi) \, e^{-\frac{1}{2}\xi^2} \, d\xi| < \varepsilon.$$

Restricting this function g_n to the interval $[0, 2n[$ and considering it as an element of $\mathfrak{L}_2(0, 2n)$ we choose a trigonometric polynomial $q = \sum_{k=-m}^{m} \alpha_k \, e^{i\pi kx/n}$ such that

$$\int_{0}^{2n} |g_n(\xi) - q(\xi)|^2 \, d\xi < \varepsilon^2 / 2 \sum_{k=0}^{\infty} e^{-4k^2 n^2}.$$

This is possible since $\{e^{i\pi kx/n}/\sqrt{2n}\}_{k=-\infty}^{+\infty}$ is a basis of $\mathfrak{L}_2(0, 2n)$ (cf. § 2 example 4, § 8 example 3). Taking into account that g_n and q are periodic functions with period $2n$ we obtain

$$(4) \qquad \|(g_n - q) \, e^{-\frac{1}{2}x^2}\|^2 = \int_{-\infty}^{+\infty} |g_n(\xi) - q(\xi)|^2 \, e^{-\xi^2} \, d\xi$$

$$= \sum_{k=-\infty}^{\infty} \int_{2nk}^{2n(k+1)} |g_n(\xi) - q(\xi)|^2 \, e^{-\xi^2} \, d\xi$$

$$\leqslant 2 \sum_{k=0}^{\infty} e^{-4n^2 k^2} \int_{0}^{2n} |g_n(\xi) - q(\xi)|^2 \, d\xi < \varepsilon^2.$$

Furthermore from (2) we conclude

$$(5) \qquad \int_{-\infty}^{+\infty} f(\xi) q(\xi) \, e^{-\frac{1}{2}\xi^2} \, d\xi = \sum_{k=-m}^{m} \alpha_k \langle f, e^{-i\pi kx/n} \, e^{-\frac{1}{2}x^2} \rangle = 0.$$

Combining (3), (4) and (5) we finally obtain

$$\int_{-\infty}^{+\infty} |f(\xi)| \, e^{-\frac{1}{2}\xi^2} \, d\xi \leqslant |\int_{-\infty}^{+\infty} |f(\xi)| \, e^{-\frac{1}{2}\xi^2} - \int_{-\infty}^{+\infty} f(\xi) g_n(\xi) \, e^{-\frac{1}{2}\xi^2} \, d\xi| +$$

$$+ |\int_{-\infty}^{+\infty} f(\xi) [g_n(\xi) - q(\xi)] \, e^{-\frac{1}{2}\xi^2} \, d\xi|$$

$$\leqslant \varepsilon + \|f\| \cdot \|(g_n - q) \, e^{-\frac{1}{2}x^2}\| \leqslant \varepsilon(1 + \|f\|).$$

Since $\varepsilon > 0$ was arbitrary we conclude

$$\int_{-\infty}^{+\infty} |f(\xi)| \, e^{-\frac{1}{2}\xi^2} \, d\xi = 0,$$

$$f(\xi) \, e^{-\frac{1}{2}\xi^2} = 0 \quad \text{for almost all } \xi \in \mathbf{R},$$
$$f(\xi) = 0 \quad \text{for almost all } \xi \in \mathbf{R},$$
$$f = o.$$

This proves the assertion. □

THEOREM 6. *For $0 \leqslant n < \infty$ let*

$$h_n = e^{\frac{1}{2}x^2} \frac{d^n \, e^{-x^2}}{dx^n}.$$

Then

$$H_n = e^{x^2} \frac{d^n \, e^{-x^2}}{dx^n} = h_n \, e^{\frac{1}{2}x^2}$$

is a polynomial of degree n in x and the sequence $\{h_n / \|h_n\|\}_{n=0}^{\infty} = \{H_n e^{-\frac{1}{2}x^2} / \|h_n\|\}_{n=0}^{\infty}$ is a basis of $\mathfrak{L}_2(-\infty, +\infty)$ obtained by applying Gram-Schmidt orthonormalization to the sequence $\{x^n \, e^{-\frac{1}{2}x^2}\}_{n=0}^{\infty}$.

PROOF. The first assertion follows from a repeated application of the formula

$$\frac{dx^m \, e^{-x^2}}{dx} = (-2x^{m+1} + mx^{m-1}) \, e^{-x^2} \quad \text{for all } m \geqslant 1.$$

In order to show the second assertion it again suffices to prove

$$\int_{-\infty}^{+\infty} \xi^m \frac{d^n \, e^{-\xi^2}}{d\xi^n} \, d\xi = 0 \quad \text{for } 0 \leqslant m < n < \infty.$$

Indeed by repeated integration by parts we obtain

$$\int_{-\infty}^{+\infty} \xi^m \frac{d^n \, e^{-\xi^2}}{d\xi^n} \, d\xi = \xi^m \frac{d^{n-1} \, e^{-\xi^2}}{d\xi^{n-1}} \Big|_{-\infty}^{+\infty} - m \int_{-\infty}^{+\infty} \xi^{m-1} \frac{d^{n-1} \, e^{-\xi^2}}{d\xi^{n-1}} \, d\xi$$

$$= -m \int_{-\infty}^{+\infty} \xi^{m-1} \frac{d^{n-1} \, e^{-\xi^2}}{d\xi^{n-1}} \, d\xi = \cdots$$

$$= (-1)^m m! \int\limits_{-\infty}^{+\infty} \frac{d^{n-m} e^{-\xi^2}}{d\xi^{n-m}} \, d\xi$$

$$= (-1)^m m! \left. \frac{d^{n-m-1} e^{-\xi^2}}{d\xi^{n-m-1}} \right|_{-\infty}^{+\infty} = 0.$$

It follows that the sequence $\{h_n/\|h_n\|\}_{n=0}^\infty$ is orthonormal and obtained by Gram-Schmidt orthonormalization from the sequence $\{x^n e^{-\frac{1}{2}x^2}\}_{n=0}^\infty$ (cf. § 8 corollary 1.1 c), d)). Since this sequence spans $\mathfrak{L}_2(-\infty, +\infty)$ (cf. theorem 5), the sequence $\{h_n/\|h_n\|\}_{n=0}^\infty$ is a basis of $\mathfrak{L}_2(-\infty, +\infty)$ (cf. § 8 corollary 1.1 b), § 8 theorem 3 a), c)). $\square$

The function $h_n (0 \leqslant n < \infty)$ is called *Hermite function*, the polynomial $H_n = h_n \, e^{\frac{1}{2}x^2}$ is called the *Hermite polynomial* of degree n. Sometimes this name is also reserved for the polynomial

$$2^{-\frac{1}{2}n} H_n(-y/\sqrt{2}) = (-1)^n \, e^{\frac{1}{2}y^2} \frac{d^n \, e^{-\frac{1}{2}y^2}}{dy^n} \qquad (0 \leqslant n < \infty)$$

which is obtained from the polynomial $H_n(x)$ essentially by the substitution $x = -y/\sqrt{2}$. In this formula the coefficients are designed in such a way that the factor -2 turning up as soon as the exponential function is differentiated is immediately cancelled out against another factor $-\frac{1}{2}$. The function $2^{-\frac{1}{2}n} H_n(-y/\sqrt{2})$ is therefore a polynomial of degree n in y with integral coefficients and highest term y^n (cf. [12] VI § 3).

Let us now consider the Hilbert space $\mathfrak{L}_2(0, \infty)$ (the corresponding results for $\mathfrak{L}_2(\alpha, +\infty)$ or $\mathfrak{L}_2(-\infty, \beta)$ are obtained by applying the substitution $x = y - \alpha$ or $x = \beta - y$ respectively).

THEOREM 7. *The sequence* $\{x^n e^{-\frac{1}{2}x}\}_{n=0}^\infty$ *is a linearly independent family in* $\mathfrak{L}_2(0, \infty)$.
PROOF. Replace $e^{-\frac{1}{2}x^2}$ in the proof of theorem 4 by $e^{-\frac{1}{2}x}$ and $-\infty$ by 0. $\square$

The following theorem is stated without proof. We shall only apply it to assert that the orthonormal family in $\mathfrak{L}_2(0, \infty)$ exhibited in theorem 9 below is actually a basis, a fact which we shall not need in the sequel. In fact the proof of theorem 5 can be modified so as to apply to the functions $x^n e^{-\frac{1}{2}x}$ (in place of $x^n e^{-\frac{1}{2}x^2}$). However, in order to secure $\langle f, e^{i\eta x - \frac{1}{2}x} \rangle = 0$ for all $\eta \in \mathbf{R}$ one has first to establish and then to use the fact that this inner product is an analytic function in η for $|\text{Re } \eta| < \frac{1}{2}$. The interested reader is referred to [10] 7.3.2, 7.3.3 or to the exercise below.

THEOREM 8. $\bigvee \{x^n e^{-\frac{1}{2}x}\}_{n=0}^{\infty} = \mathfrak{L}_2(0, \infty)$.

THEOREM 9. For $0 \leqslant n < \infty$ let

$$l_n = e^{\frac{1}{2}x} \frac{d^n x^n e^{-x}}{dx^n}.$$

Then

$$L_n = e^x \frac{d^n x^n e^{-x}}{dx^n} = l_n e^{\frac{1}{2}x}$$

is a polynomial of degree n in x and the sequence $\{l_n/\|l_n\|\}_{n=0}^{\infty} = \{L_n e^{-\frac{1}{2}x}/\|l_n\|\}_{n=0}^{\infty}$ is a basis of $\mathfrak{L}_2(0, \infty)$ obtained by applying Gram-Schmidt orthonormalization to the sequence $\{x^n e^{-\frac{1}{2}x}\}_{n=0}^{\infty}$.

PROOF. The first assertion follows from a repeated application of the formula

$$\frac{dx^m e^{-x}}{dx} = (-x^m + mx^{m-1}) e^{-x} \quad \text{for all } m \geqslant 1.$$

In order to show the second assertion it again suffices to prove

$$\int_0^{\infty} \xi^m \frac{d^n \xi^n e^{-\xi}}{d\xi^n} d\xi = 0 \quad \text{for } 0 \leqslant m < n < \infty.$$

By repeated integration by parts we again obtain

$$\int_0^{\infty} \xi^m \frac{d^n \xi^n e^{-\xi}}{d\xi^n} d\xi = \xi^m \frac{d^{n-1} \xi^n e^{-\xi}}{d\xi^{n-1}} \Big|_0^{\infty} - m \int_0^{\infty} \xi^{m-1} \frac{d^{n-1} \xi^n e^{-\xi}}{d\xi^{n-1}} d\xi$$

$$= - m \int_0^{\infty} \xi^{m-1} \frac{d^{n-1} \xi^n e^{-\xi}}{d\xi^{n-1}} d\xi = \cdots$$

$$= (-1)^m m! \int_0^{\infty} \frac{d^{n-m} \xi^n e^{-\xi}}{d\xi^{n-m}} d\xi$$

$$= (-1)^m m! \frac{d^{n-m-1} \xi^n e^{-\xi}}{d\xi^{n-m-1}} \Big|_0^{\infty} = 0.$$

It follows that $\{l_n/\|l_n\|\}_{n=0}^{\infty}$ is an orthonormal sequence and in fact obtained from the sequence $\{x^n e^{-\frac{1}{2}x}\}_{n=0}^{\infty}$ by Gram-Schmidt orthonormalization (cf. § 8 corollary 1.1 c), d)). Since the last sequence spans $\mathfrak{L}_2(0, \infty)$ (cf. theorem 8), so does the sequence $\{l_n/\|l_n\|\}_{n=0}^{\infty}$ which is therefore a basis of $\mathfrak{L}_2(0, \infty)$ (cf. § 8 corollary 1.1 b), § 8 theorem 3 a), c)). □

The function $l_n (0 \leqslant n < \infty)$ is called *Laguerre function*, the polynomial $L_n = l_n\, e^{\pm x}$ (or a convenient scalar multiple) is called the *Laguerre polynomial* of degree n (cf. [12] VI § 1).

Finally we consider the Hilbert space $\mathfrak{L}_2(X)$ where X is the "square" $X =]\alpha, \beta[\times]\alpha, \beta[(-\infty \leqslant \alpha < \beta \leqslant +\infty)$ in the Euclidean plane $\mathbf{R}^2$ (cf. § 5). Any polynomial basis of $\mathfrak{L}_2(\alpha, \beta)$ gives rise to a polynomial basis of $\mathfrak{L}_2(X)$ by the following general theorem.

THEOREM 10. *Let* $X =]\alpha, \beta[\times]\alpha, \beta[\subset \mathbf{R}^2$ *(α and β may be replaced by $-\infty$ and $+\infty$ respectively) and let* $\{e_k\}_{k=1}^\infty$ *be a basis of* $\mathfrak{L}_2(\alpha, \beta)$. *If the functions* $e_{k,l} \in \mathfrak{L}_2(X)$ *are defined on* X *by*

$$e_{k,l}(\xi, \eta) = e_k(\xi)\, \overline{e_l(\eta)},$$

then the family $\mathfrak{E} = \{e_{k,l}\}_{1 \leqslant k, l < \infty}$ *is a basis of* $\mathfrak{L}_2(X)$.

PROOF. The fact that $\mathfrak{E}$ is an orthonormal family in $\mathfrak{L}_2(X)$ has already been shown in § 5. For every function $f \subset \mathfrak{L}_2(X)$ and every $\eta \in]\alpha, \beta[$ define the function $f^\eta = f(x, \eta)$ on $]\alpha, \beta[$ by

$$f^\eta(\xi) = f(\xi, \eta)$$

whenever $f(\xi, \eta)$ is defined for $(\xi, \eta) \in X$. For almost all $\eta \in]\alpha, \beta[$ the function f^η is Lebesgue measurable on X (cf. appendix B5) and by FUBINI's theorem we have

$$\infty > \|f\|^2 = \int_X |f(\xi, \eta)|^2\, d(\xi, \eta) = \int_\alpha^\beta \int_\alpha^\beta |f(\xi, \eta)|^2\, d\xi\, d\eta$$

$$= \int_\alpha^\beta \int_\alpha^\beta |f^\eta(\xi)|^2\, d\xi\, d\eta$$

(cf. appendix B8). We conclude $f^\eta \in \mathfrak{L}_2(\alpha, \beta)$ for almost all $\eta \in]\alpha, \beta[$. Denoting the norm of f^η in $\mathfrak{L}_2(\alpha, \beta)$ by $\|f^\eta\|$ we get

(6)
$$\|f\|^2 = \int_\alpha^\beta \|f^\eta\|^2\, d\eta = \int_\alpha^\beta \sum_{k=1}^\infty |\langle f^\eta, e_k \rangle|^2\, d\eta$$

$$= \int_\alpha^\beta \sum_{k=1}^\infty |\int_\alpha^\beta f(\xi, \eta)\, \overline{e_k(\xi)}\, d\xi|^2\, d\eta$$

$$= \sum_{k=1}^\infty \int_\alpha^\beta |\int_\alpha^\beta f(\xi, \eta)\, \overline{e_k(\xi)}\, d\xi|^2\, d\eta.$$

We conclude that the function g_k defined on $]\alpha, \beta[$ by

$$g_k(\eta) = \int_\alpha^\beta f(\xi, \eta)\,\overline{e_k(\xi)}\,d\xi$$

(which is Lebesgue measurable by FUBINI's theorem) belongs to $\mathfrak{L}_2(\alpha, \beta)$. Another application of PARSEVAL's identity, this time to the basis $\{e_l\}_{l=1}^\infty$ of $\mathfrak{L}_2(\alpha, \beta)$, gives

$$\int_\alpha^\beta |\int_\alpha^\beta f(\xi, \eta)\,\overline{e_k(\xi)}\,d\xi|^2\,d\eta = \int_\alpha^\beta |g_k(\eta)|^2\,d\eta$$

$$= \sum_{l=1}^\infty |\langle g_k, \overline{e_l}\rangle|^2 = \sum_{l=1}^\infty |\int_\alpha^\beta g_k(\eta)\,e_l(\eta)\,d\eta|^2$$

$$= \sum_{l=1}^\infty |\int_\alpha^\beta\int_\alpha^\beta f(\xi, \eta)\,\overline{e_k(\xi)}\,e_l(\eta)\,d\xi\,d\eta|^2$$

$$= \sum_{l=1}^\infty |\int_X f(\xi, \eta)\,\overline{e_{k,l}(\xi, \eta)}\,d(\xi, \eta)|^2$$

$$= \sum_{l=1}^\infty |\langle f, e_{k,l}\rangle|^2.$$

Combining this with (6) we get

$$\|f\|^2 = \sum_{k=1}^\infty \sum_{l=1}^\infty |\langle f, e_{k,l}\rangle|^2.$$

By § 8 theorem 3 a), f) this equation implies that the family $\{e_{k,l}\}_{1 \leqslant k, l < \infty}$ is a basis of $\mathfrak{L}_2(X)$. $\square$

Remark 1. For later reference note the following fact, also demonstrated in the foregoing proof: if $f \in \mathfrak{L}_2(X)$ is given, then $f^\eta = f(x, \eta) \in \mathfrak{L}_2(\alpha, \beta)$ for almost all $\eta \in]\alpha, \beta[$. Similarly we have $f_\xi = f(\xi, y) \in \mathfrak{L}_2(\alpha, \beta)$ for almost all $\xi \in]\alpha, \beta[$.

COROLLARY 10.1. *Let X be as in theorem 10. Then $\mathfrak{L}_2(X)$ is separable.*
PROOF. Apply theorem 10 and § 8 theorem 2. $\square$

COROLLARY 10.2. *Let X be as in theorem 10 and let $\mathfrak{A}$ be the set of all functions* *f defined on X by*

$$f\,(\xi, \eta) = g\,(\xi)\, h\,(\eta)$$

where $g \in \mathfrak{L}_2(\alpha, \beta)$ and $h \in \mathfrak{L}_2(\alpha, \beta)$. Then $\mathfrak{A} \subset \mathfrak{L}_2(X)$ and $\bigvee \mathfrak{A} = \mathfrak{L}_2(X)$.
PROOF. Every function $f \in \mathfrak{A}$ is Lebesgue measurable on X (in fact g and h may be considered as measurable functions on X; cf. appendix B5) and square integrable on X by FUBINI's theorem:

$$\int_X |f\,(\xi, \eta)|^2 \, \mathrm{d}(\xi, \eta) = \int_\alpha^\beta \int_\alpha^\beta |g\,(\xi)\, h\,(\eta)|^2 \, \mathrm{d}\xi \, \mathrm{d}\eta$$

$$= \int_\alpha^\beta |g\,(\xi)|^2 \, \mathrm{d}\xi \cdot \int_\alpha^\beta |h\,(\eta)|^2 \, \mathrm{d}\eta < \infty \, .$$

The set $\mathfrak{A}$ contains the basis $\{e_{k,l}\}_{1 \leqslant k, l < \infty}$ exhibited in theorem 10 which in turn spans $\mathfrak{L}_2(X)$ by § 8 theorem 3 c). $\square$

Exercise. The following chain of assertions leads up to a proof of theorem 8 which essentially deviates from the proof of theorem 5. It uses only that part of theorem 9 which asserts that the sequence $\{l_n/\|l_n\|\}_{n=0}^{\infty}$ is obtained from the sequence $\{x^n \, \mathrm{e}^{-\frac{1}{2}x}\}_{n=0}^{\infty}$ by Gram-Schmidt orthonormalization and that $L_n = l_n \, \mathrm{e}^{\frac{1}{2}x}$ is a polynomial of degree n in x.

a) $L_n = \sum_{k=0}^n \binom{n}{k} \dfrac{n!}{(n-k)!} (-x)^{n-k} .$

$\left(\text{Hint:} \dfrac{\mathrm{d}^n fg}{\mathrm{d}x^n} = \sum_{k=0}^n \binom{n}{k} \dfrac{\mathrm{d}^k f}{\mathrm{d}x^k} \dfrac{\mathrm{d}^{n-k} g}{\mathrm{d}x^{n-k}} . \right)$

b) $\langle x^n \, \mathrm{e}^{-\frac{1}{2}x}, l_n \rangle = (-1)^n (n!)^2 ,$

       $\langle x^{n+1} \, \mathrm{e}^{-\frac{1}{2}x}, l_n \rangle = (-1)^n [(n+1)!]^2 ,$

       $\|l_n\| = n! .$

(Hint: integrate by parts as in the proof of theorem 9.)

c) $L_{n+1} = (2n+1-x)L_n - n^2 L_{n-1} \quad \text{for } n \geqslant 0, \, L_{-1} \equiv 0 .$

(Hint: show that $L_{n+1} + xL_n$ is a polynomial of degree at most n and that therefore $l_{n+1} + xl_n$ is a linear combination of $l_0, \ldots, l_n$; the coefficients may be determined using the orthonormality of the sequence $\{l_n/n!\}_{n=0}^{\infty}$.)

d) $\dfrac{1}{1-\zeta}\, e^{-\xi\zeta/(1-\zeta)} = \displaystyle\sum_{n=0}^{\infty} \dfrac{1}{n!}\, L_n(\xi)\,\zeta^n$ for all $\xi \in \mathbf{R}$ and

for all $\zeta \in \mathbf{C}$ satisfying $|\zeta|^{1} < 1$.

(Hint: for every fixed $\xi \in \mathbf{R}$ the function

$$F(z) = \frac{1}{1-z}\, e^{-\xi z/(1-z)} = \frac{1}{1-z}\, e^{-\xi/(1-z)}\, e^{\xi}$$

is analytic in the open unit disk of the complex plane and satisfies

$$(1 - 2z + z^2)\,F'(z) = (1 - \xi - z)\,F(z);$$

this implies that the coefficients $a_n(\xi)$ of the Taylor expansion

$$F(z) = \sum_{k=0}^{\infty} \frac{1}{n!}\, a_n(\xi)\, z^n$$

satisfy the recursion formula c) while $a_0(x) = L_0(x)$.)

e) For every integer $m \geqslant 0$ we have

$$e^{-mx-\frac{1}{2}x} = \sum_{n=0}^{\infty} \frac{m^n}{(m+1)^{n+1}}\, \frac{l_n(x)}{n!}$$

where the series converges in $\mathfrak{L}_2(0, \infty)$ and pointwise on $\mathbf{R}$. (Hint: for $\zeta = m/(m+1)$ the Taylor series d) converges pointwise for all $\xi \in \mathbf{R}$ and the series

$$\sum_{n=0}^{\infty} \left(\frac{m}{m+1}\right)^n \frac{l_n}{n!}$$

converges in $\mathfrak{L}_2(0, \infty)$; then apply a reasoning similar as in § 5 remark 3 and § 5 exercise.)

f) If $f \in \mathfrak{L}_2(0, \infty)$ and $\varepsilon > 0$ are given, then there exists a polynomial in e^{-x}

$$p(e^{-x}) = \sum_{m=0}^{n} \alpha_m\, e^{-mx}$$

such that

$$\| f - e^{-\frac{1}{2}x} p(e^{-x}) \| < \varepsilon .$$

(Hint: if $f_1(x) = e^{\frac{1}{2}x} f(x)$, then $f_1(\log 1/y) \in \mathfrak{L}_2(0, 1)$.)

g) $\bigvee \{l_n\}_{n=0}^{\infty} = \mathfrak{L}_2(0, \infty)$.

(Hint: combine e) and f).)

§ 10. Isomorphisms

We still have to carry out a promise: to show that every separable infinite-dimensional Hilbert space is just a disguised copy of ℓ_2. To some extent this has already been done in § 8 corollary 3.1. The following theorem now clarifies what is meant by "disguised copy".

THEOREM 1. *If two separable Hilbert spaces $\mathfrak{H}$ and $\mathfrak{H}'$ have the same (finite or infinite) dimension, then there exists a one-to-one mapping $U:f \to Uf$ of $\mathfrak{H}$ onto $\mathfrak{H}'$ such that for all $f \in \mathfrak{H}$, all $g \in \mathfrak{H}$, and all $\lambda \in \mathbf{C}$ the following equations hold*:

a) $U(f+g)=Uf+Ug$;
b) $U(\lambda f)=\lambda Uf$;
c) $\langle Uf, Ug \rangle = \langle f, g \rangle$, *in particular* $\| Uf \| = \| f \|$.

PROOF. Let $\{e_k\}_{k=1}^{\kappa}$ and $\{e'_k\}_{k=1}^{\kappa}$ be bases of $\mathfrak{H}$ and $\mathfrak{H}'$ respectively ($\kappa < \infty$ or $\kappa = \infty$; cf. § 8 theorem 2). For every $f \in \mathfrak{H}$ we define

$$f' = Uf = \sum_{k=1}^{\kappa} \langle f, e_k \rangle e'_k.$$

This definition makes sense since, by § 8 theorem 3 f) we have

$$\sum_{k=1}^{\kappa} |\langle f, e_k \rangle|^2 = \| f \|^2 < \infty$$

and by § 7 theorem 3 a) the series $\sum_{k=1}^{\kappa} \langle f, e_k \rangle e'_k$ converges in $\mathfrak{H}'$.

Moreover, the mapping $f \to f'$ maps e_k into e'_k. For the mapping U defined in this manner we have

$$U(f+g) = \sum_{k=1}^{\kappa} \langle f+g, e_k \rangle e'_k = \sum_{k=1}^{\kappa} \langle f, e_k \rangle e'_k + \sum_{k=1}^{\kappa} \langle g, e_k \rangle e'_k = Uf + Ug,$$

$$U(\lambda f) = \sum_{k=1}^{\kappa} \langle \lambda , e_k \rangle e'_k = \lambda \sum_{k=1}^{\kappa} \langle f, e_k \rangle e'_k = \lambda Uf,$$

$$\langle Uf, Ug \rangle = \langle \sum_{k=1}^{\kappa} \langle f, e_k \rangle e'_k, \sum_{l=1}^{\kappa} \langle g, e_l \rangle e'_l \rangle = \sum_{k=1}^{\kappa} \langle f, e_k \rangle \overline{\langle g, e_k \rangle} = \langle f, g \rangle$$

(cf. § 8 theorem 3 e)). The mapping U is one-to-one since $Uf = Ug$ implies $\| f-g \| = \| U(f-g) \| = \| Uf-Ug \| = 0$, that is $f=g$. The mapping U maps $\mathfrak{H}$ onto $\mathfrak{H}'$ since for any vector $f' = \sum_{k=1}^{\kappa} \alpha_k e'_k \in \mathfrak{H}'$ we have $\sum_{k=1}^{\kappa} |\alpha_k|^2 < \infty$, thus the series $\sum_{k=1}^{\kappa} \alpha_k e_k$ converges to a vector $f \in \mathfrak{H}$ and we have $\alpha_k = \langle f, e_k \rangle$ for $1 \leqslant k_{(} \leqslant_{)} \kappa$, i.e. $f' = Uf$. $\square$

Intuitively, theorem 1 says that (by means of the mapping U) we may identify the elements of $\mathfrak{H}$ and $\mathfrak{H}'$ in such a way that each of these Hilbert spaces appears (algebraically and topologically) as a perfect copy of the other.

Example 1. Let $\{e_k\}_{k=1}^{\infty}$ be the standard basis of $\mathfrak{H}=\ell_2$ exhibited in § 8 example 2. Let $\{e_k'\}_{k=1}^{\infty}$ be the basis of $\mathfrak{H}' = \mathfrak{L}_2(0, 1)$ exhibited in §8 example 3, arranged into a "one-sided" sequence by defining for instance

$$e_{2l+1}' = e^{2\pi i l x} \qquad \text{for } l = 0, 1, 2, \ldots$$
$$e_{2l}' = e^{2\pi i (-l)x} \qquad \text{for } l = 1, 2, \ldots .$$

Furthermore, let $\{p_k\}_{k=0}^{\infty}$ be the sequence of Legendre polynomials for $\mathfrak{L}_2(0, 1)$ defined in § 9 theorem 3 (put $\alpha=0$, $\beta=1$). For

$$a = \{\alpha_k\}_{k=1}^{\infty} = \sum_{k=1}^{\infty} \alpha_k e_k \in \mathfrak{H} = \ell_2$$

let

$$(1) \quad a' = U_1 a = \sum_{k=1}^{\infty} \alpha_k e_k' = \sum_{l=0}^{\infty} \alpha_{2l+1} e^{2\pi i l x} + \sum_{l=1}^{\infty} \alpha_{2l} e^{-2\pi i l x} \in \mathfrak{H}' = \mathfrak{L}_2(0, 1),$$

$$(2) \quad a'' = U_2 a = \sum_{k=1}^{\infty} \alpha_k p_{k-1} \in \mathfrak{H}'' = \mathfrak{L}_2(0, 1).$$

The mappings U_1 and U_2 are two different one-to-one mappings of ℓ_2 onto $\mathfrak{L}_2(0, 1)$ with properties as stated in theorem 1. Starting from the Fourier expansion (1) of an element $a' \in \mathfrak{L}_2(0, 1)$ with respect to the basis $\{e_k'\}_{k=1}^{\infty}$ and defining $U_3 a' = a''$ we obtain a one-to-one mapping U_3 of $\mathfrak{L}_2(0, 1)$ onto itself with the same properties. Note that in (1) and (2) the Fourier series defining the elements a' and a'' of $\mathfrak{L}_2(0, 1)$ converge in norm (cf. § 3 definition 5) but need not converge pointwise to the functions a' and a'' respectively in $\mathfrak{L}_2(0, 1)$.

DEFINITION 1. A (not necessarily continuous) mapping $A: f \to Af$ of a linear manifold $\mathfrak{D}$ in a Hilbert space $\mathfrak{H}$ into a Hilbert space $\mathfrak{H}'$ is called *linear* if for all $f \in \mathfrak{D}$, all $g \in \mathfrak{D}$, and all $\lambda \in \mathbf{C}$

$$A(f + g) = Af + Ag,$$
$$A(\lambda f) = \lambda Af.$$

A mapping A of $\mathfrak{D}$ into $\mathfrak{H}'$ is called *isometric* (or an *isometry*) if for all $f \in \mathfrak{D}$ and all $g \in \mathfrak{D}$

$$\langle Af, Ag \rangle = \langle f, g \rangle.$$

A linear and isometric mapping of $\mathfrak{H}$ onto $\mathfrak{H}'$ is called an *isomorphism* of $\mathfrak{H}$ onto $\mathfrak{H}'$. An isomorphism of $\mathfrak{H}$ onto itself is called an *automorphism*.

Remark 1. A linear isometry A, in particular an isomorphism, is automatically one-to-one. Indeed, the reasoning already used in the proof of theorem 1 shows that $Af = Ag$ implies

$$\|f - g\| = \|A(f - g)\| = \|Af - Ag\| = 0.$$

Remark 2. If there exists an isomorphism A of $\mathfrak{H}$ onto $\mathfrak{H}'$ then there also exists an isomorphism B of $\mathfrak{H}'$ onto $\mathfrak{H}$: in the pairing $f \to f' = Af$ of elements of $\mathfrak{H}$ and $\mathfrak{H}'$ simply reverse the arrow and define the mapping B by $f = Bf' \leftarrow f'$. This can also be expressed by writing

$$BAf = f \quad \text{for all } f \in \mathfrak{H},$$
$$ABf' = f' \quad \text{for all } f' \in \mathfrak{H}'.$$

Since the pairs (f, f') of vectors of $\mathfrak{H}$ and $\mathfrak{H}'$ have remained the same, the mapping B is also one-to-one, onto, linear, and isometric. This possibility of reversing an isomorphism which is onto justifies the following definition.

DEFINITION 2. Two Hilbert spaces $\mathfrak{H}$ and $\mathfrak{H}'$ are called *isomorphic* if there exists an isomorphism of one space onto the other.

COROLLARY 1.1. *Every separable infinite-dimensional Hilbert space is isomorphic with ℓ_2.*
PROOF. Combine theorem 1 with the fact that every separable infinite-dimensional Hilbert space has dimension $\aleph_0$ (cf. § 8). $\square$

COROLLARY 1.2. *Two separable Hilbert spaces are isomorphic iff they have the same dimension.*
PROOF. The "if" assertion coincides with theorem 1. On the other hand, let U be an isomorphism of the Hilbert space $\mathfrak{H}$ onto the Hilbert space $\mathfrak{H}'$ and let $\{e_k\}_{k=1}^{\kappa}$ be a basis of $\mathfrak{H}$. Then $\{Ue_k\}_{k=1}^{\kappa}$ is an orthonormal family in $\mathfrak{H}'$ and therefore the dimension of $\mathfrak{H}'$ is at least κ (in fact we are mixing up $\aleph_0$ with the symbol ∞ at this place, but this will hardly cause an irreparable confusion). Turning the reasoning the other way we conclude that $\mathfrak{H}$ and $\mathfrak{H}'$ must have the same dimension. $\square$

Without proof we mention that the statements of theorem 1 and corollary 1.2 remain true if the word "separable" is omitted (cf. [2] § 16 theorem 3).

Exercise. Let A be an isometric mapping of a Hilbert space $\mathfrak{H}$ into a Hilbert space $\mathfrak{H}'$. Then the following statements hold:

a) A is linear.

b) The set $\mathfrak{M}=\{Af:f\in\mathfrak{H}\}$ is a subspace of $\mathfrak{H}'$.

c) $\mathfrak{H}$ is isomorphic with $\mathfrak{M}$.

d) If $\mathfrak{H}=\mathfrak{H}'$ is finite dimensional, then $\mathfrak{M}=\mathfrak{H}$.

e) If $\mathfrak{H}=\mathfrak{H}'$ is infinite-dimensional and separable, then the isometric mapping A can be chosen such that $\mathfrak{M}\neq\mathfrak{H}$. (Hint: cf. § 6 example 2.)

CHAPTER III

BOUNDED LINEAR OPERATORS

§ 11. Bounded linear mappings

The object of our further study will be mappings of some subset $\mathfrak{D}$ of a Hilbert space $\mathfrak{H} \neq \mathfrak{O}$ into some other Hilbert space $\mathfrak{H}'$ (we have just encountered examples when talking about isomorphisms in § 10; there we have had $\mathfrak{D} = \mathfrak{H}$). In this context we get again confronted with two familiar aspects of such a mapping: the algebraic one and the topological one.

The algebraic aspect is well taken care of if the mapping A in question is linear (cf. § 10 definition 1), i.e. if $A(f+g) = Af + Ag$ and $A(\lambda f) = \lambda Af$ for all $\lambda \in \mathbf{C}$ and for all $f \in \mathfrak{D}$ and all $g \in \mathfrak{D}$. In order that this requirement should make sense it is necessary that the subset $\mathfrak{D} \subset \mathfrak{H}$ on which A is defined (also called the *domain* of A) be a linear manifold. In particular this is the case if A is defined on a subspace or even on the entire Hilbert space $\mathfrak{H}$ (note that the linearity of A already implies $Ao = o' \in \mathfrak{H}'$). Henceforth, the domain of a linear mapping will therefore always be assumed to be a linear manifold in a given Hilbert space $\mathfrak{H}$.

In order to be able to take care of the topological aspect we define for linear mappings two concepts which will turn out to be closely related with each other: boundedness and continuity.

DEFINITION 1. A linear mapping A with domain $\mathfrak{D} \subset \mathfrak{H}$ into a Hilbert space $\mathfrak{H}'$ is called *bounded* if there exists a real number $\gamma \geqslant 0$ such that

$$(1) \qquad \|Af\| \leqslant \gamma \|f\| \quad \text{for all } f \in \mathfrak{D}.$$

If A is bounded and $\mathfrak{D} \neq \mathfrak{O}$, then the non-negative number

$$(2) \qquad \|A\| = \sup_{f \in \mathfrak{D}, f \neq o} \frac{\|Af\|}{\|f\|}$$

is called the *norm* of A.

Remark 1. For a bounded linear mapping A on a domain $\mathfrak{D} \neq \mathfrak{O}$ the chain of inequalities

$$\|A\| = \sup_{f \in \mathfrak{D}, f \neq o} \frac{\|Af\|}{\|f\|} = \sup_{f \in \mathfrak{D}, f \neq o} \left\|A\left(\frac{f}{\|f\|}\right)\right\| \leqslant \sup_{f \in \mathfrak{D}, \|f\|=1} \|Af\|$$

$$\leqslant \sup_{f \in \mathfrak{D}, \|f\| \leqslant 1} \|Af\| \leqslant \sup_{f \in \mathfrak{D}, 0 < \|f\| \leqslant 1} \frac{\|Af\|}{\|f\|} \leqslant \|A\|$$

implies that all inequalities are in fact equalities. Thus the norm of A could as well have been defined by any of the expressions in the chain.

Remark 2. If A is bounded, then (2) implies

$$\|Af\| \leqslant \|A\| \, \|f\| \quad \text{for all } f \in \mathfrak{D},$$

and $\|A\|$ is the infimum of all real numbers $\gamma \geqslant 0$ satisfying (1). Furthermore $\|A\| = 0$ iff $Af = o' \in \mathfrak{H}'$ for all $f \in \mathfrak{D}$, and A is *unbounded* (= not bounded) iff

$$\sup_{f \in \mathfrak{D}, f \neq o} \frac{\|Af\|}{\|f\|} = \sup_{f \in \mathfrak{D}, \|f\|=1} \|Af\| = \infty.$$

Example 1. Let U be an isomorphism of $\mathfrak{H}$ onto some Hilbert space $\mathfrak{H}'$ (cf. § 10 definition 1). Because $\|Uf\| = \|f\|$ for all $f \in \mathfrak{H}$ the linear mapping U is bounded and $\|U\| = 1$.

Example 2. Let $\mathfrak{M}$ be a subspace of $\mathfrak{H}$ and let the mapping $P_{\mathfrak{M}} : f \to P_{\mathfrak{M}} f$ of $\mathfrak{H}$ into itself be defined by projection upon $\mathfrak{M}$ (cf. § 7 theorem 6, § 7 theorem 8'):

(3) $\qquad f = P_{\mathfrak{M}} f + P_{\mathfrak{M}^\perp} f, \ P_{\mathfrak{M}} f \in \mathfrak{M}, P_{\mathfrak{M}^\perp} f \in \mathfrak{M}^\perp \quad \text{for all } f \in \mathfrak{H}.$

From the equations

$$f + g = (P_{\mathfrak{M}} f + P_{\mathfrak{M}} g) + (P_{\mathfrak{M}^\perp} f + P_{\mathfrak{M}^\perp} g), \ P_{\mathfrak{M}} f + P_{\mathfrak{M}} g \in \mathfrak{M},$$
$$P_{\mathfrak{M}^\perp} f + P_{\mathfrak{M}^\perp} g \in \mathfrak{M}^\perp,$$
$$\lambda f = \lambda P_{\mathfrak{M}} f + \lambda P_{\mathfrak{M}^\perp} f, \ \lambda P_{\mathfrak{M}} f \in \mathfrak{M}, \lambda P_{\mathfrak{M}^\perp} f \in \mathfrak{M}^\perp,$$

and from the uniqueness of this decomposition we conclude that $P_{\mathfrak{M}}$ is linear. From (3) we deduce

$$\|f\|^2 = \|P_{\mathfrak{M}} f\|^2 + \|P_{\mathfrak{M}^\perp} f\|^2 \geqslant \|P_{\mathfrak{M}} f\|^2$$

(cf. § 2 theorem 8). Therefore $P_{\mathfrak{M}}$ is bounded and $\|P_{\mathfrak{M}}\| \leqslant 1$. For $\mathfrak{M} = \mathfrak{O} = \{o\}$ we obviously have $\|P_{\mathfrak{M}}\| = 0$ (cf. remark 2). Otherwise, for any non-zero vector $f \in \mathfrak{M}$ we have $P_{\mathfrak{M}} f = f$ and therefore $\|P_{\mathfrak{M}}\| = 1$.

Example 3. Let $\mathfrak{D}$ be the subset of "finite" sequences in $\mathfrak{H}=\ell_2$ (cf. § 2 example 2) and let B be the mapping of $\mathfrak{D}$ into $\mathfrak{H}$ defined by

$$(4) \qquad\qquad B\{\alpha_k\}_{k=1}^\infty = \{k\alpha_k\}_{k=1}^\infty.$$

Obviously B is linear. If $\{e_k\}_{k=1}^\infty$ is the standard basis of ℓ_2 exhibited in § 8 example 2, then we have $\|Be_k\|=k=k\|e_k\|$ and $\lim_{k\to\infty}\|Be_k\|=\infty$. Thus B is unbounded (cf. remark 2). Note that (4) still makes sense for some sequences $\{\alpha_k\}_{k=1}^\infty$ in ℓ_2 outside of $\mathfrak{D}$ but certainly not anymore if $\sum_{k=1}^\infty k^2|\alpha_k|^2=\infty$, as for instance the case with $\{1/k\}_{k=1}^\infty\in\ell_2$.

DEFINITION 2. A linear mapping A with domain $\mathfrak{D}\subset\mathfrak{H}$ into some Hilbert space $\mathfrak{H}'$ is called *continuous* if for every $f_0\in\mathfrak{D}$ and for every $\varepsilon>0$ there exists a $\delta>0$ such that

$$\|Af_0 - Af\| < \varepsilon \quad \text{for all } f\in\mathfrak{D} \text{ satisfying } \|f_0 - f\| < \delta.$$

Roughly speaking a continuous linear mapping maps vectors that are close together again into vectors that are close together.

THEOREM 1. *A linear mapping A with domain $\mathfrak{D}\subset\mathfrak{H}$ into a Hilbert space $\mathfrak{H}'$ is continuous iff for every sequence $\{f_n\}_{n=1}^\infty\subset\mathfrak{D}$ converging to some limit $f_0\in\mathfrak{D}$ also*

$$Af_0 = A(\lim_{n\to\infty} f_n) = \lim_{n\to\infty} Af_n.$$

PROOF. Suppose A is continuous and $\varepsilon>0$ is given. Choose δ as in definition 2 and suppose $\|f_0-f_n\|<\delta$ for all $n\geqslant n(\delta)$. Then $\|Af_0-Af_n\|<\varepsilon$ for all $n\geqslant n(\delta)$. We conclude $Af_0=\lim_{n\to\infty} Af_n$. On the other hand, suppose A is not continuous. Then for some $f_0\in\mathfrak{D}$ and some $\varepsilon>0$ by definition 2 there must be a sequence $\{f_n\}_{n=1}^\infty\subset\mathfrak{D}$ such that $\|f_0-f_n\|<1/n$ but $\|Af_0-Af_n\|\geqslant\varepsilon$ for all $n\geqslant 1$. Thus we have $f_0=\lim_{n\to\infty} f_n$ but the sequence $\{Af_n\}_{n=1}^\infty$ does not converge to Af_0. $\square$

If one carefully looks through definition 2, theorem 1, and its proof, one notices that so far the linearity of A has not been used anywhere. Thus, we could have defined continuity for mappings which are not linear and still have saved theorem 1 for these mappings. The point here is that we shall only be interested in linear mappings and that for these boundedness and continuity imply each other.

THEOREM 2. *A linear mapping A with domain $\mathfrak{D}\subset\mathfrak{H}$ into a Hilbert space $\mathfrak{H}'$ is bounded iff it is continuous.*

PROOF. a) Suppose A is bounded and $\|A\|>0$ (for $\|A\|=0$ the assertion is trivial). Let $f_0 \in \mathfrak{D}$ and $\varepsilon > 0$ be given. Then for all $f \in \mathfrak{D}$ satisfying $\|f_0 - f\| < \varepsilon / \|A\|$ we have

$$\|Af_0 - Af\| = \|A(f - f_0)\| \leqslant \|A\| \, \|f_0 - f\| < \varepsilon$$

(cf. remark 2).

b) Suppose A is unbounded. Then by remark 2 there exists a sequence of unit vectors $\{f_n\}_{n=1}^{\infty} \subset \mathfrak{D}$ such that $\|Af_n\| \geqslant n^2$ for all $n \geqslant 1$. Thus the sequence $\{f_n/n\}_{n=1}^{\infty}$ converges to o, but $\|A(f_n/n)\| \geqslant n$ for all $n \geqslant 1$. By theorem 1 the mapping A cannot be continuous. $\square$

Remark 3. Note that in part a) of the proof of theorem 2 the statement "a continuous linear mapping A maps vectors which are close together again into vectors which are close together" is made precise in a quantitative way:

$$\|Af_0 - Af\| \leqslant \|A\| \, \|f_0 - f\| \quad \text{for} \quad _0 \in \mathfrak{D} \text{ and } f \in \mathfrak{D}.$$

It is somewhat annoying that as soon as one thinks of a linear mapping one also has to think of its particular domain. Example 3 indicates that this nuisance may have something to do with unboundedness of the mapping in question. This indication will be substantiated in chapter IV. In the next theorem we shall exhibit a consequence which boundedness has for the domain of a linear mapping. For the rest of this section (and from § 12 corollary 1.1 on for the rest of this chapter) we shall stop bothering about the domain by simply talking only about mappings which are defined on the entire Hilbert space $\mathfrak{H}$.

THEOREM 3. *A bounded linear mapping A with domain $\mathfrak{D} \subset \mathfrak{H}$ into a Hilbert space $\mathfrak{H}'$ can be extended to a bounded linear mapping $\bar{A}$ with domain $\bar{\mathfrak{D}} \subset \mathfrak{H}$ into $\mathfrak{H}'$. These requirements determine the mapping $\bar{A}$ uniquely and imply $\|\bar{A}\| = \|A\|$.*

PROOF. Note that $\bar{\mathfrak{D}}$ is a subspace (cf. § 6 theorem 1). First we make sure that such an extension $\bar{A}$, if it exists at all, must be unique. Indeed, if $f \in \bar{\mathfrak{D}}$ is given, then there exists a sequence $\{f_n\}_{n=1}^{\infty} \subset \mathfrak{D}$ such that $f = \lim_{n \to \infty} f_n$ (cf. § 3 theorem 5). Since $\bar{A}$ is continuous this implies

$$(5) \qquad \qquad \bar{A}f = \lim_{n \to \infty} \bar{A}f_n = \lim_{n \to \infty} Af_n$$

(cf. theorem 1; for $f \in \mathfrak{D}$ this is automatically true since then we have $\bar{A}f = Af$). Thus there is only one possibility of defining $\bar{A}f$ if this is possible at all: $\bar{A}f$ must be the limit of the sequence $\{Af_n\}_{n=1}^{\infty}$.

Let us therefore try to use (5) in order to define $\bar{A}$ on $\bar{\mathfrak{D}}$. Can we be sure at all that, for every sequence $\{f_n\}_{n=1}^{\infty} \subset \mathfrak{D}$ which converges to f, the sequence $\{Af_n\}_{n=1}^{\infty}$ also converges? Indeed, because of

$$\|Af_m - Af_n\| \leqslant \|A\| \, \|f_m - f_n\|$$

(cf. remark 3) convergence of $\{f_n\}_{n=1}^{\infty}$ implies

$$\lim_{m,\,n \to \infty} \|Af_m - Af_n\| = 0$$

and (since $\mathfrak{H}'$ is a Hilbert space) this in turn implies the existence of a vector $f' = \lim_{n \to \infty} Af_n \in \mathfrak{H}'$ (of course for $f \in \mathfrak{D}$ we get $f' = Af$). Can we be sure that this vector f' depends only on f and not on the particular choice of the sequence $\{f_n\}_{n=1}^{\infty}$? Indeed, if also the sequence $\{g_n\}_{n=1}^{\infty}$ converges to f and if $g' = \lim_{n \to \infty} Ag_n \in \mathfrak{H}$, then we have $\{f_n - g_n\}_{n=1}^{\infty} \subset \mathfrak{D}$, $\lim_{n \to \infty}(f_n - g_n) = o$, and

$$
\begin{aligned}
f' - g' &= \lim_{n \to \infty} Af_n - \lim_{n \to \infty} Ag_n \\
&= \lim_{n \to \infty} (Af_n - Ag_n) && \text{(by § 3 theorem 6 a))} \\
&= \lim_{n \to \infty} A(f_n - g_n) && \text{(by linearity of } A) \\
&= A \lim_{n \to \infty} (f_n - g_n) && \text{(by continuity of } A) \\
&= Ao = o' \in \mathfrak{H}',
\end{aligned}
$$

and therefore $f' = g'$.

Let us verify that the mapping $\bar{A}$ defined on $\mathfrak{D}$ by (5) is linear and bounded: for $f = \lim_{n \to \infty} f_n \in \bar{\mathfrak{D}}$, $g = \lim_{n \to \infty} g_n \in \bar{\mathfrak{D}}$ ($f_n \in \mathfrak{D}$ and $g_n \in \mathfrak{D}$ for $n \geqslant 1$), and for $\lambda \in \mathbf{C}$ we have

$$\bar{A}(f + g) = \lim_{n \to \infty} A(f_n + g_n) = \lim_{n \to \infty} Af_n + \lim_{n \to \infty} Ag_n = \bar{A}f + \bar{A}g,$$

$$\bar{A}(\lambda f) = \lim_{n \to \infty} A(\lambda f_n) = \lambda \lim_{n \to \infty} Af_n = \lambda \bar{A}f,$$

$$\|\bar{A}f\| = \lim_{n \to \infty} \|Af_n\| \leqslant \lim_{n \to \infty} \|A\| \, \|f_n\| = \|A\| \lim_{n \to \infty} \|f_n\| = \|A\| \, \|f\|$$

(cf. § 3 theorem 6). The last inequality implies $\|\bar{A}\| \leqslant \|A\| < \infty$. The reverse inequality follows from

$$\|A\| = \sup_{f \in \mathfrak{D},\, \|f\|=1} \|Af\| \leqslant \sup_{f \in \bar{\mathfrak{D}},\, \|f\|=1} \|\bar{A}f\| = \|\bar{A}\|. \quad \square$$

Depending on the choice of $\mathfrak{H}'$ there are two types of linear mappings which will be of particular interest for us. Starting from the next section we shall be concerned with linear mappings of $\mathfrak{H}$ into itself. On the other hand, taking for $\mathfrak{H}'$ the one-dimensional unitary space $\mathbf{C}$, i.e. the field of complex numbers with absolute value as norm, we may consider linear mappings of $\mathfrak{H}$ into $\mathbf{C}$.

DEFINITION 3. A linear mapping Φ of $\mathfrak{H}$ into $\mathbf{C}$ is called a *linear functional* on $\mathfrak{H}$.

Example 4. Let h be any given vector in $\mathfrak{H}$. For all $f \in \mathfrak{H}$ define $\Phi f = \langle f, h \rangle \in \mathbf{C}$. Then Φ is a bounded linear functional on $\mathfrak{H}$ and $\|\Phi\| = \|h\|$. Indeed we have

$$\Phi(f + g) = \langle f + g, h \rangle = \langle f, h \rangle + \langle g, h \rangle = \Phi f + \Phi g ,$$
$$\Phi(\lambda f) = \langle \lambda f, h \rangle = \lambda \langle f, h \rangle = \lambda \Phi f ,$$
$$\|\Phi f\| = |\Phi f| = |\langle f, h \rangle| \leqslant \|f\|\,\|h\| .$$

The last inequality implies $\|\Phi\| \leqslant \|h\|$. Because of $\|\Phi h\| = |\langle h, h \rangle| = \|h\|^2$ we even get $\|\Phi\| = \|h\|$.

It is a remarkable fact that not only every vector $h \in \mathfrak{H}$ defines a bounded linear functional on $\mathfrak{H}$ as demonstrated in example 4, but in fact every bounded linear functional on $\mathfrak{H}$ is produced in this way.

THEOREM 4 (RIESZ REPRESENTATION THEOREM). *If Φ is a bounded linear functional on $\mathfrak{H}$, then there exists a unique vector $h \in \mathfrak{H}$ such that $\Phi f = \langle f, h \rangle$ for all $f \in \mathfrak{H}$ and, as a consequence, $\|\Phi\| = \|h\|$.*
PROOF. The set $\mathfrak{M} = \{g \in \mathfrak{H} : \Phi g = 0\}$ is obviously a linear manifold. In fact, $\mathfrak{M}$ is even a subspace since for every converging sequence $\{g_n\}_{n=1}^{\infty} \subset \mathfrak{M}$ we have $\Phi(\lim_{n \to \infty} g_n) = \lim_{n \to \infty} \Phi g_n = 0$. Therefore the vector $\lim_{n \to \infty} g_n$ must belong to $\mathfrak{M}$ and $\mathfrak{M}$ is closed (cf. § 3 theorem 5).

If $\mathfrak{M} = \mathfrak{H}$ (i.e. if $\Phi f = 0$ for all $f \in \mathfrak{H}$), then the vector $h = o$ satisfies the requirements of the theorem. If $\mathfrak{M} \neq \mathfrak{H}$, then by § 7 corollary 7.1 there exists a unit vector $e \perp \mathfrak{M}$. As a consequence we have $\Phi e \neq 0$ and $h = \overline{\Phi e} \cdot e \perp \mathfrak{M}$. For every vector $f \in \mathfrak{H}$ we have

$$f = \left(f - \frac{\Phi f}{|\Phi e|^2} h \right) + \frac{\Phi f}{|\Phi e|^2} h .$$

Because of

$$\Phi\left(f - \frac{\Phi f}{|\Phi e|^2} h \right) = \Phi f - \frac{\Phi f}{|\Phi e|^2} \overline{\Phi e} \cdot \Phi e = 0$$

we have

$$f_1 = f - \frac{\Phi f}{|\Phi e|^2} h = P_{\mathfrak{M}} f \in \mathfrak{M},$$

$$f_2 = \frac{\Phi f}{|\Phi e|^2} h = P_{\mathfrak{M}^\perp} f \in \mathfrak{M}^\perp \quad \text{(cf. § 7 theorem 8')},$$

$$\langle f, h \rangle = \langle f_1, h \rangle + \langle f_2, h \rangle = \frac{\Phi f}{|\Phi e|^2} \|h\|^2 = \frac{\Phi f}{|\Phi e|^2} |\Phi e|^2 = \Phi f.$$

The equation $\|\Phi\| = \|h\|$ has already been shown in example 4.

If h' is any vector in $\mathfrak{H}$ satisfying $\Phi f = \langle f, h' \rangle$ for all $f \in \mathfrak{H}$, then we conclude

$$\langle f, h - h' \rangle = \langle f, h \rangle - \langle f, h' \rangle = \Phi f - \Phi f = 0 \quad \text{for all } f \in \mathfrak{H},$$

in particular

$$\|h - h'\|^2 = \langle h - h', h - h' \rangle = 0,$$
$$h = h'. \quad \square$$

The reader is warned that theorem 4 is not the only one which is called "Riesz representation theorem" (cf. [2] § 17 theorem 3, [6] (12.36), (20.48)).

Exercise. Let $\mathfrak{H}$ be a Hilbert space and let $\{e_k\}_{k=1}^n$ be a basis of the n-dimensional unitary space $\mathbf{C}^n$.

a) Let $h_1, \ldots, h_n$ be any n vectors in $\mathfrak{H}$. The mapping A defined for all $f \in \mathfrak{H}$ by

$$A = \sum_{k=1}^n \langle f, h_k \rangle e_k$$

is bounded and linear.

b) If A is a bounded linear mapping of $\mathfrak{H}$ into $\mathbf{C}^n$, then there exist n vectors $h_1, \ldots, h_n \in \mathfrak{H}$ such that

$$Af = \sum_{k=1}^n \langle f, h_k \rangle e_k \quad \text{for all } f \in \mathfrak{H}$$

and the vectors $h_1, \ldots, h_n$ are uniquely defined by these requirements.

§ 12. Linear operators

DEFINITION 1. A mapping with a domain $\mathfrak{D} \subset \mathfrak{H}$ into $\mathfrak{H}$ is called an *operator in* $\mathfrak{H}$. A mapping of $\mathfrak{H}$ into $\mathfrak{H}$ (i.e. $\mathfrak{D} = \mathfrak{H}$) is called an *operator on* $\mathfrak{H}$.

Actually we shall only be concerned with *linear* operators (cf. § 10 definition 1), but it will happen several times, that we shall meet an operator before knowing that it is linear, let aside bounded. For this reason we have not included these concepts in the definition of an operator, but the reader is warned again that this terminology is not universal.

Examples of bounded linear operators are provided by automorphisms (§ 10 definition 1) and by the projection upon a subspace (§ 11 example 2). A special automorphism is the *identity operator* I defined by $If=f$ for all $f\in\mathfrak{H}$. A special projection operator is the *zero operator* O, the projection upon the zero subspace, defined by $Of=o$ for all $f\in\mathfrak{H}$. An example of an unbounded linear operator has been given in § 11 example 3. Further examples of bounded linear operators will follow at the end of this section. Let us just point out here that in case of a finite-dimensional unitary space there is a one-to-one correspondence between linear operators on $\mathfrak{H}$ and matrices.

Example 1. Let A be a linear operator on $\mathfrak{H}=\mathbb{C}^n$ and let $\{e_k\}_{k=1}^n$ be a basis of $\mathfrak{H}$. The action of A on the vectors $e_k(1\leqslant k\leqslant n)$ already defines the action of A on all of $\mathfrak{H}$. Indeed, suppose

$$Ae_l = \sum_{k=1}^{n} \alpha_{k,l} e_k \quad \text{for } 1 \leqslant l \leqslant n$$

(in fact we have $\alpha_{k,l}=\langle Ae_l, e_k\rangle$). Then for every vector $\sum_{l=1}^{n} \beta_l e_l \in \mathfrak{H}$ we obtain

$$A\left(\sum_{l=1}^{n} \beta_l e_l\right) = \sum_{l=1}^{n} \beta_l Ae_l = \sum_{l=1}^{n} \beta_l \left(\sum_{k=1}^{n} \alpha_{k,l} e_k\right)$$

(1)

$$= \sum_{k=1}^{n} \left(\sum_{l=1}^{n} \alpha_{k,l} \beta_l\right) e_k.$$

The action of A on $\mathfrak{H}$ may therefore be described by the matrix

(2)
$$A' = \begin{pmatrix} \alpha_{1,1} \cdots \alpha_{1,n} \\ \vdots \quad \vdots \\ \alpha_{n,1} \cdots \alpha_{n,n} \end{pmatrix}.$$

Conversely, if a matrix A' is given as in (2), then the corresponding operator A defined on $\mathfrak{H}$ by (1) is linear and bounded. In fact, from

$$\left\|A\left(\sum_{l=1}^{n} \beta_l e_l\right)\right\|^2 = \sum_{k=1}^{n} \left|\sum_{l=1}^{n} \alpha_{k,l} \beta_l\right|^2 \leqslant \sum_{k=1}^{n} \left(\sum_{l=1}^{n} |\alpha_{k,l}|^2 \cdot \sum_{l=1}^{n} |\beta_l|^2\right)$$

$$= \left(\sum_{k=1}^{n} \sum_{l=1}^{n} |\alpha_{k,l}|^2\right)\left\|\sum_{l=1}^{n} \beta_l e_l\right\|^2$$

we conclude

(3) $$\|A\| \leqslant (\sum_{k=1}^{n} \sum_{l=1}^{n} |\alpha_{k,l}|^2)^{\frac{1}{2}}.$$

It is left to the reader to check the linearity of A.

Recalling that matrices may be added and multiplied by scalars and by each other we shall not be surprised that the same operations may be performed with linear operators. Only, in the general case we ought to be careful about the domains of these operators.

THEOREM 1. *Let A and B be linear operators in $\mathfrak{H}$ with domains $\mathfrak{D}_A$ and $\mathfrak{D}_B$ respectively. Then the mappings $A+B$, λA, and AB, defined by*

$$\begin{aligned}
(A+B)f &= Af + Bf &&\text{for all } f \in \mathfrak{D}_A \cap \mathfrak{D}_B, \\
(\lambda A)f &= \lambda(Af) &&\text{for all } f \in \mathfrak{D}_A, \\
(AB)f &= A(Bf) &&\text{for all } f \in \mathfrak{D}_B \text{ for which } Bf \in \mathfrak{D}_A,
\end{aligned}$$

are linear operators in $\mathfrak{H}$ with domains as indicated.

PROOF. Linearity of all these operators is checked easily (cf. also § 6 theorem 2). The only non-trivial point to be checked is the assertion that the set $\mathfrak{D}_{AB}$ of all $f \in \mathfrak{D}_B$ for which $Bf \in \mathfrak{D}_A$ is a linear manifold. Indeed, if f_1 and f_2 have this property, then $\alpha_1 f_1 + \alpha_2 f_2 \in \mathfrak{D}_B$ and $B(\alpha_1 f_1 + \alpha_2 f_2) = \alpha_1 Bf_1 + \alpha_2 Bf_2 \in \mathfrak{D}_A$, therefore also $\alpha_1 f_1 + \alpha_2 f_2 \in \mathfrak{D}_{AB}$. $\square$

DEFINITION 2. Two operators A and B in $\mathfrak{H}$ with domains $\mathfrak{D}_A$ and $\mathfrak{D}_B$ respectively are *equal* if $\mathfrak{D}_A = \mathfrak{D}_B$ and $Af = Bf$ for all $f \in \mathfrak{D}_A = \mathfrak{D}_B$.

It is obvious that the "equality" of operators just defined meets all requirements which one usually wants to impose on an "equality"-relation: it is *reflexive* $(A=A)$, *symmetric* $(A=B$ implies $B=A)$ and *transitive* $(A=B$ and $B=C$ together imply $A=C)$. Already from example 1 it follows that in general we have $AB \neq BA$, in fact even $\mathfrak{D}_{AB}$ may differ from $\mathfrak{D}_{BA}$. The trouble with domains, however, stops if we only consider linear operators on $\mathfrak{H}$. The situation becomes especially pleasant if all linear operators considered are bounded.

COROLLARY 1.1. *Under the operations as defined in theorem 1 the set of all linear operators on $\mathfrak{H}$ is a linear space satisfying the following further identities:*

a) $(AB)C = A(BC)$;
b) $A(B+C) = AB + AC$,
   $(A+B)C = AC + BC$;

  c) $(\alpha A) B = A(\alpha B) = \alpha(AB)$;

  d) $IA = AI = A$;

  e) $OA = AO = O$.

PROOF. All assertions follow from a straightforward application of the definitions as given in theorem 1 and of definition 2. As an example we prove the first of the assertions b): for all $f \in \mathfrak{H}$ we have

$$A(B + C)f = A(Bf + Cf) = A(Bf) + A(Cf) = (AB)f + (AC)f . \ \square$$

THEOREM 2. *Let A and B be bounded linear operators on* $\mathfrak{H}$. *Then so are* $A + B$, λA, AB, *and the following inequalities hold:*

$$(4) \hspace{4em} \|A + B\| \leqslant \|A\| + \|B\| ,$$
$$\|\lambda A\| = |\lambda| \ \|A\| ,$$
$$(5) \hspace{4em} \|AB\| \leqslant \|A\| \ \|B\| .$$

PROOF. $\|A + B\| = \sup_{\|f\| = 1} \|Af + Bf\| \leqslant \sup_{\|f\| = 1} (\|Af\| + \|Bf\|)$
$\hspace{3em} \leqslant \sup_{\|f\| = 1} \|Af\| + \sup_{\|f\| = 1} \|Bf\| = \|A\| + \|B\|,$
$\hspace{1em} \|\lambda A\| = \sup_{\|f\| = 1} \|\lambda Af\| = |\lambda| \sup_{\|f\| = 1} \|Af\| = |\lambda| \ \|A\|,$
$\hspace{1em} \|AB\| = \sup_{\|f\| = 1} \|ABf\| \leqslant \sup_{\|f\| = 1} \|A\| \ \|Bf\|$
$\hspace{3em} = \|A\| \sup_{\|f\| = 1} \|Bf\| = \|A\| \ \|B\|. \ \square$

Theorem 2 together with corollary 1.1 shows that the set $\mathscr{B}$ of all bounded linear operators on a Hilbert space $\mathfrak{H}$ is a normed linear space. In fact, even more than that is true. Algebraically the presence of a multiplication in $\mathscr{B}$ provides us with a supplementary multiplicative structure described in corollary 1.1 which makes $\mathscr{B}$ an "algebra" (even "with unit"). Topologically, the norm inequality (5) (a multiplicative analogue of the triangle inequality (4)) implies that also this multiplication is continuous, making $\mathscr{B}$ a "normed algebra". Finally, as a consequence of the completeness of $\mathfrak{H}$ it turns out that also $\mathscr{B}$ is complete, in other words, that $\mathscr{B}$ is a "Banach algebra" (cf. § 3 definition 7). We shall now make these statements precise.

DEFINITION 3. A (real or complex) linear space $\mathfrak{A}$ is called a (real or complex) *algebra* if with every ordered pair (a, b) of elements of $\mathfrak{A}$ there is associated a unique element $ab \in \mathfrak{A}$ (called the *product* of a and b) in such a way that the following conditions are satisfied (a, b, c denote arbitrary elements of $\mathfrak{A}$):

  a) $(ab) c = a(bc)$ (*associative law*);

  b) $a(b+c) = ab + ac$,
     $(a+b) c = ac + bc$ (*distributive laws*);

  c) $(\alpha a) b = a(\alpha b) = \alpha(ab)$.

The element $e \in \mathfrak{A}$ is called a *unit* if
  d) $ea = ae = a$.
The algebra $\mathfrak{A}$ is called *commutative* if
  e) $ab = ba$ (*commutative law*).

Remark 1. If a unit exists in $\mathfrak{A}$ at all, then it is uniquely determined by d). The equation $oa = ao = o$ already follows from $oa = (a - a) a = a^2 - a^2 = o$ and from the symmetric reasoning.

DEFINITION 4. A normed linear space $\mathfrak{A}$ which is also an algebra is called a *normed algebra* if for all $a \in \mathfrak{A}$ and all $b \in \mathfrak{A}$ the following inequality holds:

$$\|ab\| \leqslant \|a\| \, \|b\| \, .$$

A complete normed algebra is called a *Banach algebra*.

THEOREM 3. *Under the operations as defined in theorem 1 the set $\mathscr{B}$ of all bounded linear operators on $\mathfrak{H}$ is a Banach algebra.*
PROOF. The set $\mathscr{B}$ is a normed algebra by corollary 1.1 and theorem 2. It remains to show that $\mathscr{B}$ is complete.

Let $\{A_n\}_{n=1}^{\infty}$ be a fundamental sequence in $\mathscr{B}$, i.e.

$$\lim_{m,\,n \to \infty} \|A_m - A_n\| = 0 \, .$$

As a first consequence we have

$$\lim_{m,\,n \to \infty} \big| \, \|A_m\| - \|A_n\| \, \big| \leqslant \lim_{m,\,n \to \infty} \|A_m - A_n\| = 0$$

(cf. § 3 theorem 1) and therefore the sequence $\{\|A_n\|\}_{n=1}^{\infty}$ converges. From

$$\|A_m f - A_n f\| \leqslant \|A_m - A_n\| \, \|f\|$$

we deduce that for every vector $f \in \mathfrak{H}$ the sequence $\{A_n f\}_{n=1}^{\infty}$ is fundamental and therefore converges. Let us define an operator A on $\mathfrak{H}$ by

$$Af = \lim_{n \to \infty} A_n f \quad \text{for all } f \in \mathfrak{H} \, .$$

The operator A is linear since

$$A(\alpha f + \beta g) = \lim_{n \to \infty} A_n(\alpha f + \beta g) = \lim_{n \to \infty} (\alpha A_n f + \beta A_n g)$$
$$= \alpha \lim_{n \to \infty} A_n f + \beta \lim_{n \to \infty} A_n g = \alpha Af + \beta Ag \, .$$

The linear operator A is bounded because of

$$\|Af\| = \lim_{n \to \infty} \|A_n f\| \leqslant (\lim_{n \to \infty} \|A_n\|) \|f\| .$$

We still have to show that A is the limit of the sequence $\{A_n\}_{n=1}^{\infty}$ in the sense of the operator norm. Indeed, given any $\varepsilon > 0$ we can choose the index $n(\varepsilon)$ in such a way that

$$\|A_m - A_n\| < \varepsilon \quad \text{for all } m \geqslant n(\varepsilon) \text{ and all } n \geqslant n(\varepsilon).$$

As a consequence for every vector $f \in \mathfrak{H}$ we get

$$\|A_m f - A_n f\| \leqslant \|A_m - A_n\| \|f\| < \varepsilon \|f\| \quad \text{for all } m \geqslant n(\varepsilon) \text{ and all } n \geqslant n(\varepsilon)$$

and, passing to the limit for $m \to \infty$,

$$\|Af - A_n f\| \leqslant \varepsilon \|f\| \quad \text{for all } n \geqslant n(\varepsilon),$$
$$\|A - A_n\| \leqslant \varepsilon \qquad \text{for all } n \geqslant n(\varepsilon).$$

We obtain $A = \lim_{n \to \infty} A_n$ and, in particular, $\|A\| = \lim_{n \to \infty} \|A_n\|$. $\square$

THEOREM 4. *If $\{A_n\}_{n=1}^{\infty}$ and $\{B_n\}_{n=1}^{\infty}$ are converging sequences in $\mathscr{B}$ and*

$$\lim_{n \to \infty} A_n = A,$$

$$\lim_{n \to \infty} B_n = B,$$

then

$$\lim_{n \to \infty} A_n B_n = AB (= \lim_{n \to \infty} A_n \cdot \lim_{n \to \infty} B_n).$$

PROOF. By the inequalities (4) and (5) in theorem 2 we have

$$\|AB - A_n B_n\| \leqslant \|AB - A_n B\| + \|A_n B - A_n B_n\|$$
$$\leqslant \|A - A_n\| \|B\| + \|A_n\| \|B - B_n\| .$$

Since $\|A\| = \lim_{n \to \infty} \|A_n\|$ the sequence $\{\|A_n\|\}_{n=1}^{\infty}$ is bounded. We conclude that for $n \to \infty$ the last member tends to 0. $\square$

Note that the assertion of theorem 4 remains true if $\mathscr{B}$ is replaced by any normed algebra (cf. § 3 theorem 6).

DEFINITION 5. A bounded linear operator A on $\mathfrak{H}$ is called *invertible* if there exists a bounded linear operator A^{-1} on $\mathfrak{H}$, called the *inverse* of A, such that

$$AA^{-1} = A^{-1}A = I .$$

In § 10 remark 2 it has been pointed out that an isomorphism of $\mathfrak{H}$ onto some Hilbert space $\mathfrak{H}'$, in particular every automorphism of $\mathfrak{H}$, may be "reversed". Thus, by definition 5, every automorphism of $\mathfrak{H}$ is invertible. Conversely, however, an invertible bounded linear operator need not preserve inner products and therefore need not be an automorphism. Purely algebraically it still comes close to it (cf. § 10 remark 1).

THEOREM 5. *An invertible bounded linear operator A is one-to-one and maps $\mathfrak{H}$ onto $\mathfrak{H}$. The inverse of A is unique.*

PROOF. Suppose A^{-1} and B were inverses of A. Then we conclude

$$B = IB = (A^{-1}A) B = A^{-1}(AB) = A^{-1}I = A^{-1}.$$

From $AA^{-1}=I$ we deduce that A maps $\mathfrak{H}$ onto $\mathfrak{H}$: for every $f\in\mathfrak{H}$ we have $f=(AA^{-1})f=A(A^{-1}f)$. From $A^{-1}A=I$ we deduce that A is one-to-one: for $Af_1=Af_2$ we get

$$f_1 = (A^{-1}A)f_1 = A^{-1}(Af_1) = A^{-1}(Af_2) = (A^{-1}A)f_2 = f_2. \ \square$$

Suppose $\mathfrak{H}$ is a separable Hilbert space, $\{e_k\}_{k=1}^{\infty}$ is a basis of $\mathfrak{H}$, and A is a bounded linear operator on $\mathfrak{H}$. Then, as in the finite-dimensional case (cf. example 1), the action of A on $\mathfrak{H}$ is already determined by the action of A on the elements of the given basis. Indeed, from

$$Ae_l = \sum_{k=1}^{\infty} \alpha_{k,l} e_k \quad \text{for } 1 \leqslant l < \infty$$

we get by the continuity of A for every vector $f=\sum_{l=1}^{\infty} \beta_l e_l \in \mathfrak{H}$

$$\langle Af, e_k \rangle = \langle \sum_{l=1}^{\infty} \beta_l Ae_l, e_k \rangle = \sum_{l=1}^{\infty} \beta_l \langle Ae_l, e_k \rangle = \sum_{l=1}^{\infty} \alpha_{k,l} \beta_l,$$

$$Af = \sum_{k=1}^{\infty} \langle Af, e_k \rangle e_k = \sum_{k=1}^{\infty} \left(\sum_{l=1}^{\infty} \alpha_{k,l} \beta_l \right) e_k.$$

The action of A on $\mathfrak{H}$ can therefore also be described by the infinite matrix

(6)
$$A' = \begin{pmatrix} \alpha_{1,1} & \alpha_{1,2} & \alpha_{1,3} & \cdots \\ \alpha_{2,1} & \alpha_{2,2} & \cdots & \\ \alpha_{3,1} & \cdots & & \\ \cdots & & & \end{pmatrix}.$$

Application of the linear operator A to a vector f amounts to multiplying the matrix A' from the right with the column vector consisting of the

Fourier coefficients of f with respect to the given basis. Note that we have

$$\alpha_{k,\mathcal{U}} = \langle Ae_l, e_k \rangle \qquad\qquad \text{for all } k \geqslant 1 \text{ and all } l \geqslant 1,$$

$$\sum_{k=1}^{\infty} |\alpha_{k,l}|^2 = \|Ae_l\|^2 \leqslant \|A\|^2 < \infty \quad \text{for all } l \geqslant 1.$$

The following examples of bounded linear operators will be also of some importance later when we shall busy ourselves with spectral analysis (cf. § 22 example 3, § 23 example 2, § 25 example 1).

Example 2. Let the linear operator A on $\mathfrak{H} = \ell_2$ be given by

$$Aa = A\{\alpha_k\}_{k=1}^{\infty} = \{\alpha_{k-1}\}_{k=1}^{\infty} \quad \text{for } a = \{\alpha_k\}_{k=1}^{\infty} \in \ell_2, \alpha_0 = 0.$$

In a somewhat more illustrative notation we may write

$$A\{\alpha_1, \alpha_2, \alpha_3, ...\} = \{0, \alpha_1, \alpha_2, ...\}.$$

The operator A is also called *right shift operator* in ℓ_2. Obviously we have

$$\langle Aa, Ab \rangle = \sum_{k=1}^{\infty} \alpha_{k-1}\overline{\beta_{k-1}} = \langle a, b \rangle.$$

Thus A is an isometry of ℓ_2 into itself. In fact, A maps ℓ_2 onto a proper subspace, namely the set of all absolutely square summable sequences having first term zero (cf. § 6 example 2). Therefore A is certainly not invertible (cf. theorem 5). Describing the action of A in terms of the standard basis $\{e_k\}_{k=1}^{\infty}$ of ℓ_2 exhibited in § 8 example 2 we obtain

$$Ae_l = e_{l+1} \quad \text{for all } l \geqslant 1.$$

Thus the matrix A' associated with A according to (6) has the form

$$A' = \begin{pmatrix} 0 & 0 & \cdot & \cdot & \cdot \\ 1 & 0 & & & \\ 0 & 1 & \cdot & & \\ \cdot & & \cdot & \cdot & \\ \cdot & & & \cdot & \cdot \\ \cdot & & & & \cdot \end{pmatrix}$$

(the only non-zero elements are 1's below the diagonal).

Example 3. Let $\mathfrak{H}$ be a separable Hilbert space with basis $\{e_k\}_{k=1}^{\infty}$ and let A' be an infinite matrix as in (6) with the additional property that

$$\alpha^2 = \sum_{k=1}^{\infty} \sum_{l=1}^{\infty} |\alpha_{k,l}|^2 < \infty, \quad \alpha \geqslant 0$$

(example 2 already demonstrates a situation where this condition is not satisfied). For any vector $f = \sum_{l=1}^{\infty} \beta_l e_l \in \mathfrak{H}$ we have

(7)
$$\sum_{k=1}^{\infty} |\sum_{l=1}^{\infty} \alpha_{k,l} \beta_l|^2 \leqslant \sum_{k=1}^{\infty} (\sum_{l=1}^{\infty} |\alpha_{k,l}|^2 \cdot \sum_{l=1}^{\infty} |\beta_l|^2) = \alpha^2 \|f\|^2$$

(we have applied CAUCHY's inequality in ℓ_2). The series $\sum_{k=1}^{\infty} (\sum_{l=1}^{\infty} \alpha_{k,l} \beta_l) e_k$ then converges in $\mathfrak{H}$ by § 7 theorem 3 a). It is therefore possible to define an operator A on $\mathfrak{H}$ by

$$A(\sum_{l=1}^{\infty} \beta_l e_l) = \sum_{k=1}^{\infty} (\sum_{l=1}^{\infty} \alpha_{k,l} \beta_l) e_k \quad \text{for } f = \sum_{l=1}^{\infty} \beta_l e_l \in \mathfrak{H}.$$

The operator A is obviously linear. From (7) it follows that A is bounded and that, in fact, we have

$$\|A\| \leqslant \alpha = (\sum_{k=1}^{\infty} \sum_{l=1}^{\infty} |\alpha_{k,l}|^2)^{\frac{1}{2}}$$

(cf. (3) in example 1).

The following example is a continuous analogue of example 3.

Example 4. Let $\mathfrak{H} = \mathfrak{L}_2(\beta, \gamma)$ $(-\infty \leqslant \beta < \gamma \leqslant +\infty)$ and let X be the "square" in the real plane with "sides" $]\beta, \gamma[$:

$$X =]\beta, \gamma[\times]\beta, \gamma[= \{(\xi, \eta) \in \mathbf{R}^2 : \beta < \xi < \gamma, \beta < \eta < \gamma\}.$$

Furthermore, let $a \in \mathfrak{L}_2(X)$ be given (cf. § 5) and define

$$\alpha^2 = \int_X |a(\xi, \eta)|^2 \, d(\xi, \eta), \quad \alpha \geqslant 0.$$

If $]\beta_1, \gamma_1[$ is any finite subinterval of $]\beta, \gamma[$ and $X_1 =]\beta_1, \gamma_1[\times]\beta, \gamma[$, then we have $X_1 \subset X$ and the restriction of a to X_1, which we again denote by a, satisfies $a \in \mathfrak{L}_2(X_1)$. Furthermore, any $f \in \mathfrak{L}_2(\beta, \gamma)$ may be extended to a Lebesgue measurable function f_1 on X_1 "only depending on the second variable" by defining

$$f_1(\xi, \eta) = f(\eta) \quad \text{for all } (\xi, \eta) \in X_1$$

(cf. appendix B5). In fact we have $f_1 \in \mathfrak{L}_2(X_1)$ since

$$\int_{X_1} |f_1(\xi, \eta)|^2 \, d(\xi, \eta) = \int_{\beta_1}^{\gamma_1} \int_{\beta}^{\gamma} |f(\eta)|^2 \, d\eta \, d\xi = (\gamma_1 - \beta_1) \|f\|^2 < \infty.$$

As a consequence we get $af_1 \in \mathfrak{L}_1(X_1)$ (cf. § 5 remark 1), i.e.

$$\int_{X_1} |a(\xi, \eta) f(\eta)| \, d(\xi, \eta) = \int_{\beta_1}^{\gamma_1} \int_{\beta}^{\gamma} |a(\xi, \eta) f(\eta)| \, d\eta \, d\xi < \infty.$$

By FUBINI's theorem (cf. appendix B8) the function

$$Af = \int_{\beta}^{\gamma} a(x, \eta) f(\eta) \, d\eta$$

is defined and finite a.e. and Lebesgue measurable on every subinterval $]\beta_1, \gamma_1[\subset]\beta, \gamma[$. As a consequence Af is Lebesgue measurable on the entire interval $]\beta, \gamma[$.

In order to show that Af even belongs to $\mathfrak{L}_2(\beta, \gamma)$ recall that $a(\xi, y) \in \mathfrak{L}_2(\beta, \gamma)$ for almost all $\xi \in]\beta, \gamma[$ (cf. § 9 remark 1). We therefore have

(8)
$$\int_{\beta}^{\gamma} |Af(\xi)|^2 \, d\xi = \int_{\beta}^{\gamma} \left| \int_{\beta}^{\gamma} a(\xi, \eta) f(\eta) \, d\eta \right|^2 d\xi$$

$$\leqslant \int_{\beta}^{\gamma} \left[\int_{\beta}^{\gamma} |a(\xi, \eta)|^2 \, d\eta \cdot \int_{\beta}^{\gamma} |f(\eta)|^2 \, d\eta \right] d\xi$$

$$\leqslant \alpha^2 \|f\|^2 < \infty$$

(we have applied CAUCHY's inequality in $\mathfrak{L}_2(\beta, \gamma)$ and then FUBINI's theorem in order to get $\int_{\beta}^{\gamma} \int_{\beta}^{\gamma} |a(\xi, \eta)|^2 \, d\eta \, d\xi = \alpha^2$; cf. appendix B8).

Defining $Af \in \mathfrak{L}_2(\beta, \gamma)$ on $]\beta, \gamma[$ by

$$Af(\xi) = \int_{\beta}^{\gamma} a(\xi, \eta) f(\eta) \, d\eta \quad \text{for } f \in \mathfrak{L}_2(\beta, \gamma)$$

we therefore obtain an operator A on $\mathfrak{L}_2(\beta, \gamma)$ which is obviously linear and which is called the *Hilbert-Schmidt operator* with *kernel a*. At the same time (8) shows that A is bounded and, in fact, we have

$$\|A\| \leqslant \alpha = \left[\int_{X} |a(\xi, \eta)|^2 \, d(\xi, \eta) \right]^{\frac{1}{2}}.$$

Surprising as it may sound, example 4 is not only an analogue to but even a well disguised special case of example 3. Indeed, let $\{e_k\}_{k=1}^{\infty}$ be a basis of

$\mathfrak{L}_2(\beta, \gamma)$ and let $\{e_{k,l}\}_{1 \leqslant k,\, l < \infty}$ be the corresponding basis of $\mathfrak{L}_2(X)$ defined in § 9 theorem 10. Then for $f = \sum_{l=1}^{\infty} \beta_l e_l \in \mathfrak{L}_2(\beta, \gamma)$ we have

$$Af = A\left(\sum_{l=1}^{\infty} \beta_l e_l\right) = \sum_{l=1}^{\infty} \beta_l A e_l = \sum_{l=1}^{\infty} \beta_l \left(\sum_{k=1}^{\infty} \langle A e_l, e_k \rangle\, e_k\right)$$

$$= \sum_{k=1}^{\infty} \left(\sum_{l=1}^{\infty} \alpha_{k,l} \beta_l\right) e_k$$

where

$$\alpha_{k,l} = \langle A e_l, e_k \rangle = \int_{\beta}^{\gamma} \int_{\beta}^{\gamma} a(\xi, \eta)\, e_l(\eta)\, \overline{e_k(\xi)}\, d\xi\, d\eta$$

$$= \int_X a(\xi, \eta)\, \overline{e_{k,l}(\xi, \eta)}\, d(\xi, \eta) = \langle a, e_{k,l} \rangle$$

and

$$\sum_{k=1}^{\infty} \sum_{l=1}^{\infty} |\alpha_{k,l}|^2 = \sum_{k=1}^{\infty} \sum_{l=1}^{\infty} |\langle a, e_{k,l} \rangle|^2$$

$$= \int_X |a(\xi, \eta)|^2\, d(\xi, \eta) = \alpha^2 < \infty.$$

Exercise. Let $\{e_k\}_{k=1}^{\infty}$ be a basis of the separable Hilbert space $\mathfrak{H}$. Let

$$A' = (\alpha_{k,l})_{1 \leqslant k,\, l < \infty},$$
$$B' = (\beta_{k,l})_{1 \leqslant k,\, l < \infty},$$
$$B'_n = (\beta_{k,l}^{(n)})_{1 \leqslant k,\, l < \infty}$$

be the infinite matrices associated with the bounded linear operators A, B, B_n on $\mathfrak{H}$ respectively according to (6).

  a) If $AB = C$, then $C' = (\gamma_{k,l})_{1 \leqslant k,\, l < \infty}$ where

$$\gamma_{k,l} = \sum_{m=1}^{\infty} \alpha_{k,m} \beta_{m,l} \quad \text{for } 1 \leqslant k < \infty \text{ and } 1 \leqslant l < \infty.$$

  b) If $B = \lim_{n \to \infty} B_n$, then

$$\beta_{k,l} = \lim_{n \to \infty} \beta_{k,l}^{(n)} \quad \text{for } 1 \leqslant k < \infty \text{ and } 1 \leqslant l < \infty.$$

  c) If $\|B_n\| \leqslant \gamma$ for some constant $\gamma > 0$ and all $n \geqslant 1$ and if

$$\beta_{k,l} = \lim_{n \to \infty} \beta_{k,l}^{(n)} \quad \text{for } 1 \leqslant k < \infty \text{ and } 1 \leqslant l < \infty,$$

then

$$\langle Bf, g \rangle = \lim_{n \to \infty} \langle B_n f, g \rangle \quad \text{for all } f \in \mathfrak{H} \text{ and all } g \in \mathfrak{H}$$

but not necessarily $B = \lim_{n \to \infty} B_n$. (Hint for a counterexample: let $B_n = A^n$ where A is the right shift operator in $\mathfrak{H} = \ell_2$; cf. example 2).

§ 13. Bilinear forms

Suppose A is a bounded linear operator on $\mathfrak{H}$. Let us for the moment use the symbol $\mathfrak{H} \times \mathfrak{H}$ to denote the set of all ordered pairs (f, g) of vectors in $\mathfrak{H}$. Define complex-valued functions φ and ψ on $\mathfrak{H} \times \mathfrak{H}$ by

$$\varphi(f, g) = \langle f, Ag \rangle,$$
$$\psi(f, g) = \langle Af, g \rangle.$$

The functions φ and ψ have several properties in common with the inner product in $\mathfrak{H}$:

(1)
$$\begin{cases} \varphi(f_1 + f_2, g) = \varphi(f_1, g) + \varphi(f_2, g), \\ \quad \varphi(\alpha f, g) = \alpha \varphi(f, g), \\ \varphi(f, g_1 + g_2) = \langle f, A(g_1 + g_2) \rangle = \langle f, Ag_1 + Ag_2 \rangle \\ \qquad = \varphi(f, g_1) + \varphi(f, g_2), \\ \varphi(f, \alpha g) = \langle f, A(\alpha g) \rangle = \langle f, \alpha Ag \rangle \\ \qquad = \bar{\alpha} \varphi(f, g), \end{cases}$$

(2)
$$|\varphi(f, g)| = |\langle f, Ag \rangle| \leqslant \|f\| \|Ag\|$$
$$\leqslant \|A\| \|f\| \cdot \|g\|,$$

and similarly for ψ. It is not surprising that the function $\hat{\varphi}$ on $\mathfrak{H}$ defined by $\hat{\varphi}(f) = \varphi(f, f)$ and the corresponding function $\hat{\psi}$ have several properties in common with the square of the norm in $\mathfrak{H}$.

Let us forget how we arrived at these functions and collect the relevant properties in two definitions.

DEFINITION 1. A complex-valued function φ on $\mathfrak{H} \times \mathfrak{H}$ is called a *bilinear form* (or a *bilinear functional*) on $\mathfrak{H}$ if for all $(f, g) \in \mathfrak{H} \times \mathfrak{H}$

$$\varphi(f_1 + f_2, g) = \varphi(f_1, g) + \varphi(f_2, g), \quad \varphi(f, g_1 + g_2) = \varphi(f, g_1) + \varphi(f, g_2),$$
$$\varphi(\alpha f, g) = \alpha \varphi(f, g), \qquad\qquad \varphi(f, \alpha g) = \bar{\alpha} \varphi(f, g).$$

The bilinear form φ is called *symmetric* if

$$\varphi(g, f) = \overline{\varphi(f, g)} \quad \text{for all } (f, g) \in \mathfrak{H} \times \mathfrak{H}.$$

The bilinear form is called *bounded* if there exists a real number $\gamma \geqslant 0$ such that

$$|\varphi(f, g)| \leqslant \gamma \|f\| \|g\| \quad \text{for all } (f, g) \in \mathfrak{H} \times \mathfrak{H}.$$

If φ is bounded, then the non-negative number

$$\|\varphi\| = \sup_{f \neq o, \, g \neq o} \frac{|\varphi(f, g)|}{\|f\| \|g\|}$$

is called the *norm* of φ.

DEFINITION 2. If φ is a bilinear form on $\mathfrak{H}$, then the function $\hat{\varphi}$ on $\mathfrak{H}$ defined by

$$\hat{\varphi}(f) = \varphi(f, f)$$

is called the *quadratic form* associated with φ. The quadratic form $\hat{\varphi}$ is called *real* if $\hat{\varphi}(f) \in \mathbf{R}$ for all $f \in \mathfrak{H}$. The quadratic form $\hat{\varphi}$ is called *bounded* if there exists a real number $\delta \geqslant 0$ such that

$$|\hat{\varphi}(f)| \leqslant \delta \|f\|^2 \quad \text{for all } f \in \mathfrak{H}.$$

If $\hat{\varphi}$ is bounded, then the non-negative number

$$\|\hat{\varphi}\| = \sup_{f \neq o} \frac{|\hat{\varphi}(f)|}{\|f\|^2}$$

is called the *norm* of $\hat{\varphi}$.

Remark 1. For a bounded bilinear form φ or a bounded quadratic form $\hat{\varphi}$ on $\mathfrak{H}$ we respectively have

$$\|\varphi\| = \sup_{f \neq o, \, g \neq o} \frac{|\varphi(f, g)|}{\|f\| \|g\|} = \sup_{f \neq o, \, g \neq o} \left| \varphi \left(\frac{f}{\|f\|}, \frac{g}{\|g\|} \right) \right| = \sup_{\|f\| = \|g\| = 1} |\varphi(f, g)|,$$

$$\|\hat{\varphi}\| = \sup_{f \neq o} \frac{|\hat{\varphi}(f)|}{\|f\|^2} = \sup_{f \neq o} \left| \hat{\varphi} \left(\frac{f}{\|f\|} \right) \right| = \sup_{\|f\| = 1} |\hat{\varphi}(f)|$$

(cf. § 11 remark 1). Conversely, if the suprema on the right hand side are finite, then the same chains of equalities show that φ or $\hat{\varphi}$ respectively are bounded.

In what follows φ will denote any fixed bilinear form and $\hat{\varphi}$ the associated quadratic form on $\mathfrak{H}$. It is an interesting and useful fact that not only $\hat{\varphi}$ is determined by φ but that it is also possible to recapture the original bilinear functional φ from the associated quadratic form $\hat{\varphi}$. Consider the following identities:

$$\hat{\varphi}(f + g) = \varphi(f + g, f + g) = \varphi(f, f) + \varphi(f, g) + \varphi(g, f) + \varphi(g, g),$$
$$\hat{\varphi}(f - g) = \varphi(f - g, f - g) = \varphi(f, f) - \varphi(f, g) - \varphi(g, f) + \varphi(g, g).$$

Subtracting the second identity from the first one we obtain

(3) $$\hat{\varphi}(f + g) - \hat{\varphi}(f - g) = 2\varphi(f, g) + 2\varphi(g, f).$$

Replacing g by ig we further obtain

(4) $$\hat{\varphi}(f + ig) - \hat{\varphi}(f - ig) = -2i\varphi(f, g) + 2i\varphi(g, f).$$

Multiplying (4) by i and adding it to (3) we finally get

$$\hat{\varphi}(f + g) - \hat{\varphi}(f - g) + i\hat{\varphi}(f + ig) - i\hat{\varphi}(f - ig) = 4\varphi(f, g).$$

Thus we have proved the following statement:

THEOREM 1 (POLAR IDENTITY).

$$\varphi(f, g) = \tfrac{1}{4}[\hat{\varphi}(f + g) - \hat{\varphi}(f - g) + i\hat{\varphi}(f + ig) - i\hat{\varphi}(f - ig)].$$

COROLLARY 1.1. *If φ and ψ are bilinear forms on $\mathfrak{H}$ and if $\hat{\varphi} = \hat{\psi}$ (i.e. $\hat{\varphi}(f) = \hat{\psi}(f)$ for all $f \in \mathfrak{H}$), then $\varphi = \psi$ (i.e. $\varphi(f, g) = \psi(f, g)$ for all $(f, g) \in \mathfrak{H} \times \mathfrak{H}$).*

COROLLARY 1.2. *A linear operator A on $\mathfrak{H}$ is isometric iff $\|Af\| = \|f\|$ for all $f \in \mathfrak{H}$.*
PROOF. Define two bilinear forms φ and ψ on $\mathfrak{H}$ by

$$\varphi(f, g) = \langle Af, Ag \rangle,$$
$$\psi(f, g) = \langle f, g \rangle.$$

If $\|Af\| = \|f\|$ for all $f \in \mathfrak{H}$, then $\hat{\varphi} = \hat{\psi}$ and, by corollary 1.1, $\varphi = \psi$. In other words, A is isometric. The converse is trivial (cf. § 10 definition 1). $\square$

The following theorems further illustrate the importance of the polar identity.

THEOREM 2. φ *is symmetric iff $\hat{\varphi}$ is real.*
PROOF. If φ is symmetric, then we have $\hat{\varphi}(f) = \varphi(f, f) = \overline{\varphi(f, f)} = \overline{\hat{\varphi}(f)}$ and therefore $\hat{\varphi}(f) \in \mathbf{R}$. Conversely, suppose $\hat{\varphi}$ is real. From the polar identity

and the identity $\hat{\varphi}(f)=\hat{\varphi}(-f)=\hat{\varphi}(if)$ we deduce

$$\varphi(g,f) = \tfrac{1}{4}[\hat{\varphi}(g+f) - \hat{\varphi}(g-f) + i\hat{\varphi}(g+if) - i\hat{\varphi}(g-if)]$$
$$= \tfrac{1}{4}[\hat{\varphi}(f+g) - \hat{\varphi}(f-g) + i\hat{\varphi}(f-ig) - i\hat{\varphi}(f+ig)]$$
$$= \overline{\varphi(f,g)}. \ \square$$

THEOREM 3. φ *is bounded iff* $\hat{\varphi}$ *is bounded. If* φ *is bounded, then*

$$\|\hat{\varphi}\| \leqslant \|\varphi\| \leqslant 2\|\hat{\varphi}\|.$$

PROOF. Suppose φ is bounded. Then we have

$$\sup_{\|f\|=1} |\hat{\varphi}(f)| = \sup_{\|f\|=1} |\varphi(f,f)| \leqslant \sup_{\|f\|=\|g\|=1} |\varphi(f,g)| = \|\varphi\|$$

and therefore $\hat{\varphi}$ is bounded and $\|\hat{\varphi}\| \leqslant \|\varphi\|$ (cf. remark 1). On the other hand, suppose $\hat{\varphi}$ is bounded. From the polar identity and the parallelogram law (§ 2 theorem 6) we deduce

$$|\varphi(f,g)| \leqslant \tfrac{1}{4}\|\hat{\varphi}\|\,(\|f+g\|^2 + \|f-g\|^2 + \|f+ig\|^2 + \|f-ig\|^2)$$
$$= \tfrac{1}{4}\|\hat{\varphi}\|\cdot 2(\|f\|^2 + \|g\|^2 + \|f\|^2 + \|g\|^2) = \|\hat{\varphi}\|\,(\|f\|^2 + \|g\|^2),$$
$$\sup_{\|f\|=\|g\|=1} |\varphi(f,g)| \leqslant 2\|\hat{\varphi}\|.$$

Thus φ is bounded and $\|\varphi\| \leqslant 2\|\hat{\varphi}\|$. $\square$

THEOREM 4. *If* φ *is bounded and symmetric, then* $\|\varphi\| = \|\hat{\varphi}\|$.
PROOF. Note that $\hat{\varphi}$ is real by theorem 2. In view of theorem 3 we only have to show $\|\varphi\| \leqslant \|\hat{\varphi}\|$, i.e. $|\varphi(f,g)| \leqslant \|\hat{\varphi}\|$ for any two unit vectors f and g in $\mathfrak{H}$. Suppose

$$\varphi(f,g) = \varrho e^{i\alpha} \quad (\alpha \in \mathbf{R}, \varrho \in \mathbf{R}, \varrho \geqslant 0).$$

Denoting the unit vector $e^{-i\alpha}f$ by f' we obtain

$$|\varphi(f,g)| = \varrho = \varphi(e^{-i\alpha}f,g) = \tfrac{1}{4}[\hat{\varphi}(f'+g) - \hat{\varphi}(f'-g)]$$

(the fact that $\varphi(f',g)$ is real implies that the purely imaginary terms in the polar identity must vanish),

$$|\varphi(f,g)| \leqslant \tfrac{1}{4}\|\hat{\varphi}\|\,(\|f'+g\|^2 + \|f'-g\|^2)$$
$$= \tfrac{1}{2}\|\hat{\varphi}\|\,(\|f'\|^2 + \|g\|^2) = \|\hat{\varphi}\|. \ \square$$

At the beginning of this section we already observed that a bounded linear operator A on $\mathfrak{H}$ gives rise to two bounded bilinear forms φ and ψ

on $\mathfrak{H}$ defined by

$$\varphi(f, g) = \langle f, Ag \rangle, \quad \psi(f, g) = \langle Af, g \rangle.$$

The relationship between bounded linear operators and bounded bilinear forms goes far beyond this statement. In fact, every bounded bilinear form on $\mathfrak{H}$ can be obtained in this way, and the norm of the bounded bilinear form coincides with the norm of the corresponding bounded linear operator.

THEOREM 5. a) *Let A be a bounded linear operator on $\mathfrak{H}$. Then the complex-valued function φ on $\mathfrak{H} \times \mathfrak{H}$ defined by*

$$(5) \qquad \qquad \varphi(f, g) = \langle f, Ag \rangle$$

is a bounded bilinear form on $\mathfrak{H}$, and $\|\varphi\| = \|A\|$.

b) *Conversely, let φ be a bounded bilinear form on $\mathfrak{H}$. Then there is a unique bounded linear operator A on $\mathfrak{H}$ such that*

$$(6) \qquad \qquad \varphi(f, g) = \langle f, Ag \rangle \quad \text{for all } (f, g) \in \mathfrak{H} \times \mathfrak{H}.$$

PROOF. a) If A is a bounded linear operator on $\mathfrak{H}$ we already know that the function φ defined by (5) is a bounded bilinear form on $\mathfrak{H}$ by (1) and (2). Moreover, by (2) we have

$$(7) \qquad \qquad \|\varphi\| \leqslant \|A\|.$$

b) Conversely, let φ be any bounded bilinear form on $\mathfrak{H}$. We shall first prove the existence of a bounded linear operator A satisfying (6). In the course of the proof we shall also show that for this operator A the inequality

$$\|A\| \leqslant \|\varphi\|$$

holds. Together with the inequality (7) obtained in part a) of the proof this will imply the desired equality $\|\varphi\| = \|A\|$ if A is shown to be unique.

To this end, for any $g \in \mathfrak{H}$, define the complex-valued function Φ_g on $\mathfrak{H}$ by

$$(8) \qquad \qquad \Phi_g(f) = \varphi(f, g).$$

Then Φ_g is a linear functional on $\mathfrak{H}$ (cf. § 11 definition 3). The linear functional Φ_g is bounded because of the inequality

$$|\Phi_g(f)| = |\varphi(f, g)| \leqslant \|\varphi\| \, \|f\| \, \|g\|.$$

In fact, we even conclude

$$\|\Phi_g\| \leqslant \|\varphi\| \, \|g\|.$$

By the Riesz representation theorem (§ 11 theorem 4) there exists a unique vector $Ag \in \mathfrak{H}$ such that

(9) $\Phi_g(f) = \langle f, Ag \rangle$ for all $f \in \mathfrak{H}$.

Furthermore, by the same theorem we have

(10) $\|Ag\| = \|\Phi_g\| \leqslant \|\varphi\| \, \|g\|$.

Let us convince ourselves that the operator $A: g \to Ag$ defined by (9) is linear: from

$$\langle f, A(\alpha g) \rangle = \Phi_{\alpha g}(f) = \varphi(f, \alpha g) = \bar{\alpha}\varphi(f, g)$$
$$= \bar{\alpha}\Phi_g(f) = \bar{\alpha}\langle f, Ag \rangle$$
$$= \langle f, \alpha Ag \rangle \quad \text{for all } f \in \mathfrak{H}$$

we conclude

$$\langle f, A(\alpha g) - \alpha Ag \rangle = 0 \quad \text{for all } f \in \mathfrak{H}$$

and (substituting $f = A(\alpha g) - \alpha Ag$)

$$A(\alpha g) = \alpha Ag .$$

Similarly, from

$$\langle f, A(g + h) \rangle = \Phi_{g+h}(f) = \varphi(f, g + h) = \varphi(f, g) + \varphi(f, h)$$
$$= \Phi_g(f) + \Phi_h(f) = \langle f, Ag \rangle + \langle f, Ah \rangle$$
$$= \langle f, Ag + Ah \rangle \quad \text{for all } f \in \mathfrak{H}$$

we conclude

$$\langle f, A(g + h) - Ag - Ah \rangle = 0 \quad \text{for all } f \in \mathfrak{H},$$
$$A(g + h) = Ag + Ah .$$

Thus A is a linear operator on $\mathfrak{H}$. The inequality (10) shows that A is bounded and in fact

$$\|A\| \leqslant \|\varphi\| .$$

From (8) and (9) we conclude

$$\varphi(f, g) = \Phi_g(f) = \langle f, Ag \rangle \quad \text{for all } (f, g) \in \mathfrak{H} \times \mathfrak{H}.$$

Finally, suppose B is any other bounded linear operator on $\mathfrak{H}$ having the property that

$$\varphi(f, g) = \langle f, Bg \rangle \quad \text{for all } (f, g) \in \mathfrak{H} \times \mathfrak{H}.$$

Then for every fixed $g \in \mathfrak{H}$ we conclude

$$\langle f, Ag - Bg \rangle = \langle f, Ag \rangle - \langle f, Bg \rangle = 0 \quad \text{for all } f \in \mathfrak{H},$$

therefore

$$Ag = Bg \quad \text{for all } g \in \mathfrak{H}$$

and $A = B$. This completes the proof. $\square$

COROLLARY 5.1. a) *Let A be a bounded linear operator on $\mathfrak{H}$. Then the complex-valued function ψ on $\mathfrak{H} \times \mathfrak{H}$ defined by*

$$\psi(f, g) = \langle Af, g \rangle$$

is a bounded bilinear form on $\mathfrak{H}$ and $\|\psi\| = \|A\|$.

   b) *Conversely, let ψ be a bounded bilinear form on $\mathfrak{H}$. Then there is a unique bounded linear operator A on $\mathfrak{H}$ such that*

$$\psi(f, g) = \langle Af, g \rangle \quad \text{for all } (f, g) \in \mathfrak{H} \times \mathfrak{H}.$$

PROOF. a) Define the function φ on $\mathfrak{H} \times \mathfrak{H}$ by

$$\varphi(f, g) = \overline{\psi(g, f)} = \langle f, Ag \rangle.$$

By theorem 5 a) φ is a bounded bilinear form on $\mathfrak{H}$ and $\|\varphi\| = \|A\|$. Since we have

$$\psi(f, g) = \overline{\varphi(g, f)}$$

also ψ is a bounded bilinear form on $\mathfrak{H}$ and

$$\|\psi\| = \sup_{\|f\| = \|g\| = 1} |\psi(f, g)| = \sup_{\|f\| = \|g\| = 1} |\varphi(g, f)| = \|\varphi\| = \|A\|.$$

   b) If ψ is given define again a bounded bilinear form φ on $\mathfrak{H}$ by

$$\varphi(f, g) = \overline{\psi(g, f)}.$$

By theorem 5 b) there is a bounded linear operator A on $\mathfrak{H}$ such that

$$\varphi(f, g) = \langle f, Ag \rangle \quad \text{for all } (f, g) \in \mathfrak{H} \times \mathfrak{H}.$$

Therefore we have

$$\psi(f, g) = \overline{\varphi(g, f)} = \langle Af, g \rangle \quad \text{for all } (f, g) \in \mathfrak{H} \times \mathfrak{H}.$$

Uniqueness of A follows from the uniqueness statement in theorem 5 b) or directly as in the proof thereof. $\square$

COROLLARY 5.2. *If A is a bounded linear operator on $\mathfrak{H}$, then*

$$\|A\| = \sup_{\|f\| = \|g\| = 1} |\langle f, Ag \rangle| = \sup_{\|f\| = \|g\| = 1} |\langle Af, g \rangle|.$$

Exercise. The following chain of assertions is motivated by the polar identity (theorem 1) and leads to the conclusion that in a normed linear space the norm is induced by an inner product iff it satisfies the parallelogram law (cf. § 2 theorem 6, § 3 example 1).

Let $\mathfrak{L}$ be a normed linear space satisfying the parallelogram law

$$\|f + g\|^2 + \|f - g\|^2 = 2(\|f\|^2 + \|g\|^2)$$

and define a function φ on $\mathfrak{L} \times \mathfrak{L}$ by

$$\varphi(f, g) = \tfrac{1}{4}(\|f + g\|^2 - \|f - g\|^2 + i\|f + ig\|^2 - i\|f - ig\|^2)$$
$$\text{for all } (f, g) \in \mathfrak{H} \times \mathfrak{H}.$$

Then the following assertions hold.

  a) If $f = \lim_{n \to \infty} f_n$ and $g = \lim_{n \to \infty} g_n$, then $\varphi(f, g) = \lim_{n \to \infty} \varphi(f_n, g_n)$.
  b) $\varphi(g, f) = \overline{\varphi(f, g)}$, $\varphi(f, f) \geq 0$,
     $\varphi(f, o) = 0$, $\varphi(f, f) = 0 \Rightarrow f = o$.
  c) $\|f + g + h\|^2 + \|f\|^2 + \|g\|^2 + \|h\|^2 = \|f + g\|^2 + \|f + h\|^2 + \|g + h\|^2$.
(Note that this is the only place where the parallelogram law is needed.)
  d) $\varphi(f + g, h) = \varphi(f, h) + \varphi(g, h)$.
  e) $\varphi(\alpha f, g) = \alpha\varphi(f, g)$ for all $\alpha \in \mathbf{R}$.
(Hint: prove the assertion first for $\alpha = k/m$, where k and m are integers, and then apply a).)
  f) $\varphi(\alpha f, g) = \alpha\varphi(f, g)$ for all $\alpha \in \mathbf{C}$.
(Hint: prove the assertion first for $\alpha = i$ and then write $\alpha = \beta + i\gamma$ where $\beta = \operatorname{Re} \alpha \in \mathbf{R}$ and $\gamma = \operatorname{Im} \alpha \in \mathbf{R}$.)
  g) φ is an inner product on $\mathfrak{L}$ which induces the given norm $\| \ \|$.

§ 14. Adjoint operators

Not every bounded linear operator A on $\mathfrak{H}$ has an inverse, but A always has a twin brother A^* of some other sort, connected with A by the equation

$$\langle Af, g \rangle = \langle f, A^*g \rangle.$$

This statement and further ones about A^* are consequences of what has been said about bilinear forms in the preceding section.

THEOREM 1. *Let A be a bounded linear operator on $\mathfrak{H}$. Then there exists a unique bounded linear operator A^* on $\mathfrak{H}$, called the adjoint of A, such that*

(1) $\langle Af, g \rangle = \langle f, A^*g \rangle$ *for all $f \in \mathfrak{H}$ and all $g \in \mathfrak{H}$.*

In addition the equality $\|A^*\| = \|A\|$ holds.

PROOF. The equation $\varphi(f, g) = \langle Af, g \rangle$ defines a bounded bilinear form φ on $\mathfrak{H}$ by § 13 corollary 5.1 a). By § 13 theorem 5 b) there is a unique bounded linear operator A^* on $\mathfrak{H}$ such that $\varphi(f, g) = \langle f, A^*g \rangle$ for all $f \in \mathfrak{H}$ and all $g \in \mathfrak{H}$. Moreover, by the same two theorems, we have $\|A\| = \|\varphi\| = \|A^*\|$. $\square$

Example 1. Let $\mathfrak{H}$ be the n-dimensional unitary space $\mathbf{C}^n$ with basis $\{e_k\}_{k=1}^n$. Let the linear operator A be given by a matrix $A' = (\alpha_{k,l})_{1 \leqslant k, l \leqslant n}$ as in § 12 example 1. If $b = \sum_{k=1}^n \beta_k e_k$ and $c = \sum_{k=1}^n \gamma_k e_k$ are any vectors in $\mathfrak{H}$, then we have

$$\langle Ab, c \rangle = \langle \sum_{k=1}^n (\sum_{l=1}^n \alpha_{k,l} \beta_l) e_k, \sum_{k=1}^n \gamma_k e_k \rangle$$

$$= \sum_{k=1}^n \sum_{l=1}^n \bar{\gamma}_k \alpha_{k,l} \beta_l .$$

We may also imagine to have obtained this number by multiplying the matrix A' from the right by the column vector with components β_l and from the left by the row vector with components $\bar{\gamma}_k$. Let $A^{*'} = (\alpha^*_{k,l})_{1 \leqslant k, l \leqslant n}$ be the matrix with the elements $\alpha^*_{k,l} = \overline{\alpha_{l,k}}$ ($A^{*'}$ is the complex conjugate transpose of A' and commonly called the *adjoint matrix* of A'). Then we have

$$\langle b, A^*c \rangle = \langle \sum_{k=1}^n \beta_k e_k, \sum_{k=1}^n (\sum_{l=1}^n \alpha^*_{k,l} \gamma_l) e_k \rangle$$

$$= \sum_{k=1}^n \sum_{l=1}^n \beta_k \overline{\alpha^*_{k,l} \gamma_l} = \sum_{k=1}^n \sum_{l=1}^n \bar{\gamma}_l \alpha_{l,k} \beta_k = \langle Ab, c \rangle .$$

The operator A^* corresponding to the matrix $A^{*'}$ is therefore indeed the adjoint of A, and adjoint operators correspond to adjoint matrices.

Example 2. Let A be the right shift operator in $\mathfrak{H} = \ell_2$ as in § 12 example 2, defined by

$$A \{\alpha_1, \alpha_2, \alpha_3, \ldots\} = \{0, \alpha_1, \alpha_2, \ldots\} .$$

It turns out that the adjoint operator A^* is the "*left shift operator*" given by

$$A^* \{\beta_1, \beta_2, \beta_3, \ldots\} = \{\beta_2, \beta_3, \beta_4, \ldots\} .$$

Indeed, for $a = \{\alpha_k\}_{k=1}^\infty$ and $b = \{\beta_k\}_{k=1}^\infty$ in ℓ_2 we have

$$\langle Aa, b \rangle = \sum_{k=1}^\infty \alpha_k \overline{\beta_{k+1}} = \langle a, A^*b \rangle .$$

With respect to the standard basis $\{e_k\}_{k=1}^\infty$ of ℓ_2 (cf. § 8 example 2) the action of A^* is described by

$$A^*e_k = e_{k-1} \quad \text{for } k \geqslant 2,$$
$$A^*e_1 = o .$$

In the corresponding matrix $A^{*'}$ the only non-zero elements are 1's above the diagonal. Thus $A^{*'}$ is again the adjoint of the matrix A'. Observe that

$$(2) \qquad\qquad A^*A = I \neq AA^*$$

(AA^* simply annihilates the first term of every sequence in ℓ_2). While A is isometric, A^* certainly isn't since $A^*e_1 = o$. Still we have $\|A\| = \|A^*\| = 1$.

Example 3. Let A be a Hilbert-Schmidt operator on $\mathfrak{H} = \mathfrak{L}_2(\beta, \gamma)$ with kernel a (cf. § 12 example 4). Define the function a^* on $X =]\beta, \gamma[\times]\beta, \gamma[$ by $a^*(\xi, \eta) = \overline{a(\eta, \xi)}$. Then we have

$$\int_X |a^*(\xi, \eta)|^2 \, \mathrm{d}(\xi, \eta) = \int_\beta^\gamma \int_\beta^\gamma |a(\eta, \xi)|^2 \, \mathrm{d}\xi \, \mathrm{d}\eta = \int_X |a(\xi, \eta)|^2 \, \mathrm{d}(\xi, \eta).$$

Therefore a^* belongs to $\mathfrak{L}_2(X)$ and is the kernel of another Hilbert-Schmidt operator A^* which, as it turns out, is the adjoint of A. Indeed we have

$$\langle Af, g \rangle = \int_\beta^\gamma \left[\int_\beta^\gamma a(\xi, \eta) f(\eta) \, \mathrm{d}\eta \right] \overline{g(\xi)} \, \mathrm{d}\xi$$

$$= \int_\beta^\gamma f(\eta) \left[\int_\beta^\gamma \overline{a^*(\eta, \xi) g(\xi)} \, \mathrm{d}\xi \right] \mathrm{d}\eta = \langle f, A^*g \rangle$$

(in the middle equality we have applied FUBINI's theorem).

In what follows we shall denote the operator $(A^*)^*$ by A^{**}.

THEOREM 2. *Let A and B be bounded linear operators on $\mathfrak{H}$. Then*

a) $A^{**} = A$;
b) $(\lambda A)^* = \bar{\lambda} A^*$;
c) $(A + B)^* = A^* + B^*$;
d) $(AB)^* = B^*A^*$;
e) *if A is invertible, then so is A^* and $(A^*)^{-1} = (A^{-1})^*$.*

PROOF.

a) $\langle A^*f, g \rangle = \overline{\langle g, A^*f \rangle} = \overline{\langle Ag, f \rangle} = \langle f, Ag \rangle$.
b) $\langle \lambda Af, g \rangle = \langle Af, \bar{\lambda}g \rangle = \langle f, A^*(\bar{\lambda}g) \rangle = \langle f, \lambda A^*g \rangle$.
c) $\langle (A + B)f, g \rangle = \langle Af, g \rangle + \langle Bf, g \rangle = \langle f, A^*g \rangle + \langle f, B^*g \rangle$
$\qquad\qquad\qquad = \langle f, (A^* + B^*) g \rangle$.

d) $\langle ABf, g \rangle = \langle Bf, A^*g \rangle = \langle f, B^*A^*g \rangle$.

e) Obviously we have $I^* = I$. From $AA^{-1} = A^{-1}A = I$ (cf. § 12 definition 5) and d) we conclude

$$(A^{-1})^*A^* = A^*(A^{-1})^* = I^* = I.$$

According to § 12 definition 5, A^* is invertible and its inverse is $(A^{-1})^*$. $\square$

THEOREM 3. *If A is a bounded linear operator on $\mathfrak{H}$, then*

$$\|A^*A\| = \|A\|^2.$$

PROOF. From

$$\|Af\|^2 = \langle Af, Af \rangle = \langle f, A^*Af \rangle \leqslant \|A^*A\| \|f\|^2$$

we conclude

$$\|Af\| \leqslant \|A^*A\|^{\frac{1}{2}} \|f\|,$$
$$\|A\|^2 \leqslant \|A^*A\|.$$

The converse inequality follows from

$$\|A^*A\| \leqslant \|A^*\| \|A\| = \|A\|^2. \ \square$$

There are some special types of bounded linear operators on $\mathfrak{H}$ that are worthwhile to be considered in some detail.

DEFINITION 1. A bounded linear operator A on $\mathfrak{H}$ is called *selfadjoint* (or *Hermitian*) if $A^* = A$. A is called *unitary* if $A^* = A^{-1}$. A is called *normal* if $AA^* = A^*A$.

Obviously selfadjoint bounded linear operators as well as unitary ones are normal. The right shift operator A in ℓ_2 (cf. example 2) is not normal by (2). By the same token, also the left shift operator in ℓ_2 is not normal. The operator $2iI$ in any Hilbert space $\mathfrak{H}$ is normal but neither selfadjoint nor unitary: we have $(2iI)^* = -2iI$ and

$$(2iI)(2iI)^* = (2iI)^*(2iI) = 4I.$$

From what has been said in example 1 it follows that a linear operator A in n-dimensional unitary space $\mathbf{C}^n$ is selfadjoint iff, with respect to any given basis, the corresponding matrix A' is *selfadjoint* (or *Hermitian*) in the traditional sense, i.e. $\alpha_{k,l} = \overline{\alpha_{l,k}}$. Furthermore, A is unitary iff the corresponding matrix A' is *unitary* in the usual sense, in other words iff the adjoint matrix coincides with the inverse matrix of A' (a *real* unitary matrix is also commonly called *orthogonal*).

Let us look for corresponding statements in case of a Hilbert-Schmidt operator.

Example 4. Let A be a Hilbert-Schmidt operator on $\mathfrak{L}_2(\beta, \gamma)$ $(-\infty \leqslant \beta < \gamma \leqslant +\infty)$ with kernel a. In § 25 example 3 we shall show that A cannot possibly be unitary. The operator A is selfadjoint iff $a(\xi, \eta) = \overline{a(\eta, \xi)}$ almost everywhere on $X =]\beta, \gamma[\times]\beta, \gamma[$. Indeed, the equation $\langle Af, g \rangle = \langle f, Ag \rangle$ is equivalent with

$$\int_\beta^\gamma \int_\beta^\gamma a(\xi, \eta) \, f(\eta) \, \overline{g(\xi)} \, \mathrm{d}\eta \, \mathrm{d}\xi = \int_\beta^\gamma \int_\beta^\gamma f(\eta) \, \overline{a(\eta, \xi) \, g(\xi)} \, \mathrm{d}\xi \, \mathrm{d}\eta$$

and

(3) $$\int_X [a(\xi, \eta) - \overline{a(\eta, \xi)}] \, \overline{g(\xi)} \, f(\eta) \, \mathrm{d}(\xi, \eta) = 0.$$

This equation is satisfied for all f and g in $\mathfrak{L}_2(\beta, \gamma)$ iff the function $a(x, y) - \overline{a(y, x)} \in \mathfrak{L}_2(X)$ is orthogonal to every function $g(x)\overline{f(y)} \in \mathfrak{L}_2(X)$ where $g(x) = g \in \mathfrak{L}_2(\beta, \gamma)$ and $f(y) = f \in \mathfrak{L}_2(\beta, \gamma)$. Since these functions span $\mathfrak{L}_2(X)$ by § 9 corollary 10.2 this in turn is true iff $a(x, y) - \overline{a(y, x)} = o \in \mathfrak{L}_2(X)$ (cf. § 7 corollary 7.2) or, equivalently, iff $a(\xi, \eta) = \overline{a(\eta, \xi)}$ almost everywhere on X (cf. example 3).

Remark 1. If we replace in (3) the function $\overline{a(y, x)} \in \mathfrak{L}_2(X)$ by the kernel $b(x, y) \in \mathfrak{L}_2(X)$ of some other Hilbert-Schmidt operator B on $\mathfrak{L}_2(\beta, \gamma)$, then the reasoning just described shows that we have

(4) $\langle Af, g \rangle = \langle Bf, g \rangle$ for all $f \in \mathfrak{H}$ and all $g \in \mathfrak{H}$

if and only if $a(\xi, \eta) = b(\xi, \eta)$ a.e. on X. On the other hand, (4) is equivalent with each of the following assertions:

$$\langle Af - Bf, g \rangle = 0 \quad \text{for all } f \in \mathfrak{H} \text{ and all } g \in \mathfrak{H},$$
$$Af = Bf \qquad\qquad \text{for all } f \in \mathfrak{H},$$
$$A = B.$$

The conclusion is that two Hilbert-Schmidt operators A, B on $\mathfrak{L}_2(\beta, \gamma)$ coincide if and only if their kernels a, b coincide a.e. on X. In particular it follows that a Hilbert-Schmidt operator (essentially) uniquely determines its kernel.

The following three theorems characterize selfadjoint, unitary, and normal bounded linear operators in some other way. Observe that, by definition 1, a bounded linear operator A is selfadjoint iff

$$\langle Af, g \rangle = \langle f, Ag \rangle \quad \text{for all } f \in \mathfrak{H} \text{ and all } g \in \mathfrak{H}$$

and A is unitary iff it is invertible and

$$(5) \qquad \langle Af, g \rangle = \langle f, A^{-1}g \rangle \quad \text{for all } f \in \mathfrak{H} \text{ and all } g \in \mathfrak{H}.$$

THEOREM 4. *A bounded linear operator on $\mathfrak{H}$ is unitary iff it is an automorphism.*
PROOF.
   a) Suppose the operator $A \in \mathscr{B}$ is unitary (cf. § 12 theorem 3). Then

$$\langle Af, Ag \rangle = \langle f, A^{-1}Ag \rangle = \langle f, g \rangle$$

and therefore A is an automorphism (cf. § 10 definition 1).
   b) Let A be an automorphism. Then we have

$$\langle Af, g \rangle = \langle Af, AA^{-1}g \rangle = \langle f, A^{-1}g \rangle. \ \square$$

Remark 2. A second look at part a) of the proof of theorem 4 reveals that there in fact the following statement has been proved: if A is a linear one-to-one mapping of $\mathfrak{H}$ onto $\mathfrak{H}$ satisfying (5) (A^{-1} now simply denotes the reversed mapping), then A is an automorphism and therefore bounded. This is a special case of a theorem which we shall come across later and which asserts the boundedness of a linear operator A possessing an adjoint operator A^* in the sense of (1) (cf. § 17 corollary 8.1).

THEOREM 5. *Let A be a bounded linear operator on $\mathfrak{H}$. The following statements are equivalent:*
   a) *A is selfadjoint.*
   b) *The bilinear form φ on $\mathfrak{H}$ defined by $\varphi(f, g) = \langle Af, g \rangle$ is symmetric.*
   c) *The quadratic form $\hat{\varphi}$ on $\mathfrak{H}$ defined by $\hat{\varphi}(f) = \langle Af, f \rangle$ is real.*

PROOF.
   a)$\Rightarrow$c): $\overline{\hat{\varphi}(f)} = \overline{\langle Af, f \rangle} = \langle f, Af \rangle = \langle Af, f \rangle = \hat{\varphi}(f)$.
   c)$\Rightarrow$b): cf. § 13 theorem 2.
   b)$\Rightarrow$a): $\langle Af, g \rangle = \varphi(f, g) = \overline{\varphi(g, f)} = \overline{\langle Ag, f \rangle} = \langle f, Ag \rangle. \ \square$

Properly we would always have to repeat the attribute "linear" when speaking of normal (in particular selfadjoint or unitary) bounded operators on $\mathfrak{H}$. However, since adjoints are only defined for linear operators (cf. § 17 definition 1) and since a unitary operator on $\mathfrak{H}$ is necessarily bounded (cf. remark 2) we shall from now on simply talk about bounded selfadjoint operators on $\mathfrak{H}$, bounded normal operators on $\mathfrak{H}$, and unitary operators on $\mathfrak{H}$. The following corollary should be compared with § 13 corollary 5.2.

COROLLARY 5.1. *If A is a bounded selfadjoint operator on $\mathfrak{H}$, then*

$$\|A\| = \sup_{\|f\|=1} |\langle Af, f \rangle|.$$

PROOF. Defining $\varphi(f, g) = \langle Af, g \rangle$ as in the proof of theorem 1 we obtain by § 13 corollary 5.1a) and § 13 theorem 4

$$\|A\| = \|\varphi\| = \|\hat{\varphi}\| = \sup_{\|f\|=1} |\hat{\varphi}(f)| = \sup_{\|f\|=1} |\langle Af, f \rangle|. \ \Box$$

THEOREM 6. *A bounded linear operator A on $\mathfrak{H}$ is normal iff*

$$\|Af\| = \|A^*f\| \quad \textit{for all } f \in \mathfrak{H}.$$

PROOF. Define the bilinear forms φ and ψ on $\mathfrak{H}$ by

$$\varphi(f, g) = \langle A^*Af, g \rangle,$$
$$\psi(f, g) = \langle AA^*f, g \rangle.$$

We obtain

$$\hat{\varphi}(f) = \langle A^*Af, f \rangle = \langle Af, Af \rangle = \|Af\|^2$$
$$\hat{\psi}(f) = \langle AA^*f, f \rangle = \langle A^*f, A^*f \rangle \doteq \|A^*f\|^2.$$

By § 13 corollary 1.1 we have $\hat{\varphi} = \hat{\psi}$ iff $\varphi = \psi$, in other words iff

$$\langle A^*Af, g \rangle = \langle AA^*f, g \rangle \quad \text{for all } f \in \mathfrak{H} \text{ and all } g \in \mathfrak{H}.$$

This again is the case iff $A^*A = AA^*$. $\Box$

Why should selfadjoint and normal operators be so interesting? There is a large class of very special and simple bounded selfadjoint operators, namely the projection operators which we shall investigate in the following section. Every bounded selfadjoint operator on $\mathfrak{H}$ (in fact even every unbounded selfadjoint operator in $\mathfrak{H}$; cf. § 17 definition 1 and § 17 definition 4) can be built up in some sense from projections (cf. § 28 theorem 2, § 34 theorem 1, § 39 theorem 5). This is the central result of spectral theory which will occupy us essentially beginning with § 21. The point here is that every bounded linear operator A on $\mathfrak{H}$ is a linear combination of two bounded selfadjoint operators B and C which even commute ($BC = CB$) if A is normal.

THEOREM 7. *Let A be a bounded linear operator on $\mathfrak{H}$. Then there are two bounded selfadjoint operators B and C on $\mathfrak{H}$ such that $A = B + iC$, and this*

requirement uniquely determines B and C. The operator A is normal iff BC = CB.
PROOF. Suppose there exist bounded selfadjoint operators B and C on $\mathfrak{H}$ as described in the theorem. Then we have

$$
\begin{aligned}
A &= B + \mathrm{i}C, \\
A^* &= B^* - \mathrm{i}C^* = B - \mathrm{i}C, \\
A + A^* &= 2B, \\
A - A^* &= 2\mathrm{i}C.
\end{aligned}
\tag{6}
$$

We therefore necessarily arrive at the formulas

$$
\begin{cases}
B = \tfrac{1}{2}(A + A^*), \\
C = \dfrac{1}{2\mathrm{i}}(A - A^*).
\end{cases}
\tag{7}
$$

Conversely, for every bounded linear operator A on $\mathfrak{H}$, the bounded linear operators B and C defined by (7) are selfadjoint and satisfy (6). If A is normal, then we moreover have

$$
BC = \frac{1}{4\mathrm{i}}(A^2 + A^*A - AA^* - A^{*2}) = \frac{1}{4\mathrm{i}}(A^2 - A^{*2}) = CB.
$$

Conversely, from $BC = CB$ we conclude

$$
AA^* = (B + \mathrm{i}C)(B - \mathrm{i}C) = (B - \mathrm{i}C)(B + \mathrm{i}C) = A^*A. \quad \square
$$

Exercise. Let A be a bounded linear operator on $\mathfrak{H}$.
a) A is an isometry iff $A^*A = I$.
b) A is unitary iff A and A^* are isometries.
c) Let A be an isometry and define $\mathfrak{M} = A\mathfrak{H} = \{Af : f \in \mathfrak{H}\}$ (cf. § 10 exercise). Then

$$
\begin{aligned}
\langle A^*f, A^*g \rangle &= \langle f, g \rangle \quad \text{for all } f \in \mathfrak{M} \text{ and all } g \in \mathfrak{M}, \\
A^*f &= o \qquad\quad \text{for all } f \in \mathfrak{M}^\perp.
\end{aligned}
$$

d) If A is an isometry, then $Af = f$ iff $A^*f = f$. (Hint: if $A^*f = f$ compute $\|Af - f\|^2$.)

§ 15. Projection operators

We already know that every subspace $\mathfrak{M} \subset \mathfrak{H}$ gives rise to a bounded linear operator $P = P_{\mathfrak{M}}$ on $\mathfrak{H}$ which maps every vector $f \in \mathfrak{H}$ onto its projection upon $\mathfrak{M}$ (cf. § 11 example 2). As we have already done in § 12, we shall call P the *projection operator* corresponding to $\mathfrak{M}$ or, more briefly, the *projection* upon

$\mathfrak{M}$ (there is hardly a chance for confusion with the projection of a single vector upon $\mathfrak{M}$). If we already know that P is a projection operator, then the corresponding subspace $\mathfrak{M}$ can be recovered in various ways as indicated in the following theorem.

THEOREM 1. *If P is the projection upon a subspace $\mathfrak{M}$, then*

$$\mathfrak{M} = \{f : Pf = f\} = \{f : \|Pf\| = \|f\|\} = \{Pg : g \in \mathfrak{H}\}.$$

PROOF. We already know that $Pg \in \mathfrak{M}$ for all $g \in \mathfrak{H}$ and $Pf = f$ for all $f \in \mathfrak{M}$. Hence $\mathfrak{M}$ is contained in all sets mentioned above and even coincides with the last one. On the other hand, for $f \notin \mathfrak{M}$ we have $f = Pf + P^{\perp}f$ ($P^{\perp}$ denoting the projection upon $\mathfrak{M}^{\perp}$) and $P^{\perp}f \neq o$. Therefore $\|f\|^2 = \|Pf\|^2 + \|P^{\perp}f\|^2 > \|Pf\|^2$ and f is not contained in any of the sets mentioned above. $\square$

Remark 1. It seems reasonable to denote the set $\{Pg : g \in \mathfrak{H}\}$ by $P\mathfrak{H}$. In general, if A is any linear operator in $\mathfrak{H}$ with domain $\mathfrak{D}_A$ (cf. § 12 definition 1), then for any subset $\mathfrak{A} \subset \mathfrak{D}_A$ we shall write

$$A\mathfrak{A} = \{Af : f \in \mathfrak{A}\}.$$

In particular, the set $A\mathfrak{D}_A$ is called the *range* of A. According to theorem 1 the range of a projection coincides with the corresponding subspace.

Example 1. Let $]\gamma, \delta[$ be a subinterval of the interval $]\alpha, \beta[\subset \mathbf{R}$ and let

$$\mathfrak{M} = \{f \in \mathfrak{L}_2(\alpha, \beta) : f(\xi) = 0 \text{ for almost all } \xi \in]\alpha, \beta[\setminus]\gamma, \delta[\}.$$

(Here and in the sequel the symbol $\setminus$ denotes set-theoretic subtraction.) Then $\mathfrak{M}$ is a subspace of $\mathfrak{L}_2(\alpha, \beta)$ which may be identified with $\mathfrak{L}_2(\gamma, \delta)$ (cf. § 6 example 4, § 7 example 1). If P is the projection upon $\mathfrak{M}$, then

$$Pf(\xi) = \begin{cases} f(\xi) & \text{for } \xi \in]\gamma, \delta[, \\ 0 & \text{for } \xi \in]\alpha, \beta[\setminus]\gamma, \delta[. \end{cases}$$

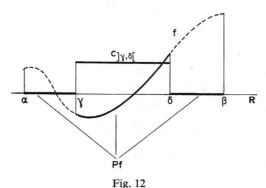

Fig. 12

The transition from $f \in \mathfrak{L}_2(\alpha, \beta)$ to Pf may also be realized by a multiplication of the function f with the function $c_{]\gamma, \delta[}$ (the "*characteristic function*"
of the interval $]\gamma, \delta[$) defined by

$$c_{]\gamma, \delta[}(\xi) = \begin{cases} 1 & \text{for } \xi \in]\gamma, \delta[, \\ 0 & \text{for } \xi \in]\alpha, \beta[\setminus]\gamma, \delta[\end{cases}$$

(fig. 12). In other words, we have (with pointwise multiplication)

$$Pf = c_{]\gamma, \delta[} f \quad \text{for all } f \in \mathfrak{L}_2(\alpha, \beta).$$

The assertions of theorem 1 may then be formulated in the following way:

$$\mathfrak{M} = \{f \in \mathfrak{L}_2(\alpha, \beta): c_{]\gamma, \delta[} f = f\}$$

$$= \{f \in \mathfrak{L}_2(\alpha, \beta): \int_\alpha^\beta |f(\xi)|^2 \, d\xi = \int_\gamma^\delta |f(\xi)|^2 \, d\xi\}$$

$$= \{c_{]\gamma, \delta[} g: g \in \mathfrak{L}_2(\alpha, \beta)\}.$$

Note that since the values of the functions $f \in \mathfrak{L}_2(\alpha, \beta)$ in the points γ and
δ are irrelevant we might in place of $c_{]\gamma, \delta[}$ as well have used the characteristic
function $c_{[\gamma, \delta]}$ of the closed interval $[\gamma, \delta]$.

How should one recognize that a given bounded linear operator P is a
projection? There is a simple algebraic answer to this question.

THEOREM 2. *A bounded linear operator P on $\mathfrak{H}$ is a projection iff $P = P^* = P^2$.*
PROOF.
   a) Suppose P is the projection upon the subspace $\mathfrak{M}$ and $P^\perp$ is the projection upon $\mathfrak{M}^\perp$. Then because of $Pf \in \mathfrak{M}$ we have

$$P^2 f = P(Pf) = Pf$$

and because of $\langle Pf, P^\perp g \rangle = \langle P^\perp f, Pg \rangle = 0$ we have

$$\langle Pf, g \rangle = \langle Pf, Pg + P^\perp g \rangle = \langle Pf, Pg \rangle = \langle Pf + P^\perp f, Pg \rangle = \langle f, Pg \rangle$$

which proves $P^2 = P$ and $P^* = P$.
   b) Suppose $P = P^* = P^2$. The set $\mathfrak{M} = \{f: Pf = f\}$ is obviously a linear
manifold. In fact it is even closed and therefore a subspace, since for any
sequence $\{f_n\}_{n=1}^\infty \subset \mathfrak{M}$ converging to a vector $f = \lim_{n \to \infty} f_n \in \mathfrak{M}$ we have

$$Pf = \lim_{n \to \infty} Pf_n = \lim f_n = f.$$

Now every vector $g \in \mathfrak{H}$ can be split into a sum of the form

$$g = Pg + (g - Pg)$$

where $Pg \in \mathfrak{M}$ since $P(Pg) = P^2g = Pg$. In order to show that P is the projection upon $\mathfrak{M}$ we only have to verify that $g - Pg \in \mathfrak{M}^\perp$. This is indeed the case since for every $h \in \mathfrak{M}$ we have

$$\langle g - Pg, h \rangle = \langle g - Pg, Ph \rangle = \langle Pg - P^2g, h \rangle = \langle Pg - Pg, h \rangle = 0. \ \square$$

Remark 2. A linear operator A in $\mathfrak{H}$ is called *idempotent* if $A^2 = A$. By theorem 2, among all bounded linear operators on $\mathfrak{H}$ the projections are characterized as being selfadjoint and idempotent. In particular also the zero operator and the identity operator are projections.

While the set $\mathscr{B}$ of all bounded linear operators on $\mathfrak{H}$ is a Banach algebra (cf. § 12 theorem 3), the subset of all projections does not nearly possess a similarly nice algebraic or topological structure. For example, if P is a projection different from the zero operator, then λP is a projection iff $\lambda = 1$ or $\lambda = 0$ as follows from the relations $\lambda P = \bar\lambda P = \lambda^2 P$ which have to be satisfied by theorem 2. There remains the question under which circumstances the sum, difference, or product of two projections again is a projection. A topological question of this sort will be treated in § 30 theorem 5.

THEOREM 3. *If P is the projection upon the subspace $\mathfrak{M}$ and $P^\perp$ is the projection upon $\mathfrak{M}^\perp$, then $P^\perp = I - P$ and $\mathfrak{M}^\perp = \{f : Pf = o\}$.*
PROOF. By § 7 theorem 8′ we have

$$P^\perp g = g - Pg = (I - P) g \,.$$

By theorem 1 we conclude

$$\mathfrak{M}^\perp = \{f : (I - P) f = f\} = \{f : Pf = o\}. \ \square$$

THEOREM 4. *Let P and Q be the projections upon the subspaces $\mathfrak{M}$ and $\mathfrak{N}$ respectively. The following statements are equivalent:*
  a) *$PQ = QP$;*
  b) *PQ is a projection;*
  c) *QP is a projection.*

PROOF. Because of the symmetry of the statements b) and c) it suffices to prove a)⇔b). We use theorem 2.
  a)⇒b) $(PQ)^* = Q^*P^* = QP = PQ$,
        $(PQ)^2 = PQPQ = PPQQ = PQ$.
  b)⇒a) $QP = Q^*P^* = (PQ)^* = PQ$. $\square$

COROLLARY 4.1. *If PQ is a projection, then it is the projection upon* $\mathfrak{M}\cap\mathfrak{N}$.
PROOF. By theorem 1 we have to show $\mathfrak{M}\cap\mathfrak{N}=PQ\mathfrak{H}$. Indeed, since $PQ=QP$
from $PQ\mathfrak{H}\subset P\mathfrak{H}\subset\mathfrak{M}$ and $PQ\mathfrak{H}=QP\mathfrak{H}\subset Q\mathfrak{H}\subset\mathfrak{N}$ we conclude $PQ\mathfrak{H}\subset\mathfrak{M}\cap\mathfrak{N}$.
On the other hand, for $f\in\mathfrak{M}\cap\mathfrak{N}$ we have $f=Qf=PQf$ and therefore
$\mathfrak{M}\cap\mathfrak{N}\subset PQ\mathfrak{H}$. $\square$

THEOREM 5. *Let P and Q be the projections upon the subspaces* $\mathfrak{M}$ *and* $\mathfrak{N}$ *respectively. The following statements are equivalent:*

a) $\mathfrak{M}\perp\mathfrak{N}$;
b) $P\mathfrak{N}=\mathfrak{O}$;
c) $Q\mathfrak{M}=\mathfrak{O}$;
d) $PQ=0$;
e) $QP=0$;
f) $P+Q$ is a projection.

PROOF. Because of the symmetry of the statements b), c) and d), e) it suffices
to prove the equivalence of a), b), d), and f).

a)$\Rightarrow$b) Since $\mathfrak{N}\subset\mathfrak{M}^\perp$ we have $P\mathfrak{N}=\mathfrak{O}$ by theorem 3.
b)$\Rightarrow$d) Since $Qf\in\mathfrak{N}$ we have $PQf=o$ for all $f\in\mathfrak{H}$.
d)$\Rightarrow$f) Since $QP=(PQ)^*=O$ we have

$$(P+Q)^2 = P^2 + PQ + QP + Q^2 = P+Q,$$
$$(P+Q)^* = P^* + Q^* = P+Q.$$

f)$\Rightarrow$a) For $f\in\mathfrak{M}$ we get

$$\|f\|^2 \geqslant \|(P+Q)f\|^2 = \langle (P+Q)f, (P+Q)f \rangle$$
$$= \langle (P+Q)f, f \rangle = \langle Pf, f \rangle + \langle Qf, f \rangle$$
$$= \|f\|^2 + \|Qf\|^2,$$

therefore $Qf=o$ and, for all $g\in\mathfrak{N}$,

$$\langle f, g \rangle = \langle f, Qg \rangle = \langle Qf, g \rangle = 0. \square$$

COROLLARY 5.1. *If $P+Q$ is a projection, then it is the projection upon* $\mathfrak{M}+\mathfrak{N}$.
PROOF. $\mathfrak{M}+\mathfrak{N}$ is a subspace since $\mathfrak{M}\perp\mathfrak{N}$ (cf. § 7 theorem 2). For $f\in\mathfrak{M}+\mathfrak{N}$
we have $f=Pf+Qf=(P+Q)f$ (cf. § 7 corollary 5.1.), and conversely this
equation implies $f\in\mathfrak{M}+\mathfrak{N}$. Thus we have $\mathfrak{M}+\mathfrak{N}=\{f:(P+Q)f=f\}$. $\square$

Remark 3. In view of the equivalence of a), d), and e), the projections P and
Q are called *orthogonal* (notation: $P\perp Q$) if $PQ=O$ or, equivalently, $QP=O$.

If $\{P_k\}_{k=1}^n$ is a sequence of mutually orthogonal projections upon the subspaces $\mathfrak{M}_k$ respectively, then an inductive argument shows that $\sum_{k=1}^n P_k$ is the projection upon the subspace $\sum_{k=1}^n \mathfrak{M}_k$.

THEOREM 6. *Let P and Q be the projections upon the subspaces $\mathfrak{M}$ and $\mathfrak{N}$ respectively. The following statements are equivalent:*

  a) $\mathfrak{M} \subset \mathfrak{N}$;
  b) $QP = P$;
  c) $PQ = P$;
  d) $\|Pf\| \leqslant \|Qf\|$ *for all $f \in \mathfrak{H}$*;
  e) $\langle (Q-P)f, f \rangle \geqslant 0$ *for all $f \in \mathfrak{H}$*;
  f) $Q - P$ *is a projection.*

PROOF.

  a)$\Rightarrow$b) Since $Pf \in \mathfrak{M} \subset \mathfrak{N}$ we have $QPf = Pf$ for all $f \in \mathfrak{H}$.
  b)$\Rightarrow$c) $PQ = P^*Q^* = (QP)^* = P^* = P$.
  c)$\Rightarrow$f) $(Q-P)^* = Q^* - P^* = Q - P$,
          $(Q-P)^2 = Q^2 - QP - PQ + P^2 = Q - (PQ)^* - P + P = Q - P$.
  f)$\Rightarrow$e) $\langle (Q-P)f, f \rangle = \langle (Q-P)f, (Q-P)f \rangle = \|(Q-P)f\|^2 \geqslant 0$.
  e)$\Rightarrow$d) $\|Qf\|^2 - \|Pf\|^2 = \langle Qf, f \rangle - \langle Pf, f \rangle = \langle (Q-P)f, f \rangle \geqslant 0$.
  d)$\Rightarrow$a) For $f \in \mathfrak{M}$ we have $\|f\| = \|Pf\| \leqslant \|Qf\| \leqslant \|f\|$ which implies $f \in \mathfrak{N}$ by theorem 1. $\square$

COROLLARY 6.1. *If $Q - P$ is a projection, then it is the projection upon $\mathfrak{N} - \mathfrak{M} = \mathfrak{N} \cap \mathfrak{M}^\perp$ (cf. § 7 definition 1).*
PROOF. $Q - P = Q(I - P)$ is the projection upon $\mathfrak{N} \cap \mathfrak{M}^\perp$ by corollary 4.1 and theorem 3. $\square$

Exercise. Let $\{P_k\}_{k=1}^\infty$ be a sequence of projections upon the subspaces $\mathfrak{M}_k$ $(1 \leqslant k < \infty)$ respectively. Let $\mathfrak{D}$ be the linear manifold of all vectors $f \in \mathfrak{H}$ for which the series $\sum_{k=1}^\infty P_k f$ converges and define the linear operator P in $\mathfrak{H}$ by

$$Pf = \sum_{k=1}^\infty P_k f \quad \text{for all } f \in \mathfrak{D}.$$

  a) The series $\sum_{k=1}^\infty P_k$ converges in $\mathscr{B}$ iff $P_k = O$ for all $k \geqslant k_0$. (Hint: use § 3 remark 3.)
  b) If $P_k \perp P_l$ for $k \neq l$, $1 \leqslant k < \infty$, $1 \leqslant l < \infty$, then $\mathfrak{D} = \mathfrak{H}$ and P is the projection upon the subspace $\mathfrak{M} = \sum_{k=1}^\infty \mathfrak{M}_k$ (cf. § 7 theorem 4).
  c) If $\mathfrak{D} = \mathfrak{H}$ and if P is a projection, then $P_k \perp P_l$ for $k \neq l$, $1 \leqslant k < \infty$, $1 \leqslant l < \infty$. (Hint: imitate the proof of the implication f)$\Rightarrow$a) in the proof of theorem 5.)

§ 16. The Fourier-Plancherel operator

Recall how we have defined a Hilbert-Schmidt operator A on $\mathfrak{L}_2(-\infty, +\infty)$: if a is any function in $\mathfrak{L}_2(\mathbf{R}^2)$ (the "square" X now coincides with the real plane $\mathbf{R}^2$), then for every $f \in \mathfrak{L}_2(-\infty, +\infty)$ the function Af is defined by

$$(1) \qquad Af(\xi) = \int_{-\infty}^{+\infty} a(\xi, \eta) f(\eta) \, d\eta \quad \text{for almost all } \xi \in \mathbf{R}$$

(cf. § 12 example 4). The function Af again belongs to $\mathfrak{L}_2(-\infty, +\infty)$ and (1) thus defines a linear operator on $\mathfrak{L}_2(-\infty, +\infty)$ which is in fact bounded.

This procedure of defining an operator breaks down if we try to use in place of the function $a \in \mathfrak{L}_2(\mathbf{R}^2)$ a function not belonging to $\mathfrak{L}_2(\mathbf{R}^2)$, even if it is as well behaved and bounded on $\mathbf{R}^2$ as the function $e^{-ixy}/\sqrt{2\pi}$. In fact, if we want to stick to Lebesgue integration, then the formula

$$(2) \qquad Ff(\xi) = \frac{1}{\sqrt{2\pi}} \int_{-\infty}^{+\infty} e^{-i\xi\eta} f(\eta) \, d\eta$$

defines a number $Ff(\xi)$ if an only if we impose on the function $f \in \mathfrak{L}_2(-\infty, +\infty)$ the additional condition of being (absolutely) integrable, i.e.

$$\int_{-\infty}^{+\infty} |f(\eta)| \, d\eta < \infty$$

(the function $1/(|x|+1)$ already provides an example of a non-integrable function in $\mathfrak{L}_2(-\infty, +\infty)$). But even then it is not yet clear at all whether the function Ff, defined by (2) for all $\xi \in \mathbf{R}$, again belongs to $\mathfrak{L}_2(-\infty, +\infty)$.

The best we can hope for is a large (more precisely: everywhere dense) linear manifold of functions $f \in \mathfrak{L}_2(-\infty, +\infty)$ for which (2) defines a function Ff again belonging to $\mathfrak{L}_2(-\infty, +\infty)$. We shall see that this hope is fully justified and that a small trick (more precisely: an application of § 11 theorem 3) even provides us with a unitary operator F defined everywhere (but not necessarily any more by formula (2)) on $\mathfrak{L}_2(-\infty, +\infty)$.

Let us start slowly: we shall first show that formula (2) in fact defines a function Ff in $\mathfrak{L}_2(-\infty, +\infty)$ if we take in place of f the Hermite function

$$h_n = e^{\frac{1}{2}x^2} \frac{d^n e^{-x^2}}{dx^n} \in \mathfrak{L}_2(-\infty, +\infty) \quad (0 \leqslant n < \infty)$$

introduced in § 9 theorem 6. The formula

$$h_n = H_n e^{-\frac{1}{2}x^2} \quad \text{for } 0 \leqslant n < \infty$$

where H_n is the Hermite polynomial of degree n already shows that h_n is indeed integrable on $\mathbf{R}$.

THEOREM 1.

$$\left. \begin{aligned} \frac{1}{\sqrt{2\pi}} \int_{-\infty}^{+\infty} e^{-i\xi\eta} h_n(\eta)\, d\eta &= (-i)^n h_n(\xi) \\ \frac{1}{\sqrt{2\pi}} \int_{-\infty}^{+\infty} e^{i\xi\eta} h_n(\eta)\, d\eta &= i^n h_n(\xi) \end{aligned} \right\} \quad \begin{aligned} &\text{for all } \xi \in \mathbf{R} \text{ and} \\ &\text{for } 0 \leqslant n < \infty. \end{aligned}$$

PROOF. It suffices to show the first formula since the second one follows from it after i has been replaced by $(-i)$. Observe that

$$\frac{d^k e^{-y^2}}{dy^k} = q_k(y)\, e^{-y^2},$$

$$\frac{\partial^k e^{-ixy+\frac{1}{2}y^2}}{\partial y^k} = r_k(x, y)\, e^{-ixy+\frac{1}{2}y^2},$$

where q_k and r_k are polynomials of degree k in x and y. Therefore we have

$$\lim_{\eta \to \pm\infty} \frac{\partial^k e^{-i\xi\eta+\frac{1}{2}\eta^2}}{\partial\eta^k} \cdot \frac{d^{n-k-1} e^{-\eta^2}}{d\eta^{n-k-1}} =$$

$$= \lim_{\eta \to \pm\infty} r_k(\xi, \eta)\, q_{n-k-1}(\eta)\, e^{-i\xi\eta - \frac{1}{2}\eta^2} = 0 \quad \text{for all } \xi \in \mathbf{R}.$$

These terms appear (and fortunately vanish) if we apply integration by parts n times to the integral to be computed. We therefore obtain

$$\frac{1}{\sqrt{2\pi}} \int_{-\infty}^{+\infty} e^{-i\xi\eta+\frac{1}{2}\eta^2} \frac{d^n e^{-\eta^2}}{d\eta^n}\, d\eta = \frac{(-1)^n}{\sqrt{2\pi}} \int_{-\infty}^{+\infty} e^{-\eta^2} \frac{\partial^n e^{-i\xi\eta+\frac{1}{2}\eta^2}}{\partial\eta^n}\, d\eta$$

$$= \frac{(-1)^n}{\sqrt{2\pi}} e^{\frac{1}{2}\xi^2} \int_{-\infty}^{+\infty} e^{-\eta^2} \frac{\partial^n e^{\frac{1}{2}(\eta-i\xi)^2}}{\partial\eta^n}\, d\eta$$

$$\left(\text{since for } \eta - i\xi = \zeta \text{ we have } \frac{\partial^n e^{\frac{1}{2}(\eta - i\xi)^2}}{\partial \eta^n} = \frac{d^n e^{\frac{1}{2}\zeta^2}}{d\zeta^n} = \frac{1}{(-i)^n} \frac{\partial^n e^{\frac{1}{2}(\eta - i\xi)^2}}{\partial \xi^n} \right)$$

$$= \frac{(-i)^n}{\sqrt{2\pi}} e^{\frac{1}{2}\xi^2} \int_{-\infty}^{+\infty} \frac{\partial^n e^{\frac{1}{2}(\eta - i\xi)^2 - \eta^2}}{\partial \xi^n} \, d\eta$$

$$= \frac{(-i)^n}{\sqrt{2\pi}} e^{\frac{1}{2}\xi^2} \int_{-\infty}^{+\infty} \frac{\partial^n e^{-\frac{1}{2}(\eta^2 + 2i\xi\eta + \xi^2)}}{\partial \xi^n} \, d\eta$$

(integration and differentiation may be interchanged since for every $k \geqslant 0$ the function

$$\frac{\partial^k e^{-\frac{1}{2}(y^2 + 2ixy + x^2)}}{\partial x^k} = s_k(x, y) \, e^{-\frac{1}{2}(y^2 + 2ixy + x^2)}$$

(s_k being a polynomial of degree k in x and y) is integrable on **R** for every fixed value of x, it is the integral of its partial derivative with respect to x for every fixed value of y, and this partial derivative is integrable on every rectangular strip $[\alpha, \beta] \times \mathbf{R} \subset \mathbf{R}^2$; cf. appendix B10)

$$= \frac{(-i)^n}{\sqrt{2\pi}} e^{\frac{1}{2}\xi^2} \frac{d^n}{d\xi^n} \left[e^{-\xi^2} \int_{-\infty}^{+\infty} e^{-\frac{1}{2}(\eta + i\xi)^2} \, d\eta \right]$$

$$= (-i)^n e^{\frac{1}{2}\xi^2} \frac{d^n e^{-\xi^2}}{d\xi^n}$$

(since $\displaystyle\int_{-\infty}^{+\infty} e^{-\frac{1}{2}(\eta + i\xi)^2} \, d\eta = \int_{-\infty}^{+\infty} e^{-\frac{1}{2}\zeta^2} \, d\zeta = \sqrt{2\pi}$; the attentive reader is warned that some analytic function theory has to be used in order to justify the last change of variable). $\square$

The result is pleasant indeed: if we ever succeed in defining a linear operator F in $\mathfrak{L}_2(-\infty, +\infty)$ which on integrable functions in $\mathfrak{L}_2(-\infty, +\infty)$ works according to formula (2), then we have

(3) $$\qquad\qquad\qquad F h_n = (-i)^n h_n \quad \text{for } 0 \leqslant n < \infty.$$

Let us try to use this knowledge now in order to define the operator F.

THEOREM 2. *Let* $\mathfrak{L} \subset \mathfrak{L}_2(-\infty, +\infty)$ *be the linear manifold of all finite linear combinations of the Hermite functions* h_n $(0 \leqslant n < \infty)$. *For every* $f \in \mathfrak{L}$ *the function* $F_0 f$ *defined on* **R** *by*

(4)
$$F_0 f(\xi) = \frac{1}{\sqrt{2\pi}} \int_{-\infty}^{+\infty} e^{-i\xi\eta} f(\eta) \, d\eta$$

belongs to $\mathfrak{L}_2(-\infty, +\infty)$, *and the operator* F_0 *defined on* $\mathfrak{L}$ *by* (4) *maps* $\mathfrak{L}$ *isometrically and therefore one-to-one onto* $\mathfrak{L}$. *Its inverse* F_0^{-1} *is given on* $\mathfrak{L}$ *by*

(5)
$$F_0^{-1} f(\xi) = \frac{1}{\sqrt{2\pi}} \int_{-\infty}^{+\infty} e^{i\xi\eta} f(\eta) \, d\eta.$$

PROOF. Every function $f \in \mathfrak{L}$ may be written in the form $f = \sum_{n=0}^{m} \alpha_n h_n$ (of course some α_n's may be zero). Since every h_n is integrable so is f. Using theorem 1 we obtain

$$F_0 f(\xi) = \frac{1}{\sqrt{2\pi}} \int_{-\infty}^{+\infty} e^{-i\xi\eta} \sum_{n=0}^{m} \alpha_n h_n(\eta) \, d\eta$$

$$= \sum_{n=0}^{m} \alpha_n (-i)^n h_n(\xi) \quad \text{for all } \xi \in \mathbf{R},$$

$$F_0 f = \sum_{n=0}^{m} (-i)^n \alpha_n h_n \in \mathfrak{L}_2(-\infty, +\infty).$$

If two functions $f = \sum_{n=0}^{m} \alpha_n h_n$ and $g = \sum_{n=0}^{m} \beta_n h_n$ in $\mathfrak{L}$ are given (without loss of generality we may assume that n runs through the same index set for both functions), then we have

$$\langle F_0 f, F_0 g \rangle = \langle \sum_{n=0}^{m} (-i)^n \alpha_n h_n, \sum_{n=0}^{m} (-i)^n \beta_n h_n \rangle$$

$$= \sum_{n=0}^{m} \alpha_n \bar{\beta}_n \|h_n\|^2 = \langle f, g \rangle.$$

In particular we obtain $\|F_0 f\| = \|f\|$ for all $f \in \mathfrak{L}$. Thus F_0 is isometric and therefore one-to-one (cf. § 10 remark 1). In order to show that $\mathfrak{L}$ is mapped onto $\mathfrak{L}$ by F_0 we observe that

$$F_0 \left(\sum_{n=0}^{m} i^n \alpha_n h_n \right) = \sum_{n=0}^{m} \alpha_n h_n = f.$$

Finally, by the second formula of theorem 1 we have

$$\frac{1}{\sqrt{2\pi}} \int\limits_{-\infty}^{+\infty} e^{i\xi\eta} F_0 f(\eta) \, d\eta = \sum_{n=0}^{m} (-i)^n \alpha_n \frac{1}{\sqrt{2\pi}} \int\limits_{-\infty}^{+\infty} e^{i\xi\eta} h_n(\eta) \, d\eta$$

$$= \sum_{n=0}^{m} \alpha_n h_n(\xi) = f(\xi) \quad \text{for all } \xi \in \mathbf{R},$$

which shows that the operator F_0^{-1} as given on $\mathfrak{L}$ by formula (5) indeed reverses the action of F_0 on $\mathfrak{L}$. $\square$

THEOREM 3. *Let $\mathfrak{L}$ be defined as in theorem 2. Then there exists a unique unitary operator F on $\mathfrak{L}_2(-\infty, +\infty)$, called the Fourier-Plancherel operator, such that for every $f \in \mathfrak{L}$ the functions $Ff \in \mathfrak{L}$ and $F^{-1}f \in \mathfrak{L}$ are given by*

$$Ff(\xi) = \frac{1}{\sqrt{2\pi}} \int\limits_{-\infty}^{+\infty} e^{-i\xi\eta} f(\eta) \, d\eta \quad \text{for all } \xi \in \mathbf{R},$$

$$F^{-1}f(\xi) = \frac{1}{\sqrt{2\pi}} \int\limits_{-\infty}^{+\infty} e^{i\xi\eta} f(\eta) \, d\eta \quad \text{for all } \xi \in \mathbf{R}.$$

PROOF. The normalized Hermite functions form a basis of $\mathfrak{L}_2(-\infty, +\infty)$ by § 9 theorem 6 and therefore span $\mathfrak{L}_2(-\infty, +\infty)$ (cf. § 8 theorem 3 c)). Consequently $\mathfrak{L}$ is everywhere dense in $\mathfrak{H} = \mathfrak{L}_2(-\infty, +\infty)$ (cf. § 6 theorem 4). By § 11 theorem 3 we can extend the isometric linear operators F_0 and F_0^{-1} defined on $\mathfrak{L}$ as in theorem 2 to bounded linear operators F and F' on $\mathfrak{H}$, and these extensions are determined uniquely by F_0 and F_0^{-1}. If $\{f_n\}_{n=1}^{\infty}$ and $\{g_n\}_{n=1}^{\infty}$ are sequences in $\mathfrak{L}$ converging to given vectors $f \in \mathfrak{H}$ and $g \in \mathfrak{H}$ respectively, then by the continuity of F and F' and by § 3 theorem 6 c) we get

$$\langle Ff, Fg \rangle = \lim_{n \to \infty} \langle Ff_n, Fg_n \rangle = \lim_{n \to \infty} \langle f_n, g_n \rangle = \langle f, g \rangle,$$

$$FF'f = \lim_{n \to \infty} FF'f_n = \lim_{n \to \infty} f_n = f,$$

$$F'Ff = \lim_{n \to \infty} F'Ff_n = \lim_{n \to \infty} f_n = f,$$

$$FF' = F'F = I.$$

We find that F is isometric and invertible in the sense of § 12 definition 5, its inverse F^{-1} being the operator F'. Thus F is an automorphism (cf. § 10 definition 1) and therefore unitary (cf. § 14 theorem 4). $\square$

By means of yet another trick we can even obtain an analytic formula describing the action of F and F^{-1} on arbitrary functions in $\mathfrak{L}_2(-\infty, +\infty)$.

THEOREM 4. *If F is the Fourier-Plancherel operator, then for every $f \in \mathfrak{L}_2(-\infty, +\infty)$ the functions Ff and $F^{-1}f$ are given by*

$$Ff(\xi) = \frac{1}{\sqrt{2\pi}} \frac{d}{d\xi} \int_{-\infty}^{+\infty} \frac{e^{-i\xi\eta} - 1}{-i\eta} f(\eta) \, d\eta \quad \textit{for almost all } \xi \in \mathbf{R},$$

$$F^{-1}f(\xi) = \frac{1}{\sqrt{2\pi}} \frac{d}{d\xi} \int_{-\infty}^{+\infty} \frac{e^{i\xi\eta} - 1}{i\eta} f(\eta) \, d\eta \quad \textit{for almost all } \xi \in \mathbf{R}.$$

As a consequence, if f is also integrable on $\mathbf{R}$, then

$$Ff(\xi) = \frac{1}{\sqrt{2\pi}} \int_{-\infty}^{+\infty} e^{-i\xi\eta} f(\eta) \, d\eta \quad \textit{for almost all } \xi \in \mathbf{R},$$

$$F^{-1}f(\xi) = \frac{1}{\sqrt{2\pi}} \int_{-\infty}^{+\infty} e^{i\xi\eta} f(\eta) \, d\eta \quad \textit{for almost all } \xi \in \mathbf{R}.$$

PROOF. Let $\mathfrak{L}$ be defined as in theorem 2 and suppose $\{f_n\}_{n=1}^{\infty}$ is a sequence in $\mathfrak{L}$ converging to the given function $f \in \mathfrak{L}_2(-\infty, +\infty)$. Let the function $c_\tau \in \mathfrak{L}_2(-\infty, +\infty)$ be defined for $\tau \geq 0$ and $\tau < 0$ respectively by

$$c_\tau(\xi) = \begin{cases} 1 & \text{for } \xi \in [0, \tau] \text{ or } \xi \in [\tau, 0] \text{ respectively}, \\ 0 & \text{for } \xi \notin [0, \tau] \text{ or } \xi \notin [\tau, 0] \text{ respectively}. \end{cases}$$

(The function c_τ is the characteristic function of the interval $[0, \tau]$ or $[\tau, 0]$ respectively; cf. § 15 fig. 12. In the further computations we formally only treat the case $\tau \geq 0$. The case $\tau < 0$ is treated analogously.) Then we have $Ff = \lim_{n \to \infty} Ff_n$ and

$$\int_0^\tau Ff(\xi) \, d\xi = \langle Ff, c_\tau \rangle = \lim_{n \to \infty} \langle Ff_n, c_\tau \rangle$$

$$= \lim_{n \to \infty} \int_0^\tau \frac{1}{\sqrt{2\pi}} \int_{-\infty}^{+\infty} e^{-i\xi\eta} f_n(\eta) \, d\eta \, d\xi.$$

Observe at this point that the function $e^{-ixy}f_n(y)$ is Lebesgue measurable and integrable on the rectangular strip $[0, \tau] \times\ ]-\infty, +\infty[$ in the plane, since f_n is integrable on $\mathbf{R}$ and

$$\int\limits_0^\tau \int\limits_{-\infty}^{+\infty} |e^{-i\xi\eta}f_n(\eta)|\ d\eta\ d\xi = \tau \int\limits_{-\infty}^{+\infty} |f_n(\eta)|\ d\eta < \infty .$$

Applying FUBINI's theorem (cf. appendix B8) we therefore get

$$\int\limits_0^\tau Ff(\xi)\ d\xi = \lim_{n\to\infty} \frac{1}{\sqrt{2\pi}} \int\limits_{-\infty}^{+\infty} \left[\int\limits_0^\tau e^{-i\xi\eta}\ d\xi \right] f_n(\eta)\ d\eta$$

$$= \lim_{n\to\infty} \frac{1}{\sqrt{2\pi}} \int\limits_{-\infty}^{+\infty} \frac{e^{-i\tau\eta}-1}{-i\eta} f_n(\eta)\ d\eta .$$

Here we observe that for every $\tau \in \mathbf{R}$ the function

$$g_\tau = \frac{e^{-i\tau y}-1}{-iy}$$

belongs to $\mathfrak{L}_2(-\infty, +\infty)$. Indeed, defining

$$g_\tau(0) = \lim_{\eta\to 0} \frac{e^{-i\tau\eta}-1}{-i\eta} = \tau$$

we see that g_τ is continuous on $\mathbf{R}$, therefore in particular bounded on $[-1, +1]$, and that

$$|g_\tau(\eta)| \leqslant 2/\eta \quad \text{for } \eta \notin [-1, +1] .$$

We obtain

$$\int\limits_0^\tau Ff(\xi)\ d\xi = \lim_{n\to\infty} \frac{1}{\sqrt{2\pi}} \langle g_\tau, f_n \rangle = \frac{1}{\sqrt{2\pi}} \langle g_\tau, f \rangle$$

$$= \frac{1}{\sqrt{2\pi}} \int\limits_{-\infty}^{+\infty} \frac{e^{-i\tau\eta}-1}{-i\eta} f(\eta)\ d\eta .$$

Since the function $Ff \in \mathfrak{L}_2(-\infty, +\infty)$ is integrable on every finite interval (cf. § 5 corollary 2.1), the function $h(x) = \int_0^x Ff(\xi)\, d\xi$ is absolutely continuous on every finite interval in $\mathbf{R}$ (cf. appendix B9; for $\tau < 0$ we have $h(\tau) = -\int_\tau^0 Ff(\xi)\, d\xi$. We conclude that in the above chain of equalities the first member and therefore also the last member is almost everywhere on $\mathbf{R}$ differentiable with respect to τ and furthermore

$$Ff(\tau) = \frac{d}{d\tau} \int_0^\tau Ff(\xi)\, d\xi$$

$$= \frac{1}{\sqrt{2\pi}} \frac{d}{d\tau} \int_{-\infty}^{+\infty} \frac{e^{-i\tau\eta} - 1}{-i\eta} f(\eta)\, d\eta \quad \text{for almost all } \tau \in \mathbf{R}.$$

If f is also integrable on $\mathbf{R}$, then the function $\{(e^{-ixy} - 1)/-iy\} f(y)$ is integrable on $\mathbf{R}$ for every fixed value of x, it is the integral of its partial derivative with respect to x for almost every fixed value of y, and its partial derivative $e^{-ixy} f(y)$ is integrable on every rectangular strip $[\alpha, \beta] \times \mathbf{R} \subset \mathbf{R}^2$ (cf. appendix B10). Therefore in this case integration and differentiation may be interchanged and we get

$$Ff(\tau) = \frac{1}{\sqrt{2\pi}} \int_{-\infty}^{+\infty} \frac{\partial}{\partial \tau} \frac{e^{-i\tau\eta} - 1}{-i\eta} f(\eta)\, d\eta$$

$$= \frac{1}{\sqrt{2\pi}} \int_{-\infty}^{+\infty} e^{-i\tau\eta} f(\eta)\, d\eta \quad \text{for almost all } \tau \in \mathbf{R}.$$

The formulas for F^{-1} are obtained in the same way if i is replaced by $-i$. $\square$

For any $f \in \mathfrak{L}_2(-\infty, +\infty)$ the function Ff is called the *Fourier transform* of f. Actually formula (2) allows to define a Fourier transform also for every function in $\mathfrak{L}_1(-\infty, +\infty)$, even if it does not belong to $\mathfrak{L}_2(-\infty, +\infty)$. Of course we cannot expect then the Fourier transform to belong to $\mathfrak{L}_2(-\infty, +\infty)$ either. The interested reader is referred to e.g. [11] or [6] (21.38)–(21.64).

Exercise. Let F be the Fourier-Plancherel operator in $\mathfrak{L}_2(-\infty, +\infty)$. For every function $f \in \mathfrak{L}_2(-\infty, +\infty)$ we have

$$Ff = \lim_{n \to \infty} \int_{-n}^{+n} e^{-ix\eta} f(\eta) \, d\eta,$$

$$F^{-1}f = \lim_{n \to \infty} \int_{-n}^{+n} e^{ix\eta} f(\eta) \, d\eta$$

in the sense of the norm in $\mathfrak{L}_2(-\infty, +\infty)$. (Hint: let f_n be the projection of f upon $\mathfrak{L}_2(-n, +n)$ considered as a subspace of $\mathfrak{L}_2(-\infty, +\infty)$ (cf. § 15 example 1) and show $f = \lim_{n \to \infty} f_n$.)

CHAPTER IV

GENERAL THEORY OF LINEAR OPERATORS

§ 17. Adjoint operators (general case)

Our knowledge about bounded linear operators on $\mathfrak{H}$, as accumulated so far, is not yet sufficient for our purposes. In applications of Hilbert space theory an important role is played by linear operators which are unbounded or defined only on part of $\mathfrak{H}$. An example of an unbounded linear operator in ℓ_2 has been given in § 11 example 3. We shall study two classes of interesting unbounded linear operators in $\mathfrak{L}_2$ spaces in the next two sections.

A particularly nice feature of a bounded linear operator A on $\mathfrak{H}$ is that it has a bounded linear adjoint A^* on $\mathfrak{H}$ connected with A by the equation

$$\langle Af, g \rangle = \langle f, A^*g \rangle \quad \text{for all } f \in \mathfrak{H} \quad \text{and all } g \in \mathfrak{H}.$$

The proof of the existence of such an adjoint leans on the Riesz representation theorem for bounded linear functionals on $\mathfrak{H}$ (§ 11 theorem 4) and breaks down in case of an operator that is unbounded or not defined on all of $\mathfrak{H}$. Let us see what still can be done about adjoints in this case.

Suppose A is a linear operator in $\mathfrak{H}$ (cf. § 12 definition 1; recall that the domain of A which we shall denote by $\mathfrak{D}_A$ is always assumed to be a linear manifold). It may happen that for some vector g there still exists a vector g^* such that

(1) $$\langle Af, g \rangle = \langle f, g^* \rangle \quad \text{for all } f \in \mathfrak{D}_A$$

(in any case an example is provided by $g = g^* = o$). If, for some fixed g, there is only one vector g^* satisfying (1), then it is legitimate to write this unique vector g^* as $g^* = A^*g$ and to consider A^* as a mapping which is, at least, well defined for this vector g. The question thus arises: just when does the equation (1), for every given vector g, admit at most one solution $g^* \in \mathfrak{H}$?

THEOREM 1. *Let A be a linear operator in $\mathfrak{H}$. Suppose for the vectors g and g^* it is true that*
$$\langle Af, g \rangle = \langle f, g^* \rangle \quad \text{for all } f \in \mathfrak{D}_A.$$

Then g^* is uniquely determined by g and (1) iff $\overline{\mathfrak{D}_A} = \mathfrak{H}$ (in other words iff $\mathfrak{D}_A$ is everywhere dense in $\mathfrak{H}$).

PROOF. a) Suppose $\overline{\mathfrak{D}}_A = \mathfrak{H}$. If also

$$\langle Af, g \rangle = \langle f, g' \rangle \quad \text{for all } f \in \mathfrak{D}_A,$$

for some vector g', then we conclude

$$\langle f, g^* - g' \rangle = 0 \quad \text{for all } f \in \mathfrak{D}_A,$$
$$(g^* - g') \perp \mathfrak{D}_A,$$
$$g^* - g' = o$$

(cf. § 7 corollary 7.2).

b) Suppose $\overline{\mathfrak{D}}_A \neq \mathfrak{H}$. By § 7 corollary 7.1 there exists a non-zero vector $h \perp \mathfrak{D}_A$. Then we have $g^* + h \neq g^*$ and

$$\langle f, g^* + h \rangle = \langle f, g^* \rangle = \langle Af, g \rangle \quad \text{for all } f \in \mathfrak{D}_A. \quad \square$$

Theorem 1 also tells us that only if $\overline{\mathfrak{D}}_A = \mathfrak{H}$ there is some hope of getting an adjoint operator A^*. Theorem 2 now substantiates this hope.

THEOREM 2. *Let A be a linear operator in $\mathfrak{H}$ such that $\overline{\mathfrak{D}}_A = \mathfrak{H}$. Let*

$$\mathfrak{D}_{A^*} = \{ g \in \mathfrak{H} : \text{there exists a vector } g^* \text{ satisfying (1)} \}.$$

Then $\mathfrak{D}_{A^*}$ is a linear manifold and the operator A^* with domain $\mathfrak{D}_{A^*}$, defined by

$$A^* g = g^* \quad \text{for all } g \in \mathfrak{D}_{A^*},$$

is linear.

PROOF. From

$$\langle Af, g \rangle = \langle f, g^* \rangle \quad \text{for all } f \in \mathfrak{D}_A,$$
$$\langle Af, h \rangle = \langle f, h^* \rangle \quad \text{for all } f \in \mathfrak{D}_A$$

we conclude

$$\langle Af, \alpha g + \beta h \rangle = \langle f, \alpha g^* + \beta h^* \rangle \quad \text{for all } f \in \mathfrak{D}_A.$$

Thus we have $\alpha g + \beta h \in \mathfrak{D}_{A^*}$ and (by the uniqueness statement in theorem 1)

$$A^*(\alpha g + \beta h) = \alpha g^* + \beta h^* = \alpha A^* g + \beta A^* h. \quad \square$$

DEFINITION 1. Let A be a linear operator in $\mathfrak{H}$ such that $\overline{\mathfrak{D}}_A = \mathfrak{H}$. The linear operator A^* with domain $\mathfrak{D}_{A^*}$ as defined in theorem 2 is called the *adjoint* of A.

By the definition of A^* we have

$$\langle Af, g \rangle = \langle f, A^* g \rangle \quad \text{for all } f \in \mathfrak{D}_A \text{ and all } g \in \mathfrak{D}_{A^*}.$$

This shows that in case of a bounded linear operator A on $\mathfrak{H}$ the operator A^*

indeed coincides with the adjoint of A as defined in § 14 theorem 1. If we try to carry over to the general case some of the other statements in § 14 a peculiar difficulty arises in the fact that we always have to worry about domains. The following definition allows us to deal more efficiently with this difficulty.

DEFINITION 2. A linear operator A in $\mathfrak{H}$ is *contained* in (or *extended* by) a linear operator B in $\mathfrak{H}$ (notation: $A \subset B$) if $\mathfrak{D}_A \subset \mathfrak{D}_B$ and

$$Af = Bf \quad \text{for all } f \in \mathfrak{D}_A.$$

The operator B is then called an *extension* of A.

THEOREM 3. *If A and B are linear operators in $\mathfrak{H}$ such that $A \subset B$ and $\overline{\mathfrak{D}}_A = \mathfrak{H}$ (which also implies $\overline{\mathfrak{D}}_B = \mathfrak{H}$), then $A^* \supset B^*$.*
PROOF. From $A \subset B$ and

$$\langle Bf, g \rangle = \langle f, B^*g \rangle \quad \text{for all } f \in \mathfrak{D}_B \text{ and all } g \in \mathfrak{D}_{B^*}$$

we conclude in particular

$$\langle Af, g \rangle = \langle f, B^*g \rangle \quad \text{for all } \in \mathfrak{D}_A \text{ and all } g \in \mathfrak{D}_{B^*}.$$

This implies $\mathfrak{D}_{B^*} \subset \mathfrak{D}_{A^*}$ and $A^*g = B^*g$ for all $g \in \mathfrak{D}_{B^*}$. $\square$

THEOREM 4. *If A is a linear operator $\mathfrak{H}$ such that $\overline{\mathfrak{D}}_A = \overline{\mathfrak{D}}_{A^*} = \mathfrak{H}$, then $A^{**} \supset A$.*
PROOF. From (1) we conclude

$$\langle A^*g, f \rangle = \langle g, Af \rangle \quad \text{for all } g \in \mathfrak{D}_{A^*} \text{ and all } f \in \mathfrak{D}_A.$$

This implies $\mathfrak{D}_A \subset \mathfrak{D}_{A^{**}}$ and $A^{**}f = Af$ for all $f \in \mathfrak{D}_A$. $\square$

DEFINITION 3. If the linear operator A in $\mathfrak{H}$ is one-to-one on $\mathfrak{D}_A$, then the linear operator A^{-1} in $\mathfrak{H}$ defined on $\mathfrak{D}_{A^{-1}} = A\mathfrak{D}_A$ by

$$A^{-1}(Af) = f \quad \text{for all } f \in \mathfrak{D}_A$$

is called the *inverse* of A.

Note that $\mathfrak{D}_{A^{-1}}$ is a linear manifold. If I_A denotes the identity operator restricted to the domain $\mathfrak{D}_A$, then the statement that A is one-to-one and that A^{-1} is its inverse is equivalent with the pair of equations

(2)
$$\begin{cases} A^{-1}A = I_A, \\ AA^{-1} = I_{A^{-1}} \end{cases}$$

(cf. § 12 theorem 5 and its proof). In fact the inverse of A in the sense of definition 3 obviously satisfies (2). On the other hand, if A^{-1} is any operator satisfying (2), then the first equation implies that A is one-to-one, $\mathfrak{D}_{A^{-1}} \supset A\mathfrak{D}_A$ and $A^{-1}(Af) = f$ for all $f \in \mathfrak{D}_A$, while the second equation shows that $\mathfrak{D}_{A^{-1}} \subset A\mathfrak{D}_A$. Thus, in case of an invertible bounded linear operator A on $\mathfrak{H}$, the operator A^{-1} coincides with the inverse of A as defined in § 12 definition 5.

THEOREM 5. *Let A be a one-to-one linear operator in $\mathfrak{H}$ such that $\overline{\mathfrak{D}}_A = \overline{\mathfrak{D}}_{A^{-1}} = = \mathfrak{H}$. Then also A^* is one-to-one and*

$$(A^*)^{-1} = (A^{-1})^* .$$

PROOF. We have to show

$$(A^{-1})^* A^* = I_{A^*},$$
$$A^* (A^{-1})^* = I_{(A^{-1})^*}.$$

In order to check the first equation take any $g \in \mathfrak{D}_{A^*}$. Then for every $f \in \mathfrak{D}_{A^{-1}}$ we have $A^{-1}f \in \mathfrak{D}_A$ and therefore

$$\langle A^{-1}f, A^*g \rangle = \langle AA^{-1}f, g \rangle = \langle f, g \rangle .$$

We conclude $A^*g \in \mathfrak{D}_{(A^{-1})^*}$ and $(A^{-1})^* A^* g = g$. In order to check the second equation take any $g \in \mathfrak{D}_{(A^{-1})^*}$. Then for every $f \in \mathfrak{D}_A$ we have $Af \in \mathfrak{D}_{A^{-1}}$ and therefore

$$\langle Af, (A^{-1})^*g \rangle = \langle A^{-1}Af, g \rangle = \langle f, g \rangle .$$

We conclude $(A^{-1})^*g \in \mathfrak{D}_{A^*}$ and $A^*(A^{-1})^*g = g$. □

Example 1. Let A be the bounded linear operator on $\mathfrak{H} = \ell_2$ defined by

$$A \{\alpha_k\}_{k=1}^\infty = \{\alpha_k/k\}_{k=1}^\infty .$$

We have

$$\|Aa\|^2 = \sum_{k=1}^\infty \frac{|\alpha_k|^2}{k^2} \leqslant \sum_{k=1}^\infty |\alpha_k|^2 = \|a\|^2$$

and $\|A\| = 1$ since $Ae_1 = A\{\delta_{1,k}\}_{k=1}^\infty = \{\delta_{1,k}\}_{k=1}^\infty = e_1$. From

$$\langle Aa, b \rangle = \sum_{k=1}^\infty \frac{\alpha_k}{k} \bar{\beta}_k = \sum_{k=1}^\infty \alpha_k \frac{\bar{\beta}_k}{k} = \langle a, Ab \rangle$$

we conclude that A is selfadjoint. The operator A is obviously one-to-one. Furthermore, the linear manifold $A\mathfrak{H} = \mathfrak{D}_{A^{-1}}$ is everywhere dense since it contains in particular all "finite" sequences in ℓ_2 (cf. § 2 example 2). Note that $\mathfrak{D}_{A^{-1}}$ is the set of all sequences $\{\beta_k\}_{k=1}^{\infty} \in \ell_2$ having the property that $\sum_{k=1}^{\infty} k^2 |\beta_k|^2 < \infty$. The inverse A^{-1} is defined on $\mathfrak{D}_{A^{-1}}$ by

$$A^{-1} \{\beta_k\}_{k=1}^{\infty} = \{k\beta_k\}_{k=1}^{\infty}.$$

The operator A^{-1} is unbounded and contains the operator B in ℓ_2 defined in § 11 example 3. Still, by theorem 5 we have

$$(A^{-1})^* = (A^*)^{-1} = A^{-1},$$

in other words the operator A^{-1} also has the familiar property of being its own adjoint.

DEFINITION 4. A linear operator A in $\mathfrak{H}$ such that $\mathfrak{D}_A = \mathfrak{H}$ is called *self-adjoint* if $A^* = A$. The operator A is called *symmetric* if $A^* \supset A$.

Again this definition extends § 14 definition 1, the concept of symmetry still allowing for some enlargement in the domain of the operator A when passing from A to A^*. Note that every selfadjoint operator is symmetric but that an operator may be symmetric without being selfadjoint. The following theorem takes away the surprising element in example 1 by claiming it for its own statement.

THEOREM 6. *Let A be a one-to-one selfadjoint operator in $\mathfrak{H}$. Then $\mathfrak{D}_{A^{-1}} = (= \overline{A\mathfrak{H}}) = \mathfrak{H}$ and A^{-1} is selfadjoint.*
PROOF. If $\mathfrak{D}_{A^{-1}} \neq \mathfrak{H}$, then by § 7 corollary 7.1 there would exist a nonzero vector $g \perp \mathfrak{D}_{A^{-1}} = A\mathfrak{H}$. The equation

$$\langle Af, g \rangle = 0 = \langle f, o \rangle \quad \text{for all } f \in \mathfrak{D}_A$$

then would imply $g \in \mathfrak{D}_{A^*} = \mathfrak{D}_A$ and $Ag = o$ in contradiction to the fact that A is one-to-one and that $g \neq o$. We therefore have $\mathfrak{D}_A = \mathfrak{D}_{A^{-1}} = \mathfrak{H}$. Applying theorem 5 we get $(A^{-1})^* = (A^*)^{-1} = A^{-1}$. $\square$

We shall round off this section by another surprising statement: a self-adjoint operator defined everywhere on $\mathfrak{H}$ is bounded. The proof requires some preparation by a renowned theorem which we shall also have to use later on.

THEOREM 7 (BANACH-STEINHAUS THEOREM). *Let* $\{f_n\}_{n=1}^{\infty}$ *be a sequence in* $\mathfrak{H}$ *such that the sequence* $\{\langle f_n, g \rangle\}_{n=1}^{\infty}$ *is bounded for every* $g \in \mathfrak{H}$. *Then the sequence* $\{\|f_n\|\}_{n=1}^{\infty}$ *is bounded.*

PROOF. Recall that

$$\|f_n\| = \sup_{\|g\|=1} |\langle f_n, g \rangle|$$

(cf. § 2 corollary 4.2) and assume the statement were false, i.e.

$$(3) \qquad \sup_{1 \leqslant n < \infty} \|f_n\| = \sup_{1 \leqslant n < \infty, \|g\|=1} |\langle f_n, g \rangle| = \infty.$$

Our first observation will be that if k is any natural number and if $\mathfrak{S}$ is any open sphere in $\mathfrak{H}$ (cf. § 3 definition 2), then there exists a vector $h_k \in \mathfrak{S}$ and an element f_{n_k} of our sequence such that

$$|\langle f_{n_k}, h_k \rangle| > k.$$

Indeed, if $\mathfrak{S}$ has center c and radius ϱ, then by hypothesis we have for some $\gamma > 0$

$$|\langle f_n, c \rangle| < \gamma \quad \text{for all } n \geqslant 1.$$

By (3) we can find a unit vector g_k and an element f_{n_k} of our sequence such that

$$|\langle f_{n_k}, g_k \rangle| > 2 \frac{k + \gamma}{\varrho}.$$

Then we have

$$h_k = c + \tfrac{1}{2}\varrho g_k \in \mathfrak{S}$$

and

$$\begin{aligned}
|\langle f_{n_k}, h_k \rangle| &= |\langle f_{n_k}, c + \tfrac{1}{2}\varrho g_k \rangle| \\
&\geqslant \tfrac{1}{2}\varrho |\langle f_{n_k}, g_k \rangle| - |\langle f_{n_k}, c \rangle| \\
&> k + \gamma - \gamma = k.
\end{aligned}$$

If we keep f_{n_k} fixed, then $\langle f_{n_k}, h \rangle$ is a continuous function in h (cf. § 3 theorem 6 c)) and the inequality

$$|\langle f_{n_k}, h \rangle| > k$$

will even hold in a sufficiently small open sphere $\mathfrak{S}_k$ about h_k. This will remain true if in addition we require $\mathfrak{S}_k$ to be contained in $\mathfrak{S}$ and to have a radius $\varrho_k < 1/k$. Starting with any open sphere $\mathfrak{S} = \mathfrak{S}_0$ and repeating the argument while k runs through all natural numbers and $\mathfrak{S}_{k-1}$ is put in place of $\mathfrak{S}$ we obtain a shrinking sequence of open spheres

$$\mathfrak{S}_0 \supset \mathfrak{S}_1 \supset \cdots \supset \mathfrak{S}_{k-1} \supset \mathfrak{S}_k \supset \cdots$$

($\mathfrak{S}_k$ has center h_k and radius ϱ_k) and a sequence of indices $\{n_k\}_{k=1}^{\infty}$ such that

$$|\langle f_{n_k}, h \rangle| > k \quad \text{for all } h \in \mathfrak{S}_k,$$
$$\varrho_k < 1/k.$$

Because of $h_l \in \mathfrak{S}_k$ for $l \geqslant k$ we have

$$\|h_k - h_l\| < 1/k \quad \text{for all } l \geqslant k.$$

The sequence $\{h_l\}_{l=1}^{\infty}$ is therefore fundamental and has a limit

$$h_\infty = \lim_{l \to \infty} h_l \in \mathfrak{H}.$$

We conclude

$$|\langle f_{n_k}, h_\infty \rangle| = \lim_{l \to \infty} |\langle f_{n_k}, h_l \rangle| \geqslant k$$

(since $h_l \in \mathfrak{S}_k$ for $l \geqslant k$), in contradiction to the hypothesis that the sequence $\{\langle f_n, h_\infty \rangle\}_{n=1}^{\infty}$ is bounded. $\square$

THEOREM 8. *If the linear operator A is defined everywhere on $\mathfrak{H}$, then A^* is bounded.*

PROOF. If A^* were not bounded, then there would exist a sequence of unit vectors $g_n \in \mathfrak{D}_{A^*}$ such that

$$(4) \qquad \lim_{n \to \infty} \|A^* g_n\| = \infty$$

(cf. § 11 remark 2). From

$$|\langle g, A^* g_n \rangle| = |\langle Ag, g_n \rangle| \leqslant \|Ag\|$$

we deduce that the sequence $\{\langle A^* g_n, g \rangle\}_{n=1}^{\infty}$ is bounded for every $g \in \mathfrak{H}$. By theorem 7 the sequence $\{\|A^* g_n\|\}_{n=1}^{\infty}$ then has to be bounded, in contradiction to (4). $\square$

COROLLARY 8.1. *If A is a linear operator in $\mathfrak{H}$ such that $\mathfrak{D}_A = \mathfrak{D}_{A^*} = \mathfrak{H}$, then A is bounded.*

PROOF. By theorem 4 we have $A \subset A^{**}$, in fact $A = A^{**}$ since $\mathfrak{D}_A = \mathfrak{H} = \mathfrak{D}_{A^{**}}$, and by theorem 8 A^{**} is bounded. $\square$

Note that this corollary contains as a special case the statement made in § 14 remark 2.

COROLLARY 8.2. *A selfadjoint operator defined everywhere on $\mathfrak{H}$ is bounded.*

Exercise. Let A be a symmetric operator in $\mathfrak{H}$.

a) If B is any other symmetric operator in $\mathfrak{H}$ which contains A, then $B \subset A^*$.

b) If A is selfadjoint, then every symmetric operator in $\mathfrak{H}$ which contains A coincides with A.

c) If $A\mathfrak{D}_A = \mathfrak{H}$, then A is selfadjoint. (Hint: for every $g \in \mathfrak{D}_{A^*}$ there exists a vector $h \in \mathfrak{D}_A$ such that $A^*g = Ah$.)

§ 18. Differentiation operators in $\mathfrak{L}_2$ spaces

In this section and in the following one we shall exhibit two types of particularly interesting unbounded linear operators in $\mathfrak{L}_2(\alpha, \beta)\,(-\infty \leqslant \alpha < \beta \leqslant +\infty)$.

Entering upon the discussion of the first type of operators, namely the differentiation operators, we shall first consider the case of a finite and closed interval $[\alpha, \beta]$. The fact that the interval is closed does not make any difference as far as the Hilbert space $\mathfrak{L}_2(\alpha, \beta)$ is concerned (two functions still will be identified if they coincide almost everywhere on $[\alpha, \beta]$; cf. § 5). It does make some difference, however, as far as the domain of the differentiation operator is concerned: this domain will consist of continuous functions and for a continuous function the values in α and β affect the behaviour of the function in the neighbourhood of α and β.

In fact, since the action of the operator to be considered will essentially amount to differentiation, we shall require of the functions in its domain more than just continuity. A complex-valued function f is called *absolutely continuous* on $[\alpha, \beta]$ if there exists an integrable function g on $[\alpha, \beta]$ and a number $\gamma \in \mathbf{C}$ such that

$$f(\xi) = \int_\alpha^\xi g(\eta)\,\mathrm{d}\eta + \gamma \quad \text{for all } \xi \in [\alpha, \beta]$$

(in short: if f is an indefinite integral). If f is absolutely continuous on $[\alpha, \beta]$, then f is not only continuous on $[\alpha, \beta]$ but also differentiable a.e. on $[\alpha, \beta]$ and its derivative there coincides with g a.e. on $[\alpha, \beta]$ (cf. appendix B9). Since the function f is bounded it certainly belongs to $\mathfrak{L}_2(\alpha, \beta)$. Still, its derivative g in general need not belong to $\mathfrak{L}_2(\alpha, \beta)$, as demonstrated by the function $f = \sqrt{x}$ in $\mathfrak{L}_2(0, 1)$. In what follows we shall denote the derivative of an absolutely continuous function f by f'.

THEOREM 1. *Suppose* $-\infty < \alpha < \beta < +\infty$ *and let the linear operator D in* $\mathfrak{L}_2(\alpha, \beta)$, *called the differentiation operator, be defined on the linear manifold*

$\mathfrak{D}_D = \{f \in \mathfrak{L}_2(\alpha, \beta): \ f \text{ is absolutely continuous on } [\alpha, \beta],$
$$f' \in \mathfrak{L}_2(\alpha, \beta), f(\alpha) = f(\beta) = 0\}$$
by
$$Df = if'.$$

Then $\overline{\mathfrak{D}}_D = \mathfrak{L}_2(\alpha, \beta)$ and D is unbounded and symmetric.

PROOF. For every integer $n \geqslant 0$ the function x^n is absolutely continuous and in fact an accumulation point of $\mathfrak{D}_D$ in $\mathfrak{L}_2(\alpha, \beta)$. Indeed, if $\varepsilon > 0$ is given, then "cutting off linearly" the function x^n in sufficiently small neighbourhoods of α and β we obtain a function $f \in \mathfrak{D}_D$ such that $\|x^n - f\| < \varepsilon$ (fig. 13).

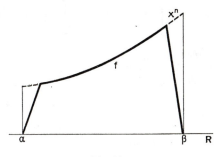

Fig. 13

Therefore $\overline{\mathfrak{D}}_D$ contains all functions $x^n(n \geqslant 0)$. Since these functions span $\mathfrak{L}_2(\alpha, \beta)$ (cf. § 9 theorem 2) and since $\overline{\mathfrak{D}}_D$ is a subspace (cf. § 6 theorem 1) we have $\overline{\mathfrak{D}}_D = \mathfrak{L}_2(\alpha, \beta)$.

In order to verify that D is unbounded we define the function $f_n \in \mathfrak{D}_D$ for $n \geqslant 2/(\beta - \alpha)$ (fig. 14) by

$$f_n(\xi) = \begin{cases} n(\xi - \alpha) & \text{for } \xi \in [\alpha, \alpha + 1/n], \\ 2 - n(\xi - \alpha) & \text{for } \xi \in [\alpha + 1/n, \alpha + 2/n], \\ 0 & \text{for } \xi \in [\alpha + 2/n, \beta]. \end{cases}$$

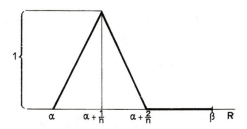

Fig. 14

We then have

$$f_n'(\xi) = \begin{cases} n & \text{for } \xi \in]\alpha, \alpha + 1/n[, \\ -n & \text{for } \xi \in]\alpha + 1/n, \alpha + 2/n[, \\ 0 & \text{for } \xi \in]\alpha + 2/n, \beta[, \end{cases}$$

$$\|f_n\|^2 = \int_\alpha^\beta |f_n(\xi)|^2 \, d\xi \leqslant 2/n,$$

$$\|if_n'\|^2 = \int_\alpha^\beta |f_n'(\xi)|^2 \, d\xi = \int_\alpha^{\alpha+2/n} n^2 d\xi = 2n,$$

$$\frac{\|Df_n\|}{\|f_n\|} \geqslant \frac{\sqrt{2n}}{\sqrt{2/n}} = n$$

and therefore

$$\sup_{f \in \mathfrak{D}_D, f \neq o} \frac{\|Df\|}{\|f\|} = \infty.$$

Finally, in order to check that D is symmetric we have to show that for all $f \in \mathfrak{D}_D$ and all $g \in \mathfrak{D}_D$ we have $\langle Df, g \rangle = \langle f, Dg \rangle$. Indeed, integrating by parts (cf. appendix B9) we obtain

$$(1) \qquad \langle if', g \rangle - \langle f, ig' \rangle = i \int_\alpha^\beta f'(\xi) \overline{g(\xi)} \, d\xi + i \int_\alpha^\beta f(\xi) \overline{g'(\xi)} \, d\xi$$

$$= if(\xi)\overline{g(\xi)} \Big|_\alpha^\beta = 0. \ \square$$

THEOREM 2. *Let D be the differentiation operator in $\mathfrak{L}_2(\alpha, \beta)$ $(-\infty < \alpha < \beta < +\infty)$ as defined in theorem 1. Then the domain of D^* is the linear manifold*

$$\mathfrak{D}^* = \{f \in \mathfrak{L}_2(\alpha, \beta): \ f \text{ is absolutely continuous on } [\alpha, \beta],$$
$$f' \in \mathfrak{L}_2(\alpha, \beta)\}$$

and D^* is defined on $\mathfrak{D}_{D^*} = \mathfrak{D}^*$ by

$$D^*f = if'.$$

PROOF. Note that (1) still holds for $f \in \mathfrak{D}_D$ and $g \in \mathfrak{D}^*$. This implies $\mathfrak{D}_{D^*} \supset \mathfrak{D}^*$ and $D^*g = ig'$ for $g \in \mathfrak{D}^*$. It only remains to show that also

$\mathfrak{D}_{D^*} \subset \mathfrak{D}^*$. Suppose therefore a function $g \in \mathfrak{D}_{D^*}$ is given. Define the absolutely continuous function h on $[\alpha, \beta]$ by

$$h(\xi) = \int\limits_\alpha^\xi D^* g(\eta) \, \mathrm{d}\eta + \delta$$

where $\delta \in C$ is determined in such a way that

(2) $$\int\limits_\alpha^\beta [g(\xi) + ih(\xi)] \, \mathrm{d}\xi = 0$$

(this is possible since for a finite interval $[\alpha, \beta]$ the functions $g \in \mathfrak{L}_2(\alpha, \beta)$, $D^* g \in \mathfrak{L}_2(\alpha, \beta)$, and therefore also $g + ih$ are integrable; cf. § 5 corollary 2.1). Applying integration by parts we obtain for every function $f \in \mathfrak{D}_D$

$$\int\limits_\alpha^\beta if'(\xi) \, \overline{g(\xi)} \, \mathrm{d}\xi = \langle Df, g \rangle = \langle f, D^* g \rangle$$

$$= \int\limits_\alpha^\beta f(\xi) \, \overline{D^* g(\xi)} \, \mathrm{d}\xi = f(\xi) \, h(\xi) \Big|_\alpha^\beta - \int\limits_\alpha^\beta f'(\xi) \, \overline{h(\xi)} \, \mathrm{d}\xi$$

$$= i \int\limits_\alpha^\beta if'(\xi) \, \overline{h(\xi)} \, \mathrm{d}\xi,$$

(3) $$\int\limits_\alpha^\beta f'(\xi) \, \overline{[g(\xi) + ih(\xi)]} \, \mathrm{d}\xi = 0.$$

We still have free choice of $f \in \mathfrak{D}_D$. Defining f on $[\alpha, \beta]$ by

$$f(\xi) = \int\limits_\alpha^\xi [g(\eta) + ih(\eta)] \, \mathrm{d}\eta$$

we obtain an absolutely continuous function f on $[\alpha, \beta]$ having the property that

$$f' = g + ih \in \mathfrak{L}_2(\alpha, \beta),$$
$$f(\alpha) = \quad (\beta) = 0 \quad (\text{by } (2)),$$

and therefore $f \in \mathfrak{D}_D$. Using (3) we obtain

$$\int_\alpha^\beta |g(\xi) + ih(\xi)|^2 \, d\xi = 0,$$

$$g(\xi) = -ih(\xi) = -i \int_\alpha^\xi D^*g(\eta) \, d\eta - i\delta \quad \text{for almost all } \xi \in [\alpha, \beta].$$

We find that g is absolutely continuous on $[\alpha, \beta]$ and $g' = -iD^*g \in \mathfrak{L}_2(\alpha, \beta)$. We conclude $g \in \mathfrak{D}^*$ which completes the proof. $\square$

THEOREM 3. *Let D be the differentiation operator in $\mathfrak{L}_2(\alpha, \beta) \, (-\infty < \alpha < < \beta < +\infty)$ as defined in theorem 1. Then $D^{**} = D$.*

PROOF. From $D \subset D^*$ we conclude $D \subset D^{**} \subset D^*$ by § 17 theorem 3 and § 17 theorem 4. It only remains to show that $\mathfrak{D}_{D^{**}} \subset \mathfrak{D}_D$. Let therefore $f \in \mathfrak{D}_{D^{**}}$ be given. We then have

$$\langle D^{**}f, g \rangle - \langle f, D^*g \rangle = 0 \quad \text{for all } g \in \mathfrak{D}_{D^*}$$

and furthermore $D^{**}f = if'$ since $D^{**} \subset D^*$. We get

$$0 = \langle if', g \rangle - \langle \ , ig' \rangle = i \int_\alpha^\beta f'(\xi) \overline{g(\xi)} \, d\xi + i \int_\alpha^\beta f(\xi) \overline{g'(\xi)} \, d\xi$$

$$= if(\xi) \overline{g(\xi)} \Big|_\alpha^\beta = i[f(\beta) \overline{g(\beta)} - f(\alpha) \overline{g(\alpha)}].$$

Choosing first $g = (x-\alpha)/(\beta-\alpha) \in \mathfrak{D}_{D^*}$ we obtain $g(\alpha) = 0$, $g(\beta) = 1$ and therefore $f(\beta) = 0$. For $g = (\beta-x)/(\beta-\alpha) \in \mathfrak{D}_{D^*}$ we obtain $g(\alpha) = 1$, $g(\beta) = 0$ and therefore $f(\alpha) = 0$. Thus we have $f \in \mathfrak{D}_D$ which completes the proof. $\square$

We next consider the case of an interval $[\alpha, +\infty[\, (-\infty < \alpha < +\infty)$. At first sight it seems unclear how we should adapt the domain of the differentiation operator in this case since the right endpoint of the interval has disappeared. A closer look, however, reveals that the situation has already taken care of itself: if $f \in \mathfrak{L}_2(\alpha, \beta)$ is absolutely continuous on every interval $[\alpha, \beta]$ $(\alpha < \beta < +\infty)$ and if $f' \in \mathfrak{L}_2(\alpha, +\infty)$, then we have $\lim_{\xi \to +\infty} f(\xi) = 0$. Indeed, since ff' is integrable on $[\alpha, +\infty[$ we have

$$\langle f, f' \rangle + \langle f', f \rangle = \lim_{\xi \to +\infty} \int_\alpha^\xi [f(\eta) \overline{f'(\eta)} + f'(\eta) \overline{f(\eta)}] \, d\eta$$

$$= \lim_{\xi \to +\infty} [|f(\xi)|^2 - |f(\alpha)|^2].$$

This implies the existence of the limit of $|f(\xi)|^2$ as $\xi \to +\infty$ and, since $f \in \mathfrak{L}_2(\alpha, +\infty)$, even $\lim_{\xi \to +\infty} f(\xi) = 0$.

THEOREM 4. *Suppose* $-\infty < \alpha < +\infty$ *and let the differentiation operator* D *in* $\mathfrak{L}_2(\alpha, +\infty)$ *be defined on the linear manifold*

$$\mathfrak{D}_D = \{f \in \mathfrak{L}_2(\alpha, +\infty): \quad f \text{ is absolutely continuous on } [\alpha, \beta] \text{ for all } \beta > \alpha,$$
$$f' \in \mathfrak{L}_2(\alpha, +\infty), f(\alpha) = 0\}$$

by

$$Df = if'.$$

Then $\overline{\mathfrak{D}}_D = \mathfrak{L}_2(\alpha, +\infty)$ *and* D *is unbounded and symmetric.*

PROOF. The fact that $\mathfrak{D}_D$ is everywhere dense in $\mathfrak{L}_2(\alpha, +\infty)$ may be proved similarly as in the proof of theorem 1 replacing the functions x^n by the functions $(x-\alpha)^n e^{-\frac{1}{2}(x-\alpha)}$ $(n \geq 0)$ if § 9 theorem 8 is available, or replacing x^n by the functions $x^n e^{-\frac{1}{2}x^2}$ $(n \geq 0)$ and using § 9 theorem 5. (Note that, since linear combinations of the functions $x^n e^{-\frac{1}{2}x^2}$ $(n \geq 0)$ are everywhere dense in $\mathfrak{L}_2(-\infty, +\infty)$, their restrictions to $[\alpha, +\infty[$ are everywhere dense in $\mathfrak{L}_2(\alpha, +\infty)$ and therefore span $\mathfrak{L}_2(\alpha, +\infty)$.)

That D is unbounded follows from the fact that D in particular contains the (unbounded!) differentiation operator in $\mathfrak{L}_2(\alpha, \beta)$ (considered as a subspace of $\mathfrak{L}_2(\alpha, +\infty)$) where $[\alpha, \beta]$ is any finite and closed subinterval of $[\alpha, +\infty[$ (cf. theorem 1). That D is symmetric again follows from (1) where β has to be replaced by $+\infty$. $\square$

THEOREM 5. *Let* D *be the differentiation operator in* $\mathfrak{L}_2(\alpha, +\infty)$ $(-\infty < \alpha < +\infty)$ *as defined in theorem 4. Then the domain of* D^* *is the linear manifold*

$$\mathfrak{D}^* = \{f \in \mathfrak{L}_2(\alpha, +\infty): \quad f \text{ is absolutely continuous in } [\alpha, \beta] \text{ for all } \beta > \alpha,$$
$$f' \in \mathfrak{L}_2(\alpha, +\infty)\}$$

and D^* *is defined on* $\mathfrak{D}_{D^*} = \mathfrak{D}^*$ *by*

$$D^*f = if'.$$

PROOF. As in the proof of theorem 2 we observe that it suffices to show that $\mathfrak{D}_{D^*} \subset \mathfrak{D}^*$. Taking any $g \in \mathfrak{D}_{D^*}$ and fixing any $\beta > \alpha$, the proof given there word for word applies to show that g is absolutely continuous on every finite interval $[\alpha, \beta]$ and $g'(\xi) = -iD^*g(\xi)$ almost everywhere on $[\alpha, \beta]$. Therefore we have $g'(\xi) = -iD^*g(\xi)$ a.e. on $[\alpha, +\infty[$ which implies that $g' = -iD^*g$ belongs to $\mathfrak{L}_2(\alpha, +\infty)$. Thus we obtain $g \in \mathfrak{D}^*$ which completes the proof. $\square$

THEOREM 6. *Let D be the differentiation operator in* $\mathfrak{L}_2(\alpha, +\infty)$ $(-\infty < \alpha < +\infty)$ *as defined in theorem 4. Then* $D^{**} = D$.

PROOF. Again as in the proof of theorem 3 we observe that it suffices to show that $\mathfrak{D}_{D^{**}} \subset \mathfrak{D}_D$. The same reasoning as applied there shows that for any $f \in \mathfrak{D}_{D^{**}}$ and any $g \in \mathfrak{D}_{D^*}$ we have

$$f(\alpha)\,\overline{g(\alpha)} = \lim_{\xi \to +\infty} f(\xi)\,\overline{g(\xi)} = 0.$$

Choosing now $g = e^{-(x-\alpha)}$ we obtain $f(\alpha) = 0$. Thus we have $f \in \mathfrak{D}_D$ which completes the proof. $\square$

It remains to consider the differentiation operator D in $\mathfrak{L}_2(-\infty, +\infty)$. Applying the reasoning preceding theorem 4 first to the interval $[0, +\infty[$ and then to the interval $]-\infty, 0]$ we find that for any function $f \in \mathfrak{L}_2(-\infty, +\infty)$ which is absolutely continuous on every finite interval $[\alpha, \beta]$ and which has the property that $f' \in \mathfrak{L}_2(-\infty, +\infty)$ we necessarily have

$$\lim_{\xi \to +\infty} f(\xi) = \lim_{\xi \to -\infty} f(\xi) = 0.$$

This already suggests that in this case under the transition from D to D^* the domain cannot get much larger.

THEOREM 7. *Let the differentiation operator D in* $\mathfrak{L}_2(-\infty, +\infty)$ *be defined on the linear manifold*

$\mathfrak{D}_D = \{f \in \mathfrak{L}_2(-\infty, +\infty)$:   *f is absolutely continuous on every*

finite interval $[\alpha, \beta] (-\infty < \alpha < \beta < +\infty)$,

$f' \in \mathfrak{L}_2(-\infty, +\infty)\}$

by

$$Df = if'.$$

Then $\overline{\mathfrak{D}}_D = \mathfrak{L}_2(-\infty, +\infty)$ *and D is unbounded and selfadjoint.*

PROOF. The linear manifold $\mathfrak{D}_D$ certainly contains all functions $x^n e^{-\frac{1}{2}x^2}$ $(n \geqslant 0)$. Since these functions span $\mathfrak{L}_2(-\infty, +\infty)$ (cf. § 9 theorem 5) the domain $\mathfrak{D}_D$ is everywhere dense in $\mathfrak{L}_2(-\infty, +\infty)$. That D is unbounded again follows from the fact that D contains the differentiation operator in $\mathfrak{L}_2(\alpha, \beta)$ (considered as a subspace of $\mathfrak{L}_2(-\infty, +\infty)$) if $[\alpha, \beta]$ is any finite interval $(\alpha < \beta)$.

That D is symmetric again follows from (1) where α has to be replaced by $-\infty$ and β has to be replaced by $+\infty$. In order to verify that D is selfadjoint it only remains to show that $\mathfrak{D}_{D^*} \subset \mathfrak{D}_D$. To this end, take any $g \in \mathfrak{D}_{D^*}$ and choose any finite interval $[\alpha, \beta]$. The proof of theorem 2 then applies word for word to show that g is absolutely continuous on $[\alpha, \beta]$ and $g'(\xi) =$ $= -iD^*g(\xi)$ a.e. on $[\alpha, \beta]$. Since $[\alpha, \beta]$ was chosen arbitrarily this implies $g'(\xi) = -iD^*g(\xi)$ for almost all $\xi \in \mathbf{R}$ and $g' = -iD^*g \in \mathfrak{L}_2(-\infty, +\infty)$. Thus we obtain $g \in \mathfrak{D}_D$ which completes the proof. $\square$

The following exercise connects with the exercise in § 17, in particular with part a) thereof: in all three parts the fact is used that if D_0 is a symmetric operator extending (=containing) a differentiation operator D, then $D \subset D_0 \subset D_0^* \subset D^*$.

Exercise. Suppose $-\infty < \alpha < \beta < +\infty$.

a) Let $\theta \in \mathbf{C}$ be given such that $|\theta| = 1$ and let D_θ be the linear operator in $\mathfrak{L}_2(\alpha, \beta)$ defined on the linear manifold

$$\mathfrak{D}_{D_\theta} = \{f \in \mathfrak{L}_2(\alpha, \beta): \; f \text{ is absolutely continuous on } [\alpha, \beta],$$
$$f' \in \mathfrak{L}_2(\alpha, \beta), f(\beta) = \theta f(\alpha)\}$$

by

$$D_\theta f = if'.$$

Then D_θ is a selfadjoint extension of the differentiation operator D in $\mathfrak{L}_2(\alpha, \beta)$. (Hint: show that D_θ is symmetric and use $D_\theta^* \subset D^*$.)

b) If D_0 is any symmetric extension of the differentiation operator D in $\mathfrak{L}_2(\alpha, \beta)$, then $D_0 = D$ or $D_0 = D_\theta$ for some $\theta \in \mathbf{C}$, $|\theta| = 1$, as in a). (Hint: use $D_0 \subset D^*$ and show that for any $f \in \mathfrak{D}_{D_0}$, $g \in \mathfrak{D}_{D_0}$ we have

$$f(\beta) \overline{g(\beta)} - f(\alpha) \overline{g(\alpha)} = 0, \quad \text{in particular } |f(\beta)|^2 = |f(\alpha)|^2.)$$

c) If D_0 is any symmetric extension of the differentiation operator D in $\mathfrak{L}_2(\alpha, +\infty)$, then $D_0 = D$. (Hint: use $D_0 \subset D^*$ and assume $\mathfrak{D}_D$ contained a function f such that $f(\alpha) \neq 0$.)

§ 19. Multiplication operators in $\mathfrak{L}_2$ spaces

The second type of operators which we now shall consider provides a continuous analogue of the operator B discussed in § 11 example 3 (cf. also § 17 example 1).

THEOREM 1. *Suppose* $-\infty \leqslant \alpha < \beta \leqslant +\infty$ *and let the linear operator E in* $\mathfrak{L}_2(\alpha, \beta)$, *called the multiplication operator (or operator of multiplication by the independent variable), be defined on the linear manifold*

$$\mathfrak{D}_E = \{f \in \mathfrak{L}_2(\alpha, \beta): xf \in \mathfrak{L}_2(\alpha, \beta)\}$$

by

$$Ef = xf.$$

Then $\mathfrak{D}_E = \mathfrak{L}_2(\alpha, \beta)$ *and* E *is selfadjoint. The operator* E *is bounded and* $\mathfrak{D}_E = \mathfrak{L}_2(\alpha, \beta)$ *iff the interval* $]\alpha, \beta[$ *is finite.*

PROOF. Let us show the last statement first. If the interval $]\alpha, \beta[$ is finite and if $\gamma = \max\{|\alpha|, |\beta|\}$, then for every $f \in \mathfrak{L}_2(\alpha, \beta)$ we have $xf \in \mathfrak{L}_2(\alpha, \beta)$ since

$$\int_\alpha^\beta |\xi f(\xi)|^2 \, d\xi \leqslant \gamma^2 \int_\alpha^\beta |f(\xi)|^2 \, d\xi = \gamma^2 \|f\|^2 < \infty.$$

At the same time this shows that in this case the operator E is bounded.

If the interval $]\alpha, \beta[$ is infinite, then $\mathfrak{D}_E$ at least contains all functions in $\mathfrak{L}_2(\alpha, \beta)$ that vanish outside finite intervals. Since these functions form a set which is everywhere dense in $\mathfrak{L}_2(\alpha, \beta)$ (cf. the set $\mathfrak{L}'$ in § 5 p. 34) so is $\mathfrak{D}_E$. (Observe that $\mathfrak{D}_E$ still is a linear manifold since if f, xf, g, xg belong to $\mathfrak{L}_2(\alpha, \beta)$ then so does $x(\lambda f + \mu g) = \lambda x f + \mu x g$.) In order to show that in this case the operator is unbounded we assume without loss of generality $\alpha \leqslant 0$ and $\beta = +\infty$ and we define $f_n \in \mathfrak{L}_2(\alpha, \beta)$ for $n \geqslant 0$ by

$$f_n(\xi) = c_{[n, n+1[}(\xi) = \begin{cases} 1 & \text{for } \xi \in [n, n+1[, \\ 0 & \text{for } \xi \notin [n, n+1[. \end{cases}$$

Then we have $\|f_n\| = 1$ and

$$\|Ef_n\|^2 = \int_n^{n+1} \xi^2 d\xi \geqslant n^2 \quad \text{for } n \geqslant 0,$$

thus showing that E is unbounded. Furthermore, the function $c_{[1, +\infty[}/x$ belongs to $\mathfrak{L}_2(\alpha, +\infty)$ but not to $\mathfrak{D}_E$.

In order to show that E is symmetric, observe that for $f \in \mathfrak{D}_E$ and $g \in \mathfrak{D}_E$ we have

$$\langle Ef, g \rangle = \int_\alpha^\beta \xi f(\xi) \overline{g(\xi)} \, d\xi = \int_\alpha^\beta f(\xi) \overline{\xi g(\xi)} \, d\xi = \langle f, Eg \rangle.$$

We conclude $g \in \mathfrak{D}_{E^*}$ and $E^* g = Eg$, in other words $E \subset E^*$. In order to verify that E is selfadjoint it only remains to show that $\mathfrak{D}_{E^*} \subset \mathfrak{D}_E$. Suppose therefore that $g \in \mathfrak{D}_{E^*}$ is given. Then for every $f \in \mathfrak{D}_E$ we have

$$\int_\alpha^\beta \xi f(\xi) \overline{g(\xi)} \, d\xi = \langle Ef, g \rangle = \langle f, E^* g \rangle = \int_\alpha^\beta f(\xi) \overline{E^* g(\xi)} \, d\xi,$$

$$\int_\alpha^\beta f(\xi) [\overline{\xi g(\xi)} - \overline{E^* g(\xi)}] \, d\xi = 0.$$

Taking any finite subinterval $]\gamma, \delta[\subset]\alpha, \beta[$ and defining $f \in \mathfrak{D}_E$ on $]\alpha, \beta[$ by

$$f(\xi) = \begin{cases} \xi g(\xi) - E^* g(\xi) & \text{for } \xi \in]\gamma, \delta[, \\ 0 & \text{for } \xi \notin]\gamma, \delta[\end{cases}$$

we obtain

$$\int_\gamma^\delta |\xi g(\xi) - E^* g(\xi)|^2 \, d\xi = 0,$$

$$\xi g(\xi) = E^* g(\xi) \quad \text{for almost all } \xi \in]\gamma, \delta[\, .$$

Since this holds for every finite subinterval $]\gamma, \delta[\subset]\alpha, \beta[$ we even get

$$\xi g(\xi) = E^* g(\xi) \quad \text{for almost all } \xi \in]\alpha, \beta[\, ,$$
$$xg = E^* g \in \mathfrak{L}_2(\alpha, \beta) \, .$$

We conclude $g \in \mathfrak{D}_E$ and therefore $\mathfrak{D}_{E^*} = \mathfrak{D}_E$ and $E^* = E$. $\square$

Remark 1. Observe that, regardless of the size of the interval $]\alpha, \beta[$, every $f \in \mathfrak{D}_E$ is integrable on $]\alpha, \beta[$. This statement is clear if the interval $]\alpha, \beta[$ is finite (cf. § 5 corollary 2.1). For the other cases it suffices to consider the Hilbert space $\mathfrak{L}_2(\alpha, +\infty)$ where $\alpha > 0$ (the entire line $]-\infty, +\infty[$ may be cut up into e.g. the intervals $]-\infty, -1] \cup]-1, +1[\cup [+1, +\infty[$). In this case, since $xf \in \mathfrak{L}_2(\alpha, +\infty)$ and $1/x \in \mathfrak{L}_2(\alpha, +\infty)$, the function $f = xf \cdot (1/x)$ is integrable on $[\alpha, +\infty[$ by § 5 remark 1.

Remark 2. In an analoguous way we can define a multiplication operator E in $\mathfrak{L}_2(\mathbf{R}^2)$ by

$$Ef = (x + iy) f$$

for all $f = f(x, y) \in \mathfrak{L}_2(\mathbf{R}^2)$ for which $(x + iy) f \in \mathfrak{L}_2(\mathbf{R}^2)$, i.e. for which

$$\int_{\mathbf{R}^2} |\xi + i\eta|^2 \, |f(\xi, \eta)|^2 \, d(\xi, \eta) = \int_{-\infty}^{+\infty} \int_{-\infty}^{+\infty} (\xi^2 + \eta^2) \, |f(\xi, \eta)|^2 \, d\xi d\eta < \infty \, .$$

The same reasoning as displayed above then shows that E is an unbounded linear operator with everywhere dense domain and that its adjoint E^* is defined on the same domain by

$$E^* f = (x - iy) f \, .$$

The operator E in $\mathfrak{L}_2(\mathbf{R}^2)$ is therefore not any longer selfadjoint but still satisfies $E^{**} = E$.

Remark 3. Traditionally for the differentiation operator the letter P is used and for the multiplication operator the letter Q. As these letters P and Q have already been used and still will be used on these pages to denote projection operators the author asks the reader's forgiveness for departing from this tradition for the sake of notational consistency within this book.

At first sight the multiplication operator and the differentiation operator do not seem to have anything to do with each other. A second look, how-ever, seems to indicate some sort of duality in their behaviour: the multi-plication operator is selfadjoint regardless which type of Hilbert space $\mathfrak{L}_2(\alpha, \beta)$ $(-\infty \leqslant \alpha < \beta \leqslant +\infty)$ we consider; it is bounded only in the Hilbert space $\mathfrak{L}_2(\alpha, \beta)$ where $-\infty < \alpha < \beta < +\infty$. The differentiation operator is un-bounded regardless which type of Hilbert space $\mathfrak{L}_2(\alpha, \beta)$ $(-\infty \leqslant \alpha < \beta \leqslant +\infty)$ we consider; it is selfadjoint only in the Hilbert space $\mathfrak{L}_2(-\infty, +\infty)$ (cf. § 18). The following two theorems show that there is indeed some relation-ship between the operators D and E. Recall that if A is a linear operator with domain $\mathfrak{D}_A$, then we denote by I_A the identity operator restricted to the domain $\mathfrak{D}_A$ (cf. § 17 (2)).

THEOREM 2. *Let D be the differentiation operator and let E be the multiplica-tion operator in $\mathfrak{L}_2(\alpha, \beta)$ $(-\infty \leqslant \alpha < \beta \leqslant +\infty)$. Then $\overline{\mathfrak{D}}_{DE-ED} = \mathfrak{L}_2(\alpha, \beta)$ and*

$$DE - ED = iI_{DE-ED}.$$

PROOF. Recalling how the domain of an operator as $DE - ED$ is defined (cf. § 12 theorem 1) we see that in the case of $\mathfrak{L}_2(-\infty, +\infty)$ the domain $\mathfrak{D}_{DE-ED}$ certainly contains all functions $x^n e^{-\frac{1}{2}x^2}$ $(n \geqslant 0)$. Since these functions span $\mathfrak{L}_2(-\infty, +\infty)$ (cf. § 9 theorem 5) the domain $\mathfrak{D}_{DE-ED}$ is everywhere dense in $\mathfrak{L}_2(-\infty, +\infty)$. In the case of $\mathfrak{L}_2(\alpha, +\infty)$ $(-\infty < \alpha < +\infty)$ or $\mathfrak{L}_2(\alpha, \beta)$ $(-\infty < \alpha < \beta < +\infty)$ the same reasoning applies if in addition these functions are "cut off linearly" in neighbourhoods of the endpoints α or α and β respectively (cf. § 18 fig. 13; in fact we may replace in these cases the functions $x^n e^{-\frac{1}{2}x^2}$ by $x^n e^{-\frac{1}{2}x}$ or x^n respectively by § 9 theorem 8 – if this is available – and § 9 theorem 2). For every $f \in \mathfrak{D}_{DE-ED}$ we have

$$(DE - ED)f = Dxf - xif' = xif' + if - xif' = if . \ \square$$

The connection between D and E becomes especially transparent in the case of $\mathfrak{L}_2(-\infty, +\infty)$ when both operators are unbounded and selfadjoint. Let $\mathfrak{L}$ be the linear manifold of all finite complex linear combinations of the functions $x^n e^{-\frac{1}{2}x^2}$ $(n \geqslant 0)$ or, equivalently, of the Hermite functions h_n $(n \geqslant 0)$ (cf. § 9 theorem 6, § 16 theorem 2). Obviously we have $\mathfrak{L} \subset \mathfrak{D}_D$ and $\mathfrak{L} \subset \mathfrak{D}_E$.

Recall that the Fourier-Plancherel operator F maps $\mathfrak{L}$ one-to-one onto $\mathfrak{L}$ (cf. § 16 theorem 2). For $f \in \mathfrak{L}$ we therefore have $g = Ff \in \mathfrak{L} \subset \mathfrak{D}_D$ and

(1)
$$\left\{ \begin{array}{ll} g(\xi) = \dfrac{1}{\sqrt{2\pi}} \displaystyle\int\limits_{-\infty}^{+\infty} e^{-i\xi\eta} f(\eta)\, d\eta & \text{for all } \xi \in \mathbf{R}, \\[3mm] ig'(\xi) = \dfrac{1}{\sqrt{2\pi}} \displaystyle\int\limits_{-\infty}^{+\infty} e^{-i\xi\eta} \eta f(\eta)\, d\eta & \text{for all } \xi \in \mathbf{R} \end{array} \right.$$

(differentiation and integration may be interchanged since for every fixed $\xi \in \mathbf{R}$ the function $e^{-i\xi y} f(y)$ is integrable on $\mathbf{R}$, for every fixed $\eta \in \mathbf{R}$ the function $e^{-ix\eta} f(\eta)$ is absolutely continuous on every finite interval $[\alpha, \beta]$ and its partial derivative $-iye^{-ixy} f(y)$ is integrable on the rectangular strip $[\alpha, \beta] \times \mathbf{R} \subset \mathbf{R}^2$; cf. appendix B10). Vaguely speaking an application of E to f (on the right hand side of (1)) amounts to an application of D to its Fourier transform g (on the left hand side of (1)). More precisely, by means of the operators D, E and F the second equation of (1) may be written in the form

(2) $$DFf = FEf \quad \text{for all } f \in \mathfrak{L}.$$

Since F maps $\mathfrak{L}$ one-to-one onto $\mathfrak{L}$, as f varies through $\mathfrak{L}$ so does $h = Ff$. Substituting $f = F^{-1}h$ in (2) we get

$$Dh = FEF^{-1}h \quad \text{for all } h \in \mathfrak{L}.$$

Our intention in the next theorem is to show that even more is true, namely

$$D = FEF^{-1}$$

where the products are to be taken in the sense of § 12 theorem 1.

THEOREM 3. *Let D be the differentiation operator, let E be the multiplication operator, and let F be the Fourier-Plancherel operator in $\mathfrak{L}_2(-\infty, +\infty)$. Then*

$$D = FEF^{-1}.$$

PROOF. Let us first realise what we have to show: comparing the two operators D and FEF^{-1} (the second one defined according to § 12 theorem 1) we have to verify that both of them are defined on the same domain and that for every function f in this common domain we have $Df = FEF^{-1}f$.

The domains of E and of D are well known to us (cf. theorem 1 and § 18 theorem 7), but what is the domain of FEF^{-1}? According to § 12 theorem 1 the domain of FE is the set of all $f \in \mathfrak{D}_E$ for which $Ef \in \mathfrak{D}_F$. Since

$\mathfrak{D}_F = \mathfrak{L}_2(-\infty, +\infty)$ this does not impose any additional restriction upon f. Thus we see that the domain of FE still is $\mathfrak{D}_E$. Again according to § 12 theorem 1 the domain of FEF^{-1} is the set of all $f \in \mathfrak{D}_{F^{-1}} = \mathfrak{L}_2(-\infty, +\infty)$ for which $F^{-1}f \in \mathfrak{D}_{FE} = \mathfrak{D}_E$. Since F is one-to-one the relation $F^{-1}f \in \mathfrak{D}_E$ is equivalent with $f \in F\mathfrak{D}_E$. We see that the domain of FEF^{-1} is $F\mathfrak{D}_E$ (note that we should have arrived at the same result first considering EF^{-1} and then $F(EF^{-1})$; cf. § 12 corollary 1.1 a)).

Thus we have to show

(3)
$$\begin{cases} \mathfrak{D}_D = F\mathfrak{D}_E, \\ Df = FEF^{-1}f \quad \text{for all } f \in \mathfrak{D}_D. \end{cases}$$

To this end take any $f \in \mathfrak{D}_D$. Then by § 16 theorem 4 we have

$$F^{-1}Df(\xi) = \frac{1}{\sqrt{2\pi}} \frac{d}{d\xi} \int_{-\infty}^{+\infty} \frac{e^{i\xi\eta} - 1}{i\eta} \, if'(\eta) \, d\eta$$

(since for every fixed $\xi \in \mathbf{R}$ the function $(e^{i\xi y} - 1)/y$ is absolutely continuous on every interval $[\alpha, \beta]$ we can apply integration by parts; cf. appendix B9)

$$= \frac{1}{\sqrt{2\pi}} \frac{d}{d\xi} \left[\frac{e^{i\xi\eta} - 1}{\eta} \, f(\eta) \Big|_{-\infty}^{+\infty} - \int_{-\infty}^{+\infty} \frac{i\xi\eta e^{i\xi\eta} - (e^{i\xi\eta} - 1)}{\eta^2} \, f(\eta) \, d\eta \right]$$

(the first term vanishes, the second one is split up into suitable summands)

$$= \frac{1}{\sqrt{2\pi}} \frac{d}{d\xi} \xi \int_{-\infty}^{+\infty} \frac{e^{i\xi\eta} - 1}{i\eta} \, f(\eta) \, d\eta + \frac{1}{\sqrt{2\pi}} \frac{d}{d\xi} \int_{-\infty}^{+\infty} \frac{e^{i\xi\eta} - 1 - i\xi\eta}{\eta^2} \, f(\eta) \, d\eta$$

$$= \xi \frac{d}{d\xi} \frac{1}{\sqrt{2\pi}} \int_{-\infty}^{+\infty} \frac{e^{i\xi\eta} - 1}{i\eta} \, f(\eta) \, d\eta + \frac{1}{\sqrt{2\pi}} \int_{-\infty}^{+\infty} \frac{e^{i\xi\eta} - 1}{i\eta} \, f(\eta) \, d\eta +$$

$$+ \frac{1}{\sqrt{2\pi}} \int_{-\infty}^{+\infty} \frac{i\eta e^{i\xi\eta} - i\eta}{\eta^2} \, f(\eta) \, d\eta$$

(in the last summand integration and differentiation have been interchanged, since again for every fixed $\xi \in \mathbf{R}$ the function $\{(e^{i\xi y} - 1 - i\xi y)/y^2\} f(y)$ is integrable on $\mathbf{R}$, for almost all $\eta \in \mathbf{R}$ the function $\{(e^{i x\eta} - 1 - ix\eta)/\eta^2\} f(\eta)$ is absolutely continuous on every interval $[\alpha, \beta]$, and the partial derivative (with respect to x) $\{(ie^{ixy}-i)/y\}f(y)$ is integrable on the rectangular strip $[\alpha, \beta] \times \mathbf{R} \subset \mathbf{R}^2$)

$$= \xi F^{-1}f(\xi) \quad \text{for almost all } \xi \in \mathbf{R}$$

(since the last two integrals cancel against each other; for the first see § 16 theorem 4). Since the initial function $F^{-1}Df$ in this chain of equalities belongs to $\mathfrak{L}_2(-\infty, +\infty)$ so does $xF^{-1}f$ and we have $F^{-1}f \in \mathfrak{D}_E$ and

$$F^{-1}Df = EF^{-1}f \quad \text{for all } f \in \mathfrak{D}_D,$$
$$Df = FEF^{-1}f \quad \text{for all } f \in \mathfrak{D}_D,$$
(4) $$\qquad D \subset FEF^{-1}.$$

Now we observe that the operator FEF^{-1} is certainly symmetric. Indeed, for every pair g, h of functions in $\mathfrak{D}_{FEF^{-1}} = F\mathfrak{D}_E$ we have $F^{-1}g \in \mathfrak{D}_E$ and $F^{-1}h \in \mathfrak{D}_E$. Since F is unitary and E is selfadjoint we get

$$\langle FEF^{-1}g, h \rangle = \langle EF^{-1}g, F^{-1}h \rangle = \langle F^{-1}g, EF^{-1}h \rangle = \langle g, FEF^{-1}h \rangle.$$

Using the fact that FEF^{-1} is symmetric and that D is selfadjoint we conclude from (4)

$$D = D^* \supset (FEF^{-1})^* \supset FEF^{-1} \supset D$$

(cf. § 17 theorem 3). This implies $\mathfrak{D}_D = \mathfrak{D}_{FEF^{-1}}$ and $D = FEF^{-1}$. $\square$

Remark 4. It is a matter of routine to verify that the statement $D = FEF^{-1}$ is equivalent with every one of the following three statements:

$$DF = FE,$$
$$F^{-1}DF = E$$
$$F^{-1}D = EF^{-1}.$$

Let us stop another moment to consider the meaning of theorem 3. The unitary operators F and F^{-1} are mutually inverse automorphisms of the Hilbert space $\mathfrak{H} = \mathfrak{L}_2(-\infty, +\infty)$ (cf. § 14 theorem 4). Theorem 3 states that the result of an application of D is also obtained in another way which makes a detour around differentiation but which sometimes for some reasons may be more pleasant (cf. § 39 example 3). This second way starts with an application of the automorphism F^{-1}. Abstractly this does not amount to more than just a relabeling of the elements within $\mathfrak{H}$, but by the special choice of this automorphism the linear manifold $\mathfrak{D}_D$ is transformed into $\mathfrak{D}_E$. To continue on the route apply E (which can be done if we started out with a vector

f in $\mathfrak{D}_D$) and finally make the relabeling undone by applying F: you'll arrive at the looked-for element Df (fig. 15):

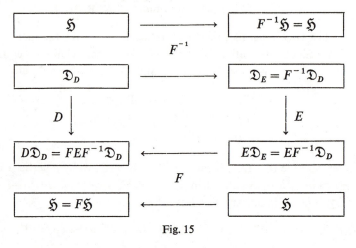

Fig. 15

Of course, according to the equation $E = F^{-1}DF$ (cf. remark 4) we would equally be justified to decompose the application of the operator E into successive applications of the operators F, D and F^{-1}. Abstractly there is no difference between the two operators D and E. They do the same things in isomorphic and even coinciding copies of the same Hilbert space. This relationship between two linear operators is important enough to be given a special name.

DEFINITION 1. A linear operator A in a Hilbert space $\mathfrak{H}$ and a linear operator A' in a Hilbert space $\mathfrak{H}'$ are called *equivalent* if there exists an isomorphism U of $\mathfrak{H}$ onto $\mathfrak{H}'$ (an automorphism U of $\mathfrak{H}$ if $\mathfrak{H} = \mathfrak{H}'$) such that

$$(5) \qquad\qquad A = U^{-1}A'U .$$

Note that in case of two Hilbert spaces $\mathfrak{H}$ and $\mathfrak{H}'$ which do not coincide equation (5) is a short notation for the combined statement

$$U\mathfrak{D}_A = \mathfrak{D}_{A'},$$
$$Af = U^{-1}[A'(Uf)] \quad \text{for all } f \in \mathfrak{D}_A .$$

In case $\mathfrak{H} = \mathfrak{H}'$ this agrees precisely with the way in which equality and products of linear operators in $\mathfrak{H}$ have been defined (cf. § 12 definition 2, § 12 theorem 1). By the same token (5) is equivalent with

$$A' = UAU^{-1} .$$

COROLLARY 3.1. *The differentiation operator and the multiplication operator in $\mathfrak{L}_2(-\infty, +\infty)$ are equivalent.*

Exercise. Let a be a continuous complex-valued function on the interval $]\alpha, \beta[\ (-\infty \leqslant \alpha < \beta \leqslant +\infty)$ and let the linear operator A in $\mathfrak{L}_2(\alpha, \beta)$ be defined on the linear manifold

$$\mathfrak{D}_A = \{f \in \mathfrak{L}_2(\alpha, \beta): af \in \mathfrak{L}_2(\alpha, \beta)\}$$

by

$$Af = af \ .$$

a) Identify A^* and show $A^{**} = A$ and $A^*A = AA^*$.

b) The operator A is bounded iff the function a is bounded on $]\alpha, \beta[$. (Hint: if $\{\xi_n\}_{n=1}^{\infty} \subset]\alpha, \beta[$ is a sequence such that $|a(\xi_n)| > n$ for all $n \geqslant 1$, then without loss of generality we may assume either $\lim_{n \to \infty} \xi_n = \alpha$ or $\lim_{n \to \infty} \xi_n = \beta$, and there exists a step function $f \in \mathfrak{L}_2(\alpha, \beta)$ such that

$$\int_\alpha^\beta |a(\xi) f(\xi)|^2 \, d\xi = \infty \ .)$$

c) The operator A is selfadjoint iff $a(\xi) \in \mathbf{R}$ for all $\xi \in]\alpha, \beta[$.

d) The operator A is unitary (and defined on all of $\mathfrak{L}_2(\alpha, \beta)$) iff $|a(\xi)| = 1$ for all $\xi \in]\alpha, \beta[$.

e) Suppose $a = e^{2\pi ix}$ and $]\alpha, \beta[=]0, 1[$. Then A is equivalent with the "right shift" operator A' on the Hilbert space $\mathfrak{H}'$ of all "two-sided" square summable complex sequences $\{\alpha_k\}_{k=-\infty}^{+\infty}$ $(\sum_{k=-\infty}^{+\infty} |\alpha_k|^2 < \infty)$, defined by

$$A' \{\alpha_k\}_{k=-\infty}^{+\infty} = \{\beta_k\}_{k=-\infty}^{+\infty}, \quad \beta_k = \alpha_{k-1} \text{ for } -\infty < k < \infty \ .$$

(Hint: use § 8 example 3.)

§ 20. Closed linear operators

Suppose A is an unbounded linear operator in $\mathfrak{H}$. Then, by § 11 theorem 2, A cannot be continuous. In fact, as shown in part b) of the proof of this theorem, there will even be a sequence $\{f_n\}_{n=1}^{\infty} \subset \mathfrak{D}_A$ such that $\lim_{n \to \infty} f_n = o$ but $\lim_{n \to \infty} \|Af_n\| = \infty$. Still, there is a chance for the operator A to save the last remains of continuity that are left for an unbounded linear operator in $\mathfrak{H}$: it might just happen, whenever both the sequence $\{f_n\}_{n=1}^{\infty} \subset \mathfrak{D}_A$ and the image sequence $\{Af_n\}_{n=1}^{\infty}$ converge, that then A in fact behaves as it should, namely $f = \lim_{n \to \infty} f_n \in \mathfrak{D}_A$ and $Af = \lim_{n \to \infty} Af_n$ (for a continuous linear operator on $\mathfrak{H}$ this goes without saying). If a linear operator in $\mathfrak{H}$ has this property, then we shall call it closed.

DEFINITION 1. A linear operator A in $\mathfrak{H}$ is called *closed* if for every sequence $\{f_n\}_{n=1}^{\infty} \subset \mathfrak{D}_A$ the existence of both limits

$$f = \lim_{n \to \infty} {}_n,$$

$$g = \lim_{n \to \infty} Af_n$$

implies $f \in \mathfrak{D}_A$ and $Af = g$.

Example 1. Let B be the unbounded linear operator in ℓ_2 defined on the linear manifold $\mathfrak{D}_B$ of all "finite" sequences in ℓ_2 by

$$B\{\alpha_k\}_{k=1}^{\infty} = \{k\alpha_k\}_{k=1}^{\infty} \quad \text{for } a = \{\alpha_k\}_{k=1}^{\infty} \in \mathfrak{D}_B$$

(cf. § 11 example 3). The operator B is not closed. Indeed, let $a_n = \{\alpha_{n,k}\}_{k=1}^{\infty} \in \mathfrak{D}_B$ be defined for $1 \leqslant n < \infty$ by

$$\alpha_{n,k} = \begin{cases} 1/k^2 & \text{for} \quad 1 \leqslant k \leqslant n, \\ 0 & \text{for } n+1 \leqslant k < \infty, \end{cases}$$

and let $b = \{1/k^2\}_{k=1}^{\infty}$, $c = \{1/k\}_{k=1}^{\infty}$. Then we have

$$\lim_{n \to \infty} a_n = b \in \ell_2,$$

$$\lim_{n \to \infty} Ba_n = c \in \ell_2,$$

but $b \notin \mathfrak{D}_B$.

It would not seem fair to reproach the operator B for not being closed since the last conclusion indicates that we have restricted its domain too drastically. Let us therefore consider the operator $\bar{B}$ defined on the linear manifold

$$\mathfrak{D}_{\bar{B}} = \{a = \{\alpha_k\}_{k=1}^{\infty} \in \ell_2 : \sum_{k=1}^{\infty} k^2 |\alpha_k|^2 < \infty\}$$

by

$$\bar{B}\{\alpha_k\}_{k=1}^{\infty} = \{k\alpha_k\}_{k=1}^{\infty}.$$

Note that $B \subset \bar{B}$ and recall that $\bar{B} = \bar{B}^*$ (cf. § 17 example 1 where the operator $\bar{B}$ has been met with under the name A^{-1}). We shall now show that $\bar{B}$ is closed. To this end, let $a_n = \{\alpha_{n,k}\}_{k=1}^{\infty} \in \mathfrak{D}_{\bar{B}}$ for $1 \leqslant n < \infty$ and suppose we have

$$\lim_{n \to \infty} a_n = a = \{\alpha_k\}_{k=1}^{\infty} \in \ell_2,$$

$$\lim_{n \to \infty} \bar{B}a_n = b = \{\beta_k\}_{k=1}^{\infty} \in \ell_2.$$

Using the standard basis $\{e_k\}_{k=1}^{\infty}$ of ℓ_2 (cf. § 8 example 2) we conclude from $\bar{B}^* = \bar{B}$

$$\beta_k = \langle b, e_k \rangle = \lim_{n \to \infty} \langle \bar{B}a_n, e_k \rangle = \lim_{n \to \infty} \langle a_n, \bar{B}e_k \rangle$$
$$= \langle a, \bar{B}e_k \rangle = k \langle a, e_k \rangle = k\alpha_k \quad \text{for } 1 \leqslant k < 0$$

(of course this could also have been shown without use of adjoints). Therefore we have $a \in \mathfrak{D}_{\bar{B}}$ and $\bar{B}a = b$.

The method of proof used above already suggests that for a linear operator in $\mathfrak{H}$ being closed might have something to do with being an adjoint. We shall substantiate this supposition after first having cleared up some connections between the properties of an operator of being closed, being bounded, and being an inverse.

THEOREM 1. *A bounded linear operator A in $\mathfrak{H}$ is closed iff $\mathfrak{D}_A$ is closed (and therefore a subspace).*
PROOF. Suppose $\mathfrak{D}_A$ is closed. For any converging sequence $\{f_n\}_{n=1}^{\infty} \subset \mathfrak{D}_A$ we have $f = \lim_{n \to \infty} f_n \in \mathfrak{D}_A$ and, since A is continuous (cf. § 11 theorem 2), $Af = \lim_{n \to \infty} Af_n$.
If $\mathfrak{D}_A$ is not closed, then we can find a converging sequence $\{f_n\}_{n=1}^{\infty} \subset \mathfrak{D}_A$ such that $f = \lim_{n \to \infty} f_n \notin \mathfrak{D}_A$. The sequence $\{Af_n\}_{n=1}^{\infty}$ still converges since A is bounded, thus showing that A is not closed. $\square$

THEOREM 2. *If A is a closed one-to-one linear operator in $\mathfrak{H}$, then A^{-1} is closed.*
PROOF. Suppose for the sequence $\{g_n\}_{n=1}^{\infty} \subset \mathfrak{D}_{A^{-1}}$ both limits

$$g = \lim_{n \to \infty} g_n,$$
$$f = \lim_{n \to \infty} A^{-1}g_n$$

exist. Writing $A^{-1}g_n = f_n$ we obtain a sequence $\{f_n\}_{n=1}^{\infty} \subset \mathfrak{D}_A$ for which both limits

$$g = \lim_{n \to \infty} Af_n,$$
$$f = \lim_{n \to \infty} f_n$$

exist. Since A is closed we have $f \in \mathfrak{D}_A$ and $Af = g$. Thus implies $g \in \mathfrak{D}_{A^{-1}}$ and $A^{-1}g = f$. $\square$

Notice that theorems 1 and 2 together supply an alternative proof of the statement in example 1 that the linear operator $\bar{B}$ in ℓ_2 is closed. In fact, the linear operator $\bar{B}$ is the inverse of the bounded linear operator A defined on $\ell_2 = \mathfrak{H}$ by

$$A\{\alpha_k\}_{k=1}^\infty = \{\alpha_k/k\}_{k=1}^\infty$$

(cf. § 17 example 1). By theorem 1, A is closed. Then $\bar{B} = A^{-1}$ is closed by theorem 2. A third proof is supplied by the following theorem and its corollary.

THEOREM 3. *If A is a linear operator in $\mathfrak{H}$ such that $\overline{\mathfrak{D}}_A = \mathfrak{H}$, then A^* is closed.*
PROOF. Suppose the sequence $\{g_n\}_{n=1}^\infty \subset \mathfrak{D}_{A^*}$ has the property that both limits

$$g = \lim_{n \to \infty} g_n,$$

$$h = \lim_{n \to \infty} A^* g_n$$

exist. We conclude that for every $f \in \mathfrak{D}_A$ we have

$$\langle Af, g \rangle = \lim_{n \to \infty} \langle Af, g_n \rangle = \lim_{n \to \infty} \langle f, A^* g_n \rangle$$
$$= \langle f, h \rangle.$$

This implies $g \in \mathfrak{D}_{A^*}$ and $A^* g = h$ which had to be proved. □

COROLLARY 3.1. *Every selfadjoint operator in $\mathfrak{H}$ is closed.*

COROLLARY 3.2. *The differentiation operator and the multiplication operator in $\mathfrak{L}_2(\alpha, \beta) (-\infty \leqslant \alpha < \beta \leqslant +\infty)$ are closed.*
PROOF. Cf. § 18 theorem 3, § 18 theorem 6, § 18 theorem 7, and § 19 theorem 1. □

The connections between closed operators and adjoints still extend beyond theorem 3. In fact, a linear operator A with everywhere dense domain in $\mathfrak{H}$ is closed iff $\overline{\mathfrak{D}}_{A^*} = \mathfrak{H}$ and $A = A^{**}$ (cf. § 17 theorem 4). A linear operator A with everywhere dense domain in $\mathfrak{H}$ admits a closed extension (as was the case with the operator B in example 1) iff $\overline{\mathfrak{D}}_{A^*} = \mathfrak{H}$, and in this case the operator $\bar{A} = A^{**}$ even is the minimal closed extension of A. If A is a closed linear operator defined everywhere on $\mathfrak{H}$, than A^* is bounded by § 17 theorem 8. By what has been said above in this paragraph we have $\overline{\mathfrak{D}}_A = \mathfrak{H}$ and since A^* is closed by theorem 3 we have $\mathfrak{D}_{A^*} = \mathfrak{H}$ by theorem 1. The adjoint A^{**} of A^* which in our case coincides with A again has to be bounded by § 17 theorem 8.

As a consequence, a closed linear operator defined everywhere on $\mathfrak{H}$ is bounded. This statement is called the *closed graph theorem*. It generalizes § 17 corollary 8.1 and § 17 corollary 8.2.

All of what just has been said depends on the crucial first proposition characterizing closed operators with domain everywhere dense in $\mathfrak{H}$. This proposition admits an elegant proof which, however, uses an apparatus which has not been introduced so far. Since the mentioned facts will not be needed in the sequel we shall not pause now to do this at this place. The interested reader is referred to appendix A (*The graph of a linear operator*) where this apparatus is introduced and where the mentioned facts are restated and proved.

What has been said above also already demonstrates why one should be interested at all in knowing whether a linear operator is closed: this information sometimes just supplies the missing link in a chain of implications ending up with the conclusion that a certain linear operator is bounded or defined everywhere on $\mathfrak{H}$. We shall actually encounter a reasoning of this type e.g. in the proof of § 23 theorem 8, § 24 theorem 3, and § 24 theorem 4.

In § 12 it has been shown that a bounded linear operator on a separable Hilbert space admits a matrix representation. The following exercise leads up to the conclusion that the same is true for a closed symmetric operator in a separable Hilbert space if one is a little bit careful in the choice of the underlying basis.

Exercise. Let A be a closed linear operator with everywhere dense domain $\mathfrak{D}_A$ in a separable infinite-dimensional Hilbert space $\mathfrak{H}$.

a) There exist an everywhere dense sequence $\{g_n\}_{n=1}^{\infty} \subset \mathfrak{D}_A$ with the following property: for every $f \in \mathfrak{D}_A$ there exists a subsequence $\{g_{n_k}\}_{k=1}^{\infty}$ of $\{g_n\}_{n=1}^{\infty}$ such that

$$f = \lim_{k \to \infty} g_{n_k},$$
$$Af = \lim_{k \to \infty} Ag_{n_k}.$$

(Hint: let $\{h_n\}_{n=1}^{\infty}$ be an everywhere dense sequence in $\mathfrak{H}$. With every ordered triplet of natural numbers (l, m, n) associate a vector $g_{l,m,n} \in \mathfrak{D}_A$ such that

$$\|h_m - g_{l,m,n}\| < \frac{1}{l},$$

$$\|h_n - Ag_{l,m,n}\| < \frac{1}{l},$$

if such a vector exists, and arrange the resulting countable infinite family of vectors $g_{l,m,n}$ arbitrarily to a sequence $\{g_n\}_{n=1}^{\infty}$.)

b) There exists a basis $\{e_k\}_{k=1}^{\infty} \subset \mathfrak{D}_A$ of $\mathfrak{H}$ with the following property: let $\mathfrak{A}$ be the linear manifold of finite complex linear combinations of basis vectors e_k $(1 \leqslant k < \infty)$; for every $f \in \mathfrak{D}_A$ there exists a sequence $\{f_n\}_{n=1}^{\infty} \subset \mathfrak{A}$ such that

$$f = \lim_{n \to \infty} f_n,$$

$$Af = \lim_{n \to \infty} Af_n.$$

(Hint: use a) and apply Gram-Schmidt orthonormalization; cf. §8 theorem 1, §8 corollary 1.1 c).)

c) Suppose A is symmetric, let $\{e_k\}_{k=1}^{\infty}$ be a basis as in b) and define

$$\alpha_{k,l} = \langle Ae_l, e_k \rangle \quad \text{for } 1 \leqslant k < \infty \text{ and } 1 \leqslant l < \infty.$$

Furthermore, define the operator A' on the linear manifold

$$\mathfrak{D}_{A'} = \{g = \sum_{l=1}^{\infty} \beta_l e_l \in \mathfrak{H}: \sum_{k=1}^{\infty} \left| \sum_{l=1}^{\infty} \alpha_{k,l} \beta_l \right|^2 < \infty\}$$

by

$$A'g = \sum_{k=1}^{\infty} \left(\sum_{l=1}^{\infty} \alpha_{k,l} \beta_l \right) e_k.$$

Then $A' = A^*$. (Hint: for $g \in \mathfrak{D}_{A^*}$ show $g \in \mathfrak{D}_{A'}$ and $A^*g = A'g$; this proves $A^* \subset A'$. For $g \in \mathfrak{D}_{A'}$ show

(1) $\langle Af, g \rangle = \langle f, A'g \rangle$

for $f = e_l$ and consequently for all $f \in \mathfrak{A}$; then apply b) to show (1) for all $f \in \mathfrak{D}_A$; this implies $g \in \mathfrak{D}_{A^*}$ and $A'g = A^*g$, at the same time proving $A' \subset A^*$.)

§ 21. Invariant subspaces of a linear operator

The idea guiding us through all of our further considerations will be, roughly speaking, to split a given linear operator A in $\mathfrak{H}$ into simpler parts in a simple way or, even better, into as simple parts as possible in a way as simple as possible. Recalling what has been said at the end of § 14 the reader will already suspect that as particularly simple we shall consider projections or at least scalar multiples of projections.

A first and reasonable step in this direction will consist, still roughly speaking, in cutting down in size the Hilbert space in which the operator A is acting. More precisely, we shall look for a proper subspace $\mathfrak{M}$ which is mapped into itself by A, and we shall then consider A as a linear operator in $\mathfrak{M}$.

DEFINITION 1. A subspace $\mathfrak{M}$ of $\mathfrak{H}$ is called *invariant* under a linear operator A in $\mathfrak{H}$ if $Af \in \mathfrak{M}$ for all $f \in \mathfrak{M} \cap \mathfrak{D}_A$.

Example 1. Let E be the multiplication operator in $\mathfrak{L}_2(\alpha, \beta)$ $(-\infty \leqslant \alpha < \beta \leqslant +\infty)$ (cf. § 19 theorem 1). Choose any $\gamma \in]\alpha, \beta[$ and let $\mathfrak{M}_1$ be the subspace defined by

$$\mathfrak{M}_1 = \{f \in \mathfrak{L}_2(\alpha, \beta): f(\xi) = 0 \quad \text{for almost all } \xi \in]\gamma, \beta[\}$$

(cf. § 6 example 4; $\mathfrak{M}_1$ may be identified with $\mathfrak{L}_2(\alpha, \gamma)$). The subspace $\mathfrak{M}_1$ is invariant since for every $f \in \mathfrak{M}_1 \cap \mathfrak{D}_E$ we have

$$\xi f(\xi) = 0 \quad \text{for almost all } \xi \in]\gamma, \beta[\,,$$
$$xf \in \mathfrak{M}_1\,.$$

By the same token also the orthogonal complement $\mathfrak{M}_2$ of $\mathfrak{M}_1$

$$\mathfrak{M}_2 = \mathfrak{M}_1^\perp = \{f \in \mathfrak{L}_2(\alpha, \beta): f(\xi) = 0 \quad \text{for almost all } \xi \in]\alpha, \gamma[\}$$

is invariant since $xf \in \mathfrak{M}_2$ for every $f \in \mathfrak{M}_2 \cap \mathfrak{D}_E$ (cf. § 7 example 1). Thus we may define operators E_1 and E_2 in $\mathfrak{M}_1$ and $\mathfrak{M}_2$ respectively by

$$E_j f = xf \quad \text{for all } f \in \mathfrak{M}_j \cap \mathfrak{D}_E \quad (j = 1, 2)\,.$$

In fact we can do more than just consider the two operators E_1, E_2 separately. Let P_1 and P_2 be the projections upon $\mathfrak{M}_1$ and $\mathfrak{M}_2$ respectively (cf. § 15 example 1). For every $f \in \mathfrak{D}_E$ we have

$$\xi P_1 f(\xi) = \begin{cases} \xi f(\xi) & \text{for } \xi \in]\alpha, \gamma[\,, \\ 0 & \text{for } \xi \notin]\alpha, \gamma[\,. \end{cases}$$

We conclude $P_1 f \in \mathfrak{M}_1 \cap \mathfrak{D}_E$ and, by the same token, $P_2 f \in \mathfrak{M} \cap \mathfrak{D}_E$. Moreover we obtain

$$f = P_1 f + P_2 f\,, \qquad P_j f \in \mathfrak{M}_j \cap \mathfrak{D}_E \qquad \text{for } j = 1, 2\,,$$
$$Ef = EP_1 f + EP_2 f$$
$$= E_1(P_1 f) + E_2(P_2 f), \quad E_j(P_j f) = EP_j f \in \mathfrak{M}_j \quad \text{for } j = 1, 2\,,$$
$$\mathfrak{D}_E = \mathfrak{D}_{EP_1} \cap \mathfrak{D}_{EP_2}\,.$$

Thus E has been split into a sum $E = EP_1 + EP_2$ where the summands EP_j may be considered as linear operators E_j in (smaller) subspaces. (The difference between EP_1 and E_1 is that EP_1 is defined on the linear manifold $(\mathfrak{M}_1 \cap \mathfrak{D}_E) + \mathfrak{M}_2$ and annihilates $\mathfrak{M}_2$, while E_1 is only defined on $\mathfrak{M}_1 \cap \mathfrak{D}_E$ and there coincides with EP_1. In particular we have $E_1 \subset EP_1$ and, analogously, $E_2 \subset EP_2$.)

The following definition collects the essential features of a splitting of a linear operator in $\mathfrak{H}$ as the one having occurred in the preceding example. Comparing this splitting with the program roughly described in the first paragraph of this section it is to be expected that we shall split further and further, giving the "parts" of A a chance to get simpler and simpler, at the same time allowing the sum to include more and more terms, eventually perhaps being glued together by an integral in place of a sum or a series (cf. § 28 theorem 2, § 34 theorem 1 d), § 39 remark 1).

DEFINITION 2. Let A be a linear operator in $\mathfrak{H}$ and let P be the projection upon a subspace $\mathfrak{M}$. The subspace $\mathfrak{M}$ (or, equivalently, the subspace $\mathfrak{M}^\perp$) *reduces* A if $P\mathfrak{D}_A \subset \mathfrak{D}_A$ and $P^\perp \mathfrak{D}_A \subset \mathfrak{D}_A$ and if there exists a linear operator A_1 in $\mathfrak{M}$ with domain $P\mathfrak{D}_A = \mathfrak{M} \cap \mathfrak{D}_A$ and a linear operator A_2 in $\mathfrak{M}^\perp$ with domain $P^\perp \mathfrak{D}_A = \mathfrak{M}^\perp \cap \mathfrak{D}_A$ such that

$$(1) \qquad\qquad Af = A_1(Pf) + A_2(P^\perp f) \quad \text{for all } f \in \mathfrak{D}_A.$$

The operators A_1 and A_2 are called the operators *induced* by A in $\mathfrak{M}$ and $\mathfrak{M}^\perp$ respectively or the *restrictions* of A to $\mathfrak{M}$ and $\mathfrak{M}^\perp$ respectively.

Note that both subspaces $\mathfrak{M}$ and $\mathfrak{M}^\perp$ in definition 2 are invariant under A and that the operators A_1 and A_2 in $\mathfrak{M}$ and $\mathfrak{M}^\perp$ are uniquely defined by (1). In fact we have

$$Af = A_1 f \quad \text{for all } f \in \mathfrak{M} \cap \mathfrak{D}_A,$$
$$Af = A_2 f \quad \text{for all } f \in \mathfrak{M}^\perp \cap \mathfrak{D}_A.$$

The next theorem points out an additional feature of this splitting which is also illustrated in example 1 (observe that there E_1 and E_2 are the multiplication operators in $\mathfrak{L}_2(\alpha, \gamma)$ and $\mathfrak{L}_2(\gamma, \beta)$ respectively).

THEOREM 1. *Let A be a unitary operator on $\mathfrak{H}$ or a selfadjoint operator in $\mathfrak{H}$ and suppose the subspace $\mathfrak{M}$ reduces A. Then the operators induced by A in $\mathfrak{M}$ and $\mathfrak{M}^\perp$ are again unitary or selfadjoint respectively.*

PROOF. It suffices to prove the assertions for the operator A_1 induced by A in $\mathfrak{M}$.

a) Suppose A is a unitary operator on $\mathfrak{H}$. From (1) we conclude $\|f\| = \|Af\| = \|A_1 f\|$ for all $f \in \mathfrak{M}$. This shows that A_1 is an isometric linear operator on $\mathfrak{M}$ (cf. § 13 corollary 1.2). Similarly A_2 is an isometric linear operator on $\mathfrak{M}^\perp$. In order to show that A_1 maps $\mathfrak{M}$ onto $\mathfrak{M}$ let any vector $g \in \mathfrak{M}$ be given. Since A maps $\mathfrak{H}$ onto $\mathfrak{H}$ there exists a vector f such that $Af = g$. Since $Af \in \mathfrak{M}$ and $A_1(Pf) \in \mathfrak{M}$, $A_2(P^\perp f) \in \mathfrak{M}^\perp$ we conclude from (1) that $A_2 P^\perp f = o$ and $\|P^\perp f\| = \|A_2 P^\perp f\| = 0$. This implies $f \in \mathfrak{M}$ and $g = Af = A_1 f$.

b) Suppose A is a selfadjoint operator in $\mathfrak{H}$ and suppose that for some g and g^* in $\mathfrak{M}$ we have

(2) $$\langle A_1 f, g \rangle = \langle f, g^* \rangle \quad \text{for all } f \in \mathfrak{M} \cap \mathfrak{D}_A.$$

Then for every $f \in \mathfrak{D}_A$ we also have

$$\begin{aligned}
\langle Af, g \rangle &= \langle A_1 Pf + A_2 P^\perp f, g \rangle \\
&= \langle A_1 Pf, g \rangle && (\text{since } A_2 P^\perp f \in \mathfrak{M}^\perp) \\
&= \langle Pf, g^* \rangle && (\text{by (2) since } Pf \in \mathfrak{M} \cap \mathfrak{D}_A) \\
&= \langle Pf + P^\perp f, g^* \rangle && (\text{since } P^\perp f \in \mathfrak{M}^\perp) \\
&= \langle f, g^* \rangle.
\end{aligned}$$

Since A is selfadjoint in $\mathfrak{H}$ this implies $g \in \mathfrak{D}_A \cap \mathfrak{M} = \mathfrak{D}_{A_1}$ and $g^* = Ag = A_1 g$. In particular the vector $g^* \in \mathfrak{M}$ is uniquely determined by the vector $g \in \mathfrak{M}$ and the relation (2). We conclude that the domain $\mathfrak{D}_{A_1} = \mathfrak{M} \cap \mathfrak{D}_A$ is everywhere dense in $\mathfrak{M}$ (cf. § 17 theorem 1) and that A_1 is a selfadjoint operator in $\mathfrak{M}$. $\square$

Which means do we have to recognize whether a given subspace reduces a given linear operator?

THEOREM 2. *Let A be a linear operator in $\mathfrak{H}$ and let P be the projection upon a subspace $\mathfrak{M}$. The subspace $\mathfrak{M}$ reduces A iff the following two conditions are satisfied:*

a) $P\mathfrak{D}_A \subset \mathfrak{D}_A$;
b) $\mathfrak{M}$ *and* $\mathfrak{M}^\perp$ *are invariant under A.*

PROOF. If $\mathfrak{M}$ reduces A, then $P\mathfrak{D}_A \subset \mathfrak{D}_A$ by definition 2. Moreover from (1) it follows that $Af \in \mathfrak{M}$ for all $f \in \mathfrak{M} \cap \mathfrak{D}_A$ and $Af \in \mathfrak{M}^\perp$ for all $f \in \mathfrak{M}^\perp \cap \mathfrak{D}_A$.

Conversely, suppose conditions a) and b) of theorem 2 are satisfied. Since for every $f \in \mathfrak{D}_A$ we have

$$P^\perp f = f - Pf \in \mathfrak{D}_A$$

we conclude $P^\perp \mathfrak{D}_A \subset \mathfrak{D}_A$. Furthermore we obtain

$$Af = APf + AP^\perp f \quad \text{for all } f \in \mathfrak{D}_A.$$

Defining the operators A_1 and A_2 in $\mathfrak{M}$ and $\mathfrak{M}^\perp$ respectively by

$$A_1 g = Ag \in \mathfrak{M} \quad \text{for all } g \in P\mathfrak{D}_A = \mathfrak{M} \cap \mathfrak{D}_A$$
$$A_2 g = Ag \in \mathfrak{M} \quad \text{for all } g \in P^\perp \mathfrak{D}_A = \mathfrak{M}^\perp \cap \mathfrak{D}_A$$

we obtain (1). $\square$

THEOREM 3. *Let A be a linear operator in $\mathfrak{H}$ and let P be the projection upon a subspace $\mathfrak{M}$. The subspace $\mathfrak{M}$ reduces A iff $PA \subset AP$.*

PROOF. Suppose $\mathfrak{M}$ reduces A and let $f \in \mathfrak{D}_A$ be given. Then we have $Pf \in \mathfrak{M} \cap \mathfrak{D}_A$, $P^\perp f \in \mathfrak{M}^\perp \cap \mathfrak{D}_A$, $APf \in \mathfrak{M}$, $AP^\perp f \in \mathfrak{M}^\perp$. Applying the projection theorem (§ 7 theorem 8') first before and then after the application of A to f we get

$$Af = APf + AP^\perp f = PAf + P^\perp Af .$$

Since the decomposition of Af into a sum of vectors in $\mathfrak{M}$ and $\mathfrak{M}^\perp$ is unique we conclude

$$PAf = APf \quad \text{for all } f \in \mathfrak{D}_A = \mathfrak{D}_{PA} .$$

This is, by definition of product and inclusion, equivalent with $PA \subset AP$ (cf. § 17 definition 2).

Conversely, suppose $PA \subset AP$. Since the domain of PA is simply $\mathfrak{D}_A$ this inclusion implies that every vector $f \in \mathfrak{D}_A$ must have the property that also $Pf \in \mathfrak{D}_A$ (cf. § 12 theorem 1), in other words we have $P\mathfrak{D}_A \subset \mathfrak{D}_A$. For every $f \in \mathfrak{M} \cap \mathfrak{D}_A$ we obtain

$$Af = APf = PAf \in \mathfrak{M}$$

which shows that $\mathfrak{M}$ is invariant under A. For every $f \in \mathfrak{M}^\perp \cap \mathfrak{D}_A$ we have

$$PAf = APf = Ao = o$$

which implies $Af \in \mathfrak{M}^\perp$. Therefore also $\mathfrak{M}^\perp$ is invariant under A. Applying theorem 2 we obtain the assertion. $\square$

The somewhat peculiar and oblique commutativity relation $PA \subset AP$ which we have just met in theorem 3 will also be of some importance in a more general form in the last chapter (cf. § 38 theorem 3, § 39 corollary 3.3). This justifies the following definition.

DEFINITION 3. Let A be a linear operator in $\mathfrak{H}$. A bounded linear operator B on $\mathfrak{H}$ is said to *commute* with A (or to be *permutable* with A) if

$$(3) \qquad\qquad\qquad BA \subset AB$$

or, equivalently, if $B\mathfrak{D}_A \subset \mathfrak{D}_A$ and

$$BAf = ABf \quad \text{for all } f \in \mathfrak{D}_A .$$

Note that while we synonymously say that B commutes with A and A commutes with B, this relation is definitely asymmetric if only B is a bounded linear operator on $\mathfrak{H}$, its less well behaved partner A playing the decisive role: on the one hand ABf must still make sense for every $f \in \mathfrak{D}_A$, on the other hand $BAf = ABf$ need only hold for all $f \in \mathfrak{D}_A$. If neither of the linear operators A and B is bounded and defined on all of $\mathfrak{H}$, then we do not try at all to define anything like commutativity of A and B. If both of them are bounded and defined on $\mathfrak{H}$, then (3) of course becomes symmetric and reduces to $BA = AB$.

Theorem 3 may now be reformulated by saying that *a subspace $\mathfrak{M}$ reduces a linear operator A in $\mathfrak{H}$ if (and only if) the corresponding projection commutes with A.* Another simple characterization of reducing subspaces can be given if the linear operator in question is bounded.

THEOREM 4. *A subspace $\mathfrak{M}$ reduces a bounded linear operator A on $\mathfrak{H}$ iff $\mathfrak{M}$ is invariant under A and A^*.*

PROOF. If $\mathfrak{M}$ reduces A, then by theorem 3 we have $PA = AP$ and (taking adjoints) $A^*P = PA^*$. Therefore (again by theorem 3) $\mathfrak{M}$ reduces not only A but also A^*.

Conversely, from $A\mathfrak{M} \subset \mathfrak{M}$ we conclude $APf \in \mathfrak{M}$ for all $f \in \mathfrak{H}$ and therefore $PAPf = APf$ for all $f \in \mathfrak{H}$ i.e. $PAP = AP$. Similarly from $A^*\mathfrak{M} \subset \mathfrak{M}$ we conclude $PA^*P = A^*P$. Combining these equalities we get

$$AP = PAP = (PA^*P)^* = (A^*P)^* = PA .$$

By theorem 3 the subspace $\mathfrak{M}$ therefore reduces A. $\square$

COROLLARY 4.1. *A bounded selfadjoint operator A on $\mathfrak{H}$ is reduced by a subspace $\mathfrak{M}$ iff $\mathfrak{M}$ is invariant under A.*

COROLLARY 4.2. *A unitary operator U on $\mathfrak{H}$ is reduced by a subspace $\mathfrak{M}$ iff $U\mathfrak{M} = \mathfrak{M}$.*
PROOF. If $U\mathfrak{M} = \mathfrak{M}$, then $\mathfrak{M}$ is invariant under U and $U^*(= U^{-1})$. Therefore $\mathfrak{M}$ reduces U. The converse assertion follows from theorem 1. $\square$

The following exercise leads up to the assertion that in the case of a finite interval $[\alpha, \beta]$ the only subspaces reducing the multiplication operator in $\mathfrak{L}_2(\alpha, \beta)$ are those of the type described in § 6 example 4.

Exercise. Suppose P is the projection upon a subspace $\mathfrak{M}$ which reduces the multiplication operator E in $\mathfrak{L}_2(\alpha, \beta)$ $(-\infty < \alpha < \beta < +\infty$; note that the finiteness of the interval $[\alpha, \beta]$ implies $x^n \in \mathfrak{L}_2(\alpha, \beta)$ for all $n \geqslant 0$).

a) $Px^n = x^n \cdot Px^0$ for all $n \geqslant 0$.

(Hint: use induction on n and theorem 3.)

b) For every $f \in \mathfrak{L}_2(\alpha, \beta)$ we have

$$Pf(\xi) = f(\xi) \cdot Px^0(\xi) \quad \text{for almost all } \xi \in [\alpha, \beta].$$

(Hint: let $\{f_n\}_{n=1}^{\infty}$ be a sequence of polynomials in x which converges in $\mathfrak{L}_2(\alpha, \beta)$ and almost everywhere on $[\alpha, \beta]$ to f; cf. § 5 remark 3, § 9 theorem 2.)

c) There exists a (Lebesgue measurable) set $Y \subset [\alpha, \beta]$ such that

$$Px^0(\xi) = \begin{cases} 0 & \text{for } \xi \in Y, \\ 1 & \text{for } \xi \in [\alpha, \beta] \backslash Y. \end{cases}$$

(Hint: choose $f = Px^0$ in b).)

d) $\mathfrak{M} = \{f \in \mathfrak{L}_2(\alpha, \beta) : f(\xi) = 0 \quad \text{for almost all } \xi \in Y\}$.

(Hint: use b) and § 7 example 1.)

§ 22. Eigenvalues of a linear operator

If we disregard the trivial subspace $\mathfrak{O}$ which is always invariant but hardly interesting, then the smallest invariant subspace we can think of would be a one-dimensional one. In this case there exists a non-zero vector f spanning this invariant subspace and having the property that $Af = \lambda f$. Note that in this invariant subspace the operator A is in fact restricted to a scalar multiple of a projection.

DEFINITION 1. A complex number λ is called an *eigenvalue* (also *proper value* or *characteristic value*) of the linear operator A in $\mathfrak{H}$ if there exists a non-zero vector $f \in \mathfrak{D}_A$, called a corresponding *eigenvector*, such that

(1) $Af = \lambda f$.

If λ is an eigenvalue of A then the set of all vectors $f \in \mathfrak{D}_A$ satisfying (1) is called the corresponding *eigenspace*.

Remark 1. Note that if A is a closed operator the eigenspace $\mathfrak{M}$ corresponding to an eigenvalue λ is indeed a subspace: for $f_1 \in \mathfrak{M}, f_2 \in \mathfrak{M}$ we have

$$A(\alpha_1 f_1 + \alpha_2 f_2) = \alpha_1 Af_1 + \alpha_2 Af_2 = \lambda(\alpha_1 f_1 + \alpha_2 f_2);$$

moreover, if for a sequence $\{f_n\}_{n=1}^{\infty} \subset \mathfrak{M}$ there exists the limit $f = \lim_{n \to \infty} f_n$, then by (1) there also exists the limit $g = \lim_{n \to \infty} Af_n = \lambda \lim_{n \to \infty} f_n$ and we have $f \in \mathfrak{D}_A$ and $Af = g = \lambda f$ since A is closed.

Remark 2. If $\mathfrak{H}$ is the n-dimensional unitary space $\mathbf{C}^n$ and if A is, with respect to a given basis $\{e_k\}_{k=1}^n$, represented by a matrix A' (cf. § 12 example 1), then the set of eigenvalues of A coincides with the set of *characteristic roots* of the matrix A' (=the roots of the equation $\det(A - \lambda I) = 0$). The Fourier coefficients of the corresponding eigenvectors of A with respect to this basis are given by the components of the corresponding eigenvectors of the matrix A'. Note, however, that sometimes the eigenspace of the matrix A' corresponding to the eigenvalue λ is defined differently from what has been said in definition 1 as the set of all vectors annihilated by some power of $(A - \lambda I)$, and not just by $(A - \lambda I)$ itself.

Example 1. Let $\{e_k\}_{k=1}^\infty$ be the standard basis of ℓ_2 and let $\bar{B}$ be the unbounded selfadjoint (and therefore closed) operator defined in § 20 example 1. Obviously every natural number n is an eigenvalue and e_n is a corresponding eigenvector. Moreover we can show that the corresponding eigenspace $\mathfrak{M}_n$ is one-dimensional (therefore also spanned by e_n) and that the natural numbers are the only eigenvalues of $\bar{B}$. Indeed, if the vector $f = \sum_{k=1}^\infty \alpha_k e_k \in \mathfrak{D}_{\bar{B}}$ satisfies

$$\bar{B}f = \lambda f \,,$$

then we conclude

$$\sum_{k=1}^\infty \alpha_k k e_k = \sum_{k=1}^\infty \alpha_k \lambda e_k \,,$$

$$(k - \lambda) \alpha_k = 0 \quad \text{for all } k \geqslant 1 \,,$$

and therefore $\alpha_k = 0$ for all $k \geqslant 1$ unless $\lambda = n$ for some natural number n. In this last case we still obtain $\alpha_k = 0$ for all $k \neq n$ and therefore $f = \alpha_n e_n$. If P_k is the projection upon the eigenspace $\mathfrak{M}_k$, then we have

$$(2) \qquad \bar{B}f = \sum_{k=1}^\infty k P_k f \quad \text{for every } f \in \mathfrak{D}_{\bar{B}} \,.$$

Example 2. Let F be the Fourier-Plancherel operator in $\mathfrak{L}_2(-\infty, +\infty)$ (cf. § 16 theorem 3). From § 16 (3) we conclude that certainly $1, -i, -1, i$ are eigenvalues of F and that for $n \geqslant 0$ the Hermite function h_n is an eigenvector corresponding to the eigenvalue $(-i)^n$. Again we can show that these are the only eigenvalues of F and that the eigenspace $\mathfrak{M}_n$ corresponding to the eigenvalue $(-i)^n$ ($n = 0, 1, 2, 3$) is given by

$$(3) \qquad \mathfrak{M}_n = \bigvee \{h_{4m+n}\}_{m=0}^\infty \quad \text{for } n = 0, 1, 2, 3 \,.$$

Indeed, if $f=\sum_{k=0}^{\infty}\alpha_k\, h_k/\|h_k\|$ satisfies $Ff = \lambda f$ (cf. § 9 theorem 6), then we conclude

$$\sum_{k=0}^{\infty}(-\mathrm{i})^k\alpha_k\,\frac{h_k}{\|h_k\|} = \sum_{k=0}^{\infty}\lambda\alpha_k\,\frac{h_k}{\|h_k\|},$$

$$[(-\mathrm{i})^k - \lambda]\,\alpha_k = 0 \quad \text{for all } k \geqslant 0,$$

and therefore $\alpha_k=0$ for all $k\geqslant0$ unless $\lambda=(-\mathrm{i})^n$ for one of the numbers $n=0, 1, 2, 3$. In this last case we still get

$$\alpha_k = 0 \quad \text{for } k \not\equiv n \;(\mathrm{mod}\;4)$$

and

(4) $$f = \sum_{m=0}^{\infty} \alpha_{4m+n}\,\frac{h_{4m+n}}{\|h_{4m+n}\|} \in \bigvee \{h_{4m+n}\}_{m=0}^{\infty}.$$

On the other hand, every non-zero function $f\in\bigvee\{h_{4m+n}\}_{m=0}^{\infty}$ admits a Fourier expansion as given in (4) (cf. § 7 theorem 5) and is therefore an eigenvector corresponding to the eigenvalue $(-\mathrm{i})^n$ of F. This proves (3). If P_n is the projection upon $\mathfrak{M}_n$ then we have

$$P_nf = \sum_{m=0}^{\infty} \langle f,\,\frac{h_{4m+n}}{\|h_{4m+n}\|} \rangle\,\frac{h_{4m+n}}{\|h_{4m+n}\|} \quad \text{for all } f \in \mathfrak{L}_2(-\infty, +\infty),$$

$$Pf = \sum_{n=0}^{3}(-\mathrm{i})^nP_nf \quad \text{for all } f \in \mathfrak{L}_2(-\infty, +\infty),$$

and therefore

(5) $$\left\{ \begin{aligned} &F = \sum_{n=0}^{3}(-\mathrm{i})^nP_n, \\ &F^{-1} = F^* = \sum_{n=0}^{3}\mathrm{i}^nP_n. \end{aligned} \right.$$

Remark 3. Note that this representation (5) of F and F^{-1} as a sum of scalar multiples of projections was not possible with the operator $\bar{B}$ in example 1 (2). In the first place the projections P_k turning up in (2) are defined on all of $\mathfrak{H}$ while the domain of $\bar{B}$ is considerably smaller (cf. § 20 example 1). In the second place, up to this point there has not been given any definition for such a thing as $\sum_{k=1}^{\infty} kP_k$. In fact this series certainly diverges in the Banach algebra $\mathscr{B}$ of bounded linear operators on $\mathfrak{H}$ (cf. § 12 theorem 3) since $\|kP_k\| =k$ (cf. § 3 remark 3). A related and legitimate but somewhat different interpretation for such a series will be given in § 39 remark 1 and § 39 example 1.

The following theorems reveal some general background for particular features of the preceding examples.

THEOREM 1. *Let* $f_1, \cdots, f_n$ *be eigenvectors corresponding to different eigen-values* $\lambda_1, \cdots, \lambda_n$ *of some linear operator* A *in* $\mathfrak{H}$. *Then* $f_1, \cdots, f_n$ *are linearly independent.*

PROOF. We proceed by induction on n. The assertion is true for $n=1$ since $f_1 \neq o$ by definition of an eigenvector. Suppose $f_1, \cdots, f_{n-1}$ are linearly independent and

$$(6) \qquad \sum_{k=1}^{n} \alpha_k f_k = o.$$

Applying the operator $(A - \lambda_n I)$ we obtain

$$o = (A - \lambda_n I) \sum_{k=1}^{n} \alpha_k f_k = \sum_{k=1}^{n} \alpha_k (A f_k - \lambda_n f_k)$$
$$= \sum_{k=1}^{n-1} \alpha_k (\lambda_k - \lambda_n) f_k.$$

Since $f_1, \cdots, f_{n-1}$ are linearly independent and $\lambda_k - \lambda_n \neq 0$ for $1 \leqslant k \leqslant n-1$ we conclude $\alpha_k = 0$ for $1 \leqslant k \leqslant n-1$. Thus (6) reduces to $\alpha_n f_n = o$ which (because of $f_n \neq o$) again implies $\alpha_n = 0$. It follows that $f_1, \cdots, f_n$ are linearly independent. $\square$

THEOREM 2. *The eigenvalues of a symmetric operator in* $\mathfrak{H}$ *are real.*

PROOF. Let A be a symmetric operator in $\mathfrak{H}$ and let $f \in \mathfrak{D}_A$ be an eigenvector corresponding to the eigenvalue λ. We obtain

$$\lambda \langle f, f \rangle = \langle \lambda f, f \rangle = \langle A f, f \rangle = \langle f, A f \rangle = \langle f, \lambda f \rangle = \bar{\lambda} \langle f, f \rangle,$$
$$\lambda = \bar{\lambda}. \square$$

THEOREM 3. *The eigenvalues of an isometric linear operator in* $\mathfrak{H}$ *have absolute value* 1.

PROOF. Let A be an isometric linear operator in $\mathfrak{H}$ and let $f \in \mathfrak{D}_A$ be an eigenvector corresponding to the eigenvalue λ. We obtain

$$\langle f, f \rangle = \langle A f, A f \rangle = \langle \lambda f, \lambda f \rangle = |\lambda|^2 \langle f, f \rangle,$$
$$|\lambda| = 1. \square$$

THEOREM 4. *Suppose* A *is a symmetric or isometric linear operator in* $\mathfrak{H}$ *and* *let f_1 and f_2 be eigenvectors corresponding to different eigenvalues λ_1 and λ_2. Then $f_1 \perp f_2$.*

PROOF. a) Suppose A is symmetric. By theorem 2 the eigenvalues λ_1 and λ_2 are real. We conclude

$$\lambda_1 \langle f_1, f_2 \rangle = \langle Af_1, f_2 \rangle = \langle f_1, Af_2 \rangle = \langle f_1, \lambda_2 f_2 \rangle = \lambda_2 \langle f_1, f_2 \rangle,$$
$$(\lambda_1 - \lambda_2) \langle f_1, f_2 \rangle = 0, \quad \lambda_1 - \lambda_2 \neq 0,$$
$$\langle f_1, f_2 \rangle = 0.$$

b) Suppose A is isometric. By theorem 3 we have $\lambda_1 \bar{\lambda}_1 = \lambda_2 \bar{\lambda}_2 = 1$. Because of $\lambda_1 \neq \lambda_2$ we conclude

$$\langle f_1, f_2 \rangle = \langle Af_1, Af_2 \rangle = \langle \lambda_1 f_1, \lambda_2 f_2 \rangle = \lambda_1 \bar{\lambda}_2 \langle f_1, f_2 \rangle,$$
$$(1 - \lambda_1 \bar{\lambda}_2) \langle f_1, f_2 \rangle = 0, \quad 1 - \lambda_1 \bar{\lambda}_2 \neq 0,$$
$$\langle f_1, f_2 \rangle = 0. \quad \square$$

Remark 4. The conclusion of theorem 4 may be stated in terms of the eigen-spaces $\mathfrak{M}_1$ and $\mathfrak{M}_2$ corresponding to λ_1 and λ_2 respectively as $\mathfrak{M}_1 \perp \mathfrak{M}_2$.

THEOREM 5. *Let U be a unitary operator on $\mathfrak{H}$ and let $\lambda \in \mathbf{C}$ be given. Then*

$$\mathfrak{M}_\lambda = \{ f \in \mathfrak{H} : Uf = \lambda f \} = [(U - \lambda I) \mathfrak{H}]^\perp.$$

As a consequence λ is an eigenvalue of U iff $\overline{(U - \lambda I) \mathfrak{H}} \neq \mathfrak{H}$.

PROOF. Suppose $Uf = \lambda f, f \neq o$. Then we have $\lambda \bar{\lambda} = 1$ by theorem 3. Applying the operator $\bar{\lambda} U^{-1}$ on both sides of the equality $Uf = \lambda f$ we get $\bar{\lambda} f = U^{-1} f$ and further

$$\langle f, (U - \lambda I) g \rangle = \langle (U^{-1} - \bar{\lambda} I) f, g \rangle = 0 \quad \text{for all } g \in \mathfrak{H}.$$

This implies $f \perp (U - \lambda I) \mathfrak{H}$. Reversing this reasoning and using the fact that also U^{-1} is unitary we see that every vector $f \in [(U - \lambda I) \mathfrak{H}]^\perp$ satisfies $Uf = \lambda f$. This proves the first assertion.

As a consequence hereof, λ is an eigenvalue of U iff there exists a non-zero vector orthogonal to the linear manifold $(U - \lambda I) \mathfrak{H}$. By § 7 corollary 7.1 this is the case iff $\overline{(U - \lambda I) \mathfrak{H}} \neq \mathfrak{H}$. $\square$

THEOREM 6. *Suppose A is a unitary operator on $\mathfrak{H}$ or a selfadjoint operator in $\mathfrak{H}$ and let $\mathfrak{M}$ be the eigenspace corresponding to an eigenvalue λ. Then $\mathfrak{M}$ reduces A.*

PROOF. If A is unitary, then $|\lambda| \neq 0$ by theorem 3 and

$$A\mathfrak{M} = \{ \lambda f : f \in \mathfrak{M} \} = \mathfrak{M}.$$

Therefore $\mathfrak{M}$ reduces A (cf. § 21 corollary 4.2).

If A is selfadjoint, then $\mathfrak{M}$ is still a subspace (cf. remark 1). If P is the projection upon $\mathfrak{M}$, then $P\mathfrak{D}_A = \mathfrak{M} \subset \mathfrak{D}_A$. Moreover, for every $g \in \mathfrak{M}^\perp \cap \mathfrak{D}_A$ we have

$$\langle f, Ag \rangle = \langle Af, g \rangle = \lambda \langle f, g \rangle = 0 \quad \text{for all } f \in \mathfrak{M}.$$

This implies $Ag \in \mathfrak{M}^\perp$. Therefore also $\mathfrak{M}^\perp$ is invariant under A and $\mathfrak{M}$ reduces A by § 21 theorem 2. $\square$

Note that an eigenspace in general need not reduce an operator (consider for instance the matrix $A' = \begin{pmatrix} 1 & 1 \\ 0 & -1 \end{pmatrix}$ with eigenvalues 1, -1 and corresponding eigenvectors $\begin{pmatrix} 1 \\ 0 \end{pmatrix}$, $\begin{pmatrix} -1 \\ 2 \end{pmatrix}$). The program which has been laid out in § 21 of splitting an operator by means of reducing subspaces into simpler parts (where scalar multiples of projections are considered to be the simplest possible ones) even encounters a more serious difficulty: for some otherwise well-behaved linear operators there are neither non-trivial reducing subspaces nor eigenvalues, as will be demonstrated in the next example (it is even still unknown whether every bounded linear operator on $\mathfrak{H}$ admits a nontrivial invariant subspace; cf. [5]). This fact motivates that in the following section we shall have to look for a concept more subtle than that of an eigenvalue and eigenspace but still allowing to recapture "part" of the operator.

Example 3. Let $\{e_k\}_{k=1}^\infty$ be the standard basis of ℓ_2 and let A be the isometric right shift operator in ℓ_2 defined by

$$(7) \qquad \qquad Ae_k = e_{k+1} \quad \text{for all } k \geqslant 1$$

(cf. § 12 example 2). For any $n \geqslant 1$ the subspace $\mathfrak{M} = \bigvee \{e_k\}_{k=n+1}^\infty$ is invariant under A but its orthogonal complement $\mathfrak{M}^\perp = \bigvee \{e_k\}_{k=1}^n$ certainly isn't, as demonstrated by (7). Therefore $\mathfrak{M}$ does not reduce A.

Suppose $\mathfrak{N}$ is any non-zero subspace which reduces A. By § 21 theorem 4 $\mathfrak{N}$ is also invariant under the left shift operator A^* in ℓ_2 defined by

$$A^* e_k = e_{k-1} \quad \text{for all } k \geqslant 2,$$
$$A^* e_1 = o$$

(cf. § 14 example 2). We choose any non-zero vector $f = \sum_{k=n}^\infty \alpha_k e_k \in \mathfrak{N}$ and suppose without loss of generality that $\alpha_n \neq 0$. Then we have

$$A^{*n} f = \sum_{k=n+1}^\infty \alpha_k e_{k-n} \in \mathfrak{N},$$
$$A^n A^{*n} f = \sum_{k=n+1}^\infty \alpha_k e_k \in \mathfrak{N},$$
$$\alpha_n e_n = f - A^n A^{*n} f \in \mathfrak{N},$$

$$e_n \in \mathfrak{N} \quad (\text{since } \alpha_n \neq 0),$$
$$e_k = A^{*n-k} e_n \in \mathfrak{N} \quad \text{for } 1 \leqslant k < n,$$
$$e_k = A^{k-n} e_n \in \mathfrak{N} \quad \text{for } n \leqslant k < \infty,$$
$$\bigvee \{e_k\}_{k=1}^\infty = \ell_2 \subset \mathfrak{N}.$$

We conclude that the only reducing subspaces of A are $\mathfrak{H} = \ell_2$ and $\mathfrak{O}$.

Similarly, if $\lambda \in \mathbf{C}$ were an eigenvalue of A and if $f = \sum_{k=n}^\infty \alpha_k e_k$ $(\alpha_n \neq 0)$ were a corresponding eigenvector, then

$$0 = \langle Af, e_n \rangle = \lambda \langle f, e_n \rangle = \lambda \alpha_n$$

in contradiction to the fact that $|\lambda| = 1 \neq 0$ by theorem 3. We conclude that A does not have any eigenvalues.

Exercise. Let D be the differentiation operator in $\mathfrak{L}_2(0, 1)$ (cf. § 18 theorem 1), let D^* be its adjoint (cf. § 18 theorem 2), and let D_1 be the self-adjoint extension of D defined in § 18 exercise (put there $\theta = 1$).

a) If $\sum_{k=1}^\infty |k\alpha_k|^2 < \infty$, then $\sum_{k=1}^\infty |\alpha_k| < \infty$. (Hint: in ℓ_2 we have

$$\sum_{k=1}^\infty |\alpha_k| = \langle \{|k\alpha_k|\}_{k=1}^\infty, \{1/k\}_{k=1}^\infty \rangle.)$$

b) Show

$$\mathfrak{D}_{D^*} = \{f = \alpha x + \sum_{k=-\infty}^{+\infty} \alpha_k e^{2\pi i k x} \in \mathfrak{L}_2(0, 1): \sum_{k=-\infty}^{+\infty} |k\alpha_k|^2 < \infty, \alpha \in \mathbf{C}\},$$

$$\mathfrak{D}_{D_1} = \{f = \sum_{k=-\infty}^{+\infty} \alpha_k e^{2\pi i k x} \in \mathfrak{L}_2(0, 1): \sum_{k=-\infty}^{+\infty} |k\alpha_k|^2 < \infty\},$$

$$\mathfrak{D}_D = \{f = \sum_{k=-\infty}^{+\infty} \alpha_k e^{2\pi i k x} \in \mathfrak{L}_2(0, 1): \sum_{k=-\infty}^{+\infty} |k\alpha_k|^2 < \infty, \sum_{k=-\infty}^{+\infty} \alpha_k = 0\}.$$

(Hint: recall that $\{e^{2\pi i k x}\}_{k=-\infty}^{+\infty}$ is a basis of $\mathfrak{L}_2(0, 1)$ (cf. § 8 example 3); for $f \in \mathfrak{D}_{D^*}$ put

$$if' = \sum_{k=-\infty}^{+\infty} \beta_k e^{2\pi i k x} \in \mathfrak{L}_2(0, 1) \subset \mathfrak{L}_1(0, 1)$$

(cf. § 5 corollary 2.1) and reconstruct f by means of the formula

$$f(\xi) = \beta + \int_0^\xi f'(\eta) \, d\eta$$

$$= \beta + \langle f', c_{[0, \xi]} \rangle \quad \text{for all } \xi \in [0, 1].)$$

c) For $f = \alpha x + \sum_{k=-\infty}^{+\infty} \alpha_k e^{2\pi ikx} \in \mathfrak{D}_{D*}$ we have

$$D^* f = i\alpha - \sum_{k=-\infty}^{+\infty} 2\pi k \alpha_k e^{2\pi ikx}$$

$$= \alpha D^* x + \sum_{k=-\infty}^{+\infty} \alpha_k D^* e^{2\pi ikx}.$$

(Hint: in the notation of b) show $\beta_0 = i\alpha$, $\beta_k = -2\pi k \alpha_k$ for $k \neq 0$.)

d) The only eigenvalues of D_1 are $-2\pi k$ ($-\infty < k < +\infty$, k an integer), the corresponding eigenspaces $\mathfrak{M}_k$ are one-dimensional, and $\mathfrak{L}_2(0, 1) = \sum_{k=-\infty}^{+\infty} \mathfrak{M}_k$.

e) The operator D does not have any eigenvalues. (Hint: use $D \subset D_1$.)

f) Every complex number λ is an eigenvalue of D^*. (Hint: consider the function $e^{-i\lambda x} \in \mathfrak{D}_{D*}$.)

§ 23. The spectrum of a linear operator

Let A be a linear operator in $\mathfrak{H}$. According to § 22 definition 1 a number $\lambda \in \mathbf{C}$ is an eigenvalue of A iff there exists a non-zero vector $f \in \mathfrak{D}_A$ such that

$$(A - \lambda I)f = o$$

or, equivalently, iff $(A - \lambda I)$ has eigenvalue 0 (note that $\mathfrak{D}_{A-\lambda I} = \mathfrak{D}_A$ by § 12 theorem 1). Since we also have

$$(A - \lambda I) o = o$$

it follows that if λ is an eigenvalue, then $A - \lambda I$ is certainly not one-to-one. Conversely, if there exist two different vectors $f_1 \in \mathfrak{D}_A$ and $f_2 \in \mathfrak{D}_A$ such that

$$(A - \lambda I) f_1 = (A - \lambda I) f_2,$$

then we have $f_1 - f_2 \in \mathfrak{D}_A$, $f_1 - f_2 \neq o$, and λ is an eigenvalue of A with corresponding eigenvector $f = f_1 - f_2$. We sum up the conclusion in a theorem.

THEOREM 1. *Let A be a linear operator in $\mathfrak{H}$. A complex number λ is an eigenvalue of A iff the operator $(A - \lambda I)$ is not one-to-one.*

If λ is not an eigenvalue of A (in fact we have just seen in § 22 example 3 that A might not have any eigenvalues), then the operator $(A - \lambda I)$ is one-to-one and therefore the inverse operator $(A - \lambda I)^{-1}$ is defined on $(A - \lambda I) \mathfrak{D}_A$ (cf. § 17 definition 3). There are still two possibilities that something goes wrong with this inverse: its domain $(A - \lambda I) \mathfrak{D}_A$ might not yet coincide with all of $\mathfrak{H}$, or $(A - \lambda I)^{-1}$ might not be bounded. If everything works out nicely, namely if $(A - \lambda I)$ is one-to-one and $(A - \lambda I)^{-1}$ is a bounded linear operator defined on all of $\mathfrak{H}$, then λ is called a *regular value* for A. In all other cases (including the case that $(A - \lambda I)$ is not one-to-one at all) λ is said to belong to the *spectrum* of A.

DEFINITION 1. Let A be a linear operator in $\mathfrak{H}$. A complex number λ is called a *regular value* for A if the operator $(A - \lambda I)$ is one-to-one and $(A - \lambda I)^{-1}$ is a bounded linear operator on $\mathfrak{H}$. The set $\Sigma(A)$ of all $\lambda \in \mathbf{C}$ that are not regular values for A is called the *spectrum* of A.

As it happens frequently, the regular values which behave nicely are not of much further interest, at least as far as our study of the operator is concerned. They leave the operator cold, if we may say so. Being tested with the expression $(A - \lambda I)^{-1}$ the operator reacts allergically, however, if λ belongs to the spectrum. It so to say recognizes λ as an integral part of its own.

This antropomorphic statement has some substantial background. Recall that we look for "parts" of A that are particularly simple or, rephrasing the problem, for "parts" of $\mathfrak{H}$ on which A approximately behaves as λI does. In fact, coincidence of A with λI on $\mathfrak{H}$ is signalized by $(A - \lambda I)$ being zero and therefore being as far from invertibility as possibly can. If $Af = \lambda If$ at least for some non-zero vectors f (making up the eigenspace $\mathfrak{M}$ corresponding to λ), then this is signalized by the fact that $(A - \lambda I)$ maps a whole non-trivial subspace into o and therefore still does not have an inverse. If $(A - \lambda I)^{-1}$ exists but is an unbounded operator, then this signalizes that, while A does not precisely act as λI does for any non-zero $f \in \mathfrak{H}$, still "somewhere in $\mathfrak{H}$" A tries hard to behave like λI:

THEOREM 2. *Let A be a linear operator in $\mathfrak{H}$. The following statements are equivalent*:

   a) *The complex number λ is an eigenvalue of A or (if λ is not an eigenvalue of A) $(A - \lambda I)^{-1}$ exists and is unbounded.*

   b) *There exists a sequence of unit vectors $\{f_n\}_{n=1}^{\infty} \subset \mathfrak{D}_A$ such that*

$$(1) \qquad \lim_{n \to \infty} (A - \lambda I) f_n = o.$$

PROOF. a)$\Rightarrow$b) If λ is an eigenvalue and if f is a corresponding eigenvector, then taking $f_n = f$ for all $n \geq 1$ we trivially get (1). If λ is not an eigenvalue and $(A - \lambda I)^{-1}$ is unbounded, then there exists a sequence of unit vectors $g_n \in \mathfrak{D}_{(A - \lambda I)^{-1}}$ such that

$$\lim_{n \to \infty} \|(A - \lambda I)^{-1} g_n\| = \infty$$

(cf. § 11 remark 2). Taking

$$f_n = \frac{(A - \lambda I)^{-1} g_n}{\|(A - \lambda I)^{-1} g_n\|} \in \mathfrak{D}_A \quad \text{for all } n \geq 1$$

we have $\| f_n \| = 1$ and

$$\lim_{n \to \infty} (A - \lambda I) f_n = \lim_{n \to \infty} \frac{g_n}{\|(A - \lambda I)^{-1} g_n\|} = o .$$

b)$\Rightarrow$a) If the sequence $\{f_n\}_{n=1}^{\infty} \subset \mathfrak{D}_A$ has the properties mentioned in b) and if λ is not an eigenvalue of A, then taking

$$g_n = \frac{(A - \lambda I) f_n}{\|(A - \lambda I) f_n\|} \in \mathfrak{D}_{(A - \lambda I)^{-1}}$$

we obtain a sequence of unit vectors $g_n \in \mathfrak{D}_{(A - \lambda I)^{-1}}$ such that

$$\lim_{n \to \infty} \|(A - \lambda I)^{-1} g_n\| = \lim_{n \to \infty} \frac{1}{\|(A - \lambda I) f_n\|} = \infty .$$

This shows that the operator $(A - \lambda I)^{-1}$ is unbounded. $\square$

DEFINITION 2. Let A be a linear operator in $\mathfrak{H}$. A complex number λ is called a *generalized* (or *approximate*) *eigenvalue* of A if there exists a sequence of unit vectors $\{f_n\}_{n=1}^{\infty} \subset \mathfrak{D}_A$ such that

$$\lim_{n \to \infty} (A - \lambda I) f_n = o .$$

By theorem 2 every generalized eigenvalue belongs to the spectrum. Conversely, it will turn out that for the classes of linear operators which interest us most, namely the unitary and selfadjoint operators, the spectrum consists entirely of generalized eigenvalues (cf. § 23 theorem 8 and § 24 theorem 4; the corresponding statement for bounded normal operators is contained in § 23 exercise e)).

Example 1. Let E be the multiplication operator in $\mathfrak{L}_2(\alpha, \beta)$ ($-\infty \leq \alpha < \beta \leq +\infty$; cf. § 19 theorem 1). We shall show that E does not have any eigenvalues, that the spectrum $\Sigma(E)$ coincides with the interval $[\alpha, \beta]$ on the real axis (of course this interval has to be taken open where an endpoint disappears at infinity), and that $\Sigma(E)$ entirely consists of generalized eigenvalues.

Let us start with any $\lambda \in \mathbf{C}$ and show that λ is not an eigenvalue. Suppose for some $f \in \mathfrak{D}_E$ we have

$$\|(E - \lambda I) f\|^2 = \int_\alpha^\beta |\xi - \lambda|^2 |f(\xi)|^2 \, d\xi = 0.$$

Since $|\xi - \lambda| > 0$ for almost all $\xi \in \mathbf{R}$ we conclude $f(\xi) = 0$ for almost all $\xi \in \mathbf{R}$ and therefore $f = o$.

Next take any $\lambda \in [\alpha, \beta]$ and for $n \geqslant 1$ let $c_n \in \mathfrak{D}_E$ be the characteristic function of the interval $[\lambda - 1/n, \lambda + 1/n] \cap [\alpha, \beta]$ (fig. 16), defined by

$$c_n(\xi) = \begin{cases} 1 & \text{for } \xi \in [\lambda - 1/n, \lambda + 1/n] \cap [\alpha, \beta], \\ 0 & \text{for } \xi \notin [\lambda - 1/n, \lambda + 1/n] \cap [\alpha, \beta]. \end{cases}$$

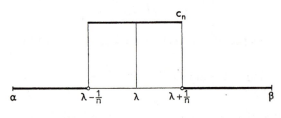

Fig. 16

Then $\|c_n\| > 0$ and $f_n = c_n / \|c_n\|$ is a unit vector in $\mathfrak{D}_E$. Moreover

$$\|(E - \lambda I) f_n\|^2 = \int_\alpha^\beta |\xi - \lambda|^2 |f_n \xi)|^2 \, d\xi$$

$$\leqslant \frac{1}{n^2} \int_\alpha^\beta |f_n(\xi)|^2 \, d\xi = \frac{1}{n^2}.$$

We see that λ is a generalized eigenvalue and therefore belongs to $\Sigma(E)$.

Finally consider any complex number $\lambda \notin [\alpha, \beta]$. Then we have $\delta = \inf_{\xi \in [\alpha, \beta]} |\xi - \lambda| > 0$. For every $g \in \mathfrak{L}_2(\alpha, \beta)$ we have $f = g/(x - \lambda) \in \mathfrak{D}_E$ since

$$\int_\alpha^\beta |f(\xi)|^2 \, d\xi = \int_\alpha^\beta \frac{|g(\xi)|^2}{|\xi - \lambda|^2} \, d\xi \leqslant \int_\alpha^\beta \frac{|g(\xi)|^2}{\delta^2} \, d\xi = \frac{1}{\delta^2} \|g\|^2 < \infty$$

and $xf = g + \lambda f \in \mathfrak{L}_2(\alpha, \beta)$. Moreover we get $(E - \lambda I) f = (x - \lambda) f = g$. We conclude $\mathfrak{D}_{(E-\lambda I)^{-1}} = (E - \lambda I)\,\mathfrak{D}_E = \mathfrak{L}_2(\alpha, \beta)$ and

$$(E - \lambda I)^{-1} g = \frac{g}{x - \lambda},$$

$$\|(E - \lambda I)^{-1} g\|^2 = \int_\alpha^\beta \frac{|g(\xi)|^2}{|\xi - \lambda|^2}\, d\xi \leqslant \frac{1}{\delta^2}\, \|g\|^2,$$

$$\|(E - \lambda I)^{-1}\| \leqslant \frac{1}{\delta}.$$

This shows that λ is a regular value for E.

Remark 1. Let E be the multiplication operator in $\mathfrak{L}_2(\mathbf{R}^2)$ as defined in § 19 remark 2. By an analogous reasoning as used above it is found that E does not have any eigenvalues, that the spectrum $\Sigma(E)$ coincides with the entire complex plane, and that $\Sigma(E)$ entirely consists of generalized eigenvalues.

Since the spectrum of a linear operator A in $\mathfrak{H}$ is defined in terms of existence and boundedness of the operator $(A - \lambda I)^{-1}$ we now look for some theorems in this direction.

THEOREM 3 (NEUMANN SERIES). *If A is a bounded linear operator on $\mathfrak{H}$ and if $\|A\| < 1$, then $(I - A)$ is invertible (in the sense of § 12 definition 5) and*

$$(I - A)^{-1} = \sum_{k=0}^{\infty} A^k.$$

PROOF. Recall that the set $\mathscr{B}$ of all bounded linear operators on $\mathfrak{H}$ is a Banach algebra (cf. § 12 theorem 3). The series $\sum_{k=0}^{\infty} A^k$ converges in $\mathscr{B}$ since for $m < n$ we have

$$\left\| \sum_{k=m}^{n} A^k \right\| \leqslant \sum_{k=m}^{n} \|A\|^k \leqslant \sum_{k=m}^{\infty} \|A\|^k = \frac{\|A\|^m}{1 - \|A\|}$$

and this is smaller than a given $\varepsilon > 0$ if m is sufficiently large (cf. § 3 remark 3). Since multiplication in $\mathscr{B}$ is continuous (cf. § 12 theorem 4) we may multiply the series $\sum_{k=0}^{\infty} A^k$ by $(I - A)$ by doing this termwise. We obtain

$$(I - A)\left(\sum_{k=0}^{\infty} A^k \right) = \sum_{k=0}^{\infty} A^k - \sum_{k=1}^{\infty} A^k = I,$$

$$\left(\sum_{k=0}^{\infty} A^k \right)(I - A) = \sum_{k=0}^{\infty} A^k - \sum_{k=1}^{\infty} A^k = I.$$

This proves the assertion. $\square$

COROLLARY 3.1. *Let A and B be bounded linear operators on $\mathfrak{H}$. If A is invertible and if*

$$\|A - B\| < \frac{1}{\|A^{-1}\|},$$

then B is invertible.

PROOF. From the hypothesis we conclude

$$\|I - A^{-1}B\| = \|A^{-1}(A - B)\| < \|A^{-1}\| \frac{1}{\|A^{-1}\|} = 1.$$

By theorem 3 the operator $C = A^{-1}B = I - (I - A^{-1}B)$ is invertible and therefore so is the operator $B = AC$, its inverse being $B^{-1} = C^{-1}A^{-1}$. $\square$

COROLLARY 3.2. *If A is a bounded linear operator on $\mathfrak{H}$ and if $|\lambda| > \|A\|$, then λ is a regular value for A and*

$$(A - \lambda I)^{-1} = -\frac{1}{\lambda} \sum_{k=0}^{\infty} \left(\frac{A}{\lambda}\right)^k.$$

Consequently $\Sigma(A)$ is bounded.

PROOF. Observe that $A - \lambda I = -\lambda(I - A/\lambda)$ and $\|A/\lambda\| < 1$ and apply theorem 3. $\square$

While in general the spectrum $\Sigma(A)$ of a linear operator A in $\mathfrak{H}$ may even coincide with the entire complex plane (cf. remark 1), for a bounded linear operator A on $\mathfrak{H}$ it is contained in a circle of radius $\|A\|$ about 0 by corollary 3.2. Next we shall show that in any case $\Sigma(A)$ is a closed subset of the complex plane by showing that if λ is a regular value for A, then all complex numbers μ in a suitable neighbourhood of λ also are regular values for A (fig. 17).

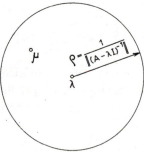

Fig. 17

THEOREM 4. *Let A be a linear operator in $\mathfrak{H}$. If λ is a regular value for A and if $|\mu - \lambda| < 1/\|(A - \lambda I)^{-1}\|$, then μ is a regular value for A.*

PROOF. In corollary 3.1 replace A and B by $(A-\lambda I)$ and $(A-\mu I)$ respectively. $\square$

COROLLARY 4.1. *The set of regular values for a linear operator A in $\mathfrak{H}$ is open. The spectrum $\Sigma(A)$ is closed.*

Example 2. Let $\{e_k\}_{k=1}^{\infty}$ be the standard basis of ℓ_2 and let A be the right shift operator in ℓ_2 (cf. § 12 example 2). Since A is isometric $\Sigma(A)$ is certainly contained in the unit disk of the complex plane by corollary 3.2. We already know that A does not have any eigenvalues (cf. § 22 example 3). We shall now show that every $\lambda \in \mathbf{C}$ satisfying $0 < |\lambda| \leqslant 1$ belongs to $\Sigma(A)$ by showing that $\mathfrak{D}_{(A-\lambda I)^{-1}} = (A-\lambda I)\,\ell_2$ does not fill up the entire Hilbert space ℓ_2. In order to do this it will suffice to show that there does not exist any $f \in \ell_2$ satisfying the equation

$$(2) \qquad\qquad (A - \lambda I)\,f = e_1.$$

In fact, if such a vector $f = \sum_{k=1}^{\infty} \alpha_k e_k$ would exist in ℓ_2, then (2) would imply

$$\sum_{k=1}^{\infty} (\alpha_{k-1} - \lambda \alpha_k)\, e_k = e_1 \quad (\alpha_0 = 0),$$

$$-\lambda \alpha_1 = 1,$$

$$\alpha_{k-1} - \lambda \alpha_k = 0 \quad \text{for } k \geqslant 2,$$

and consequently

$$\alpha_1 = -\frac{1}{\lambda},$$

$$\alpha_k = \frac{1}{\lambda}\alpha_{k-1} = -\left(\frac{1}{\lambda}\right)^k \quad \text{for } k \geqslant 2.$$

Since $|1/\lambda| \geqslant 1$ this is incompatible with the hypothesis $\sum_{k=1}^{\infty} |\alpha_k|^2 = \|f\|^2 < \infty$. Having shown that

$$\{\lambda \in \mathbf{C}: 0 < |\lambda| \leqslant 1\} \subset \Sigma(A) \subset \{\lambda \in \mathbf{C}: |\lambda| \leqslant 1\}$$

we now conclude that $\Sigma(A)$, being a closed subset of the unit disk, must in fact coincide with the unit disk in $\mathbf{C}$.

The attentive reader will perhaps also notice that 0 is a point of the spectrum but not a generalized eigenvalue of A. In fact, the inverse of $A = A - 0 \cdot I$ is isometric (and therefore bounded) but only defined on $\bigvee \{e_k\}_{k=2}^{\infty} \neq \ell_2$ (cf. definition 2 and theorem 2).

THEOREM 5 (HILBERT RELATION). *If λ and μ are regular values for a linear* *operator A in $\mathfrak{H}$, then*

  a) $(A-\lambda I)^{-1}-(A-\mu I)^{-1}=(\lambda-\mu)(A-\lambda I)^{-1}(A-\mu I)^{-1}$,
  b) $(A-\lambda I)^{-1}(A-\mu I)^{-1}=(A-\mu I)^{-1}(A-\lambda I)^{-1}$.

PROOF. a) From

$$(A-\lambda I)^{-1}=(A-\lambda I)^{-1}(A-\mu I)(A-\mu I)^{-1},$$
$$(A-\mu I)^{-1}=(A-\lambda I)^{-1}(A-\lambda I)(A-\mu I)^{-1}$$

we conclude by § 12 corollary 1.1 b) (although $\mathfrak{D}_A$ need not coincide with $\mathfrak{H}$ the distributive laws also hold in the situation considered here as may be checked directly) and § 17 (2)

$$
\begin{aligned}
(A-\lambda I)^{-1}-(A-\mu I)^{-1} &= (A-\lambda I)^{-1}[(A-\mu I)-(A-\lambda I)](A-\mu I)^{-1} \\
&= (A-\lambda I)^{-1}(\lambda-\mu)I_A(A-\mu I)^{-1} \\
&= (\lambda-\mu)(A-\lambda I)^{-1}(A-\mu I)^{-1}.
\end{aligned}
$$

  b) The assertion follows from a) after interchanging λ and μ. $\square$

Because of the important role played by the operator $(A-\lambda I)^{-1}$ it has been given a special name.

DEFINITION 3. If A is a linear operator in $\mathfrak{H}$ and if λ is not an eigenvalue, then $(A-\lambda I)^{-1}$ is called the *resolvent (operator)* of A at λ.

Remark 2. The Hilbert relation (theorem 5 a)) is used to show that, for any fixed bounded linear operator A on $\mathfrak{H}$, the resolvent is an analytic function in λ on the set of regular values in $\mathbf{C}$. This again is used to show that the spectrum of a bounded linear operator on $\mathfrak{H}$ is not empty. We shall not need and therefore not prove this statement. The interested reader is referred to [7] II § 8.4.

THEOREM 6. *If A is a closed linear operator in $\mathfrak{H}$, then the resolvent of A at λ is a closed operator for every $\lambda \in \mathbf{C}$ for which it is defined.*
PROOF. Suppose λ is not an eigenvalue. It suffices to show that the operator $A-\lambda I$ is closed and then to apply § 20 theorem 2. If $\{f_n\}_{n=1}^{\infty} \subset \mathfrak{D}_A$ is a sequence having the property that both limits

$$f = \lim_{n\to\infty} f_n,$$
$$g = \lim_{n\to\infty} (A-\lambda I)f_n$$

exist, then also the sequence $\{Af_n\}_{n=1}^{\infty} = \{(A-\lambda I)f_n+\lambda f_n\}_{n=1}^{\infty}$ converges and since A is closed we have

$$f \in \mathfrak{D}_A = \mathfrak{D}_{A-\lambda I},$$
$$g = \lim_{n \to \infty} A f_n - \lambda \lim_{n \to \infty} f_n = Af - \lambda f = (A - \lambda I) f . \;\square$$

We now apply these results to get some more information about the spectrum of a unitary operator on $\mathfrak{H}$. In the next section we shall still in more detail investigate the spectrum of a selfadjoint operator in $\mathfrak{H}$.

THEOREM 7. *The spectrum of a unitary operator U is a closed subset of the complex unit circle.*
PROOF. Note that $\|U\| = 1$ (cf. § 14 theorem 4). If $|\mu| > 1$ then μ is a regular value for U by corollary 3.2. In order to consider the case $|\mu| < 1$ we observe that $\lambda = 0$ is a regular value for U and $\|U^{-1}\| = 1$. Applying theorem 4 we find that μ is a regular value for U. $\square$

Theorem 7 extends § 22 theorem 3 in case of a unitary operator. That a similar extension is not possible for isometric operators in general follows from example 2 where we have exhibited an isometric linear operator on ℓ_2 the spectrum of which fills up the unit disk.

THEOREM 8. *The spectrum of a unitary operator U consists of generalized eigenvalues.*
PROOF. Suppose $\lambda \in \Sigma(U)$ is not an eigenvalue of U. Then $(U - \lambda I)\mathfrak{H}$ is everywhere dense in $\mathfrak{H}$ by § 22 theorem 5. By theorem 6 the operator $(U - \lambda I)^{-1}$ is closed. If $(U - \lambda I)^{-1}$ were also bounded, then $\mathfrak{D}_{(U-\lambda I)^{-1}} = (U - \lambda I)\mathfrak{H}$ would be a subspace by § 20 theorem 1 and therefore would coincide with $\mathfrak{H}$. Thus $(U - \lambda I)^{-1}$ would be bounded and defined everywhere on $\mathfrak{H}$ in contradiction to our assumption $\lambda \in \Sigma(U)$. Therefore $(U - \lambda I)^{-1}$ must be unbounded and λ is a generalized eigenvalue by theorem 2. $\square$

With an eye to the relation between the differentiation and the multiplication operator in $\mathfrak{L}_2(-\infty, +\infty)$ we finally make sure that equivalent operators in $\mathfrak{H}$ have the same spectral properties (cf. § 19 definition 1, § 19 corollary 3.1).

THEOREM 9. *Let A and B be equivalent linear operators in $\mathfrak{H}$. A complex number λ is an eigenvalue of A iff it is an eigenvalue of B. If λ is not an eigenvalue, then $(A - \lambda I)^{-1}$ is equivalent with $(B - \lambda I)^{-1}$. As a consequence, $\Sigma(A) = \Sigma(B)$ and λ is a generalized eigenvalue of A iff it is a generalized eigenvalue of B.*
PROOF. Suppose U is a unitary operator on $\mathfrak{H}$ and $A = UBU^{-1}$ (which implies

$\mathfrak{D}_A = U\mathfrak{D}_B$). If $f \in \mathfrak{D}_B$ is an eigenvector of B corresponding to the eigenvalue λ, then

$$o \neq Uf \in \mathfrak{D}_A,$$
$$A(Uf) = UBf = \lambda(Uf),$$

and Uf is an eigenvector of A corresponding to the same eigenvalue λ. Reversing this reasoning we obtain the first statement of the theorem.

If λ is not an eigenvalue, then by § 17 (2) we have

$$(A - \lambda I)^{-1} (A - \lambda I) = I_A,$$
$$(A - \lambda I)(A - \lambda I)^{-1} = I_{(A - \lambda I)^{-1}}.$$

Let $C = U^{-1}(A - \lambda I)^{-1} U$ (which implies $\mathfrak{D}_C = U^{-1}\mathfrak{D}_{(A - \lambda I)^{-1}}$). We get

$$C(B - \lambda I) = U^{-1}(A - \lambda I)^{-1}UU^{-1}(A - \lambda I) U = I_{U^{-1}AU} = I_B,$$
$$(B - \lambda I) C = U^{-1}(A - \lambda I) UU^{-1}(A - \lambda I)^{-1}U = I_C,$$

and therefore

$$(B - \lambda I)^{-1} = C = U^{-1}(A - \lambda I)^{-1}U.$$

As a consequence, the resolvent $(B - \lambda I)^{-1}$ is unbounded or not defined everywhere on $\mathfrak{H}$ if and only if so is $(A - \lambda I)^{-1}$. $\square$

Example 3. Let D and E be the differentiation operator and the multiplication operator in $\mathfrak{L}_2(-\infty, +\infty)$ respectively (cf. § 18 theorem 7, § 19 theorem 1). Since D and E are equivalent (cf. § 19 corollary 3.1) and since we already have complete information about $\Sigma(E)$ (cf. example 1) we now conclude by theorem 9 that also D does not have any eigenvalues, that $\Sigma(D)$ coincides with the real axis, and that $\Sigma(D)$ entirely consists of generalized eigenvalues.

Note that in $\mathfrak{L}_2(\alpha, \beta)$ where α or β are finite the differentiation operator D and the multiplication operator E are not anymore equivalent (this can already be deduced from the fact that E is selfadjoint while D isn't in these cases). In fact, for $-\infty < \alpha < \beta < +\infty$ every complex number λ is an eigenvalue of the operator D^* in $\mathfrak{L}_2(\alpha, \beta)$ since $e^{-i\lambda x}$ is absolutely continuous on $[\alpha, \beta]$ and

$$(3) \qquad\qquad D^*e^{-i\lambda x} = \lambda e^{-i\lambda x}$$

(cf. § 18 theorem 2, § 22 exercise f)). For the operator D^* in $\mathfrak{L}_2(\alpha, +\infty)$ $(-\infty < \alpha < +\infty)$ still every complex number $\lambda = \gamma + i\delta$ with $\gamma \in \mathbf{R}$, $\delta \in \mathbf{R}$, $\delta < 0$ is an eigenvalue. Indeed, also in this case we have $e^{-i\lambda x} = e^{-i\gamma x}e^{\delta x} \in$

$\in \mathfrak{L}_2(\alpha, +\infty)$ since .

$$\int\limits_{\alpha}^{+\infty} |e^{-i\lambda\xi}|^2 \, d\xi = \int\limits_{\alpha}^{+\infty} e^{2\delta\xi} \, d\xi = -\frac{e^{2\delta\alpha}}{2\delta} < \infty$$

and the assertion again follows from (3) (cf. § 18 theorem 5). This contrasts the fact that in either case the multiplication operator does not have any eigenvalues (cf. example 1).

The following exercise among other things contains the statement that the spectrum of a bounded normal operator on $\mathfrak{H}$ entirely consists of generalized eigenvalues.

Exercise. Let A be a bounded normal operator on $\mathfrak{H}$ and let the complex number λ be given.

a) $\lambda \in \Sigma(A)$ iff $\bar{\lambda} \in \Sigma(A^*)$. (Hint: use § 14 theorem 2 e); observe that the assertion is true even without A being normal.)

b) λ is an eigenvalue of A iff $\bar{\lambda}$ is an eigenvalue of A^*. (Hint: observe that also $(A - \lambda I)$ is normal and use § 14 theorem 6.)

c) If λ is an eigenvalue of A, then $\overline{(A - \lambda I)\,\mathfrak{H}} \neq \mathfrak{H}$ and $\overline{(A^* - \bar{\lambda} I)\,\mathfrak{H}} \neq \mathfrak{H}$. (Hint: every eigenvector of A^* corresponding to the eigenvalue $\bar{\lambda}$ is orthogonal to $(A - \lambda I)\,\mathfrak{H}$; cf. § 7 corollary 7.2.)

d) If λ is not an eigenvalue of A, then $\overline{(A - \lambda I)\,\mathfrak{H}} = \overline{(A^* - \bar{\lambda} I)\,\mathfrak{H}} = \mathfrak{H}$. (Hint: use b) to show that $g \perp (A - \lambda I)\,\mathfrak{H}$ implies $g = o$.)

e) If λ is not a generalized eigenvalue of A, then λ is a regular value for A. (Hint: use d) and imitate the proof of theorem 8.)

f) λ is a regular value for A iff $(A - \lambda I)\,\mathfrak{H} = \mathfrak{H}$. (Hint: if $(A - \lambda I)\,\mathfrak{H} = \mathfrak{H}$ use c) to show that $(A - \lambda I)^{-1}$ and $(A^* - \bar{\lambda} I)^{-1}$ exist and apply § 17 theorem 5 and § 17 theorem 8 to show that $(A^* - \bar{\lambda} I)^{-1}$ and $(A - \lambda I)^{-1}$ are bounded.)

§ 24. The spectrum of a selfadjoint operator

We already know that the eigenvalues of a selfadjoint operator in $\mathfrak{H}$ are real (if there are any eigenvalues at all; cf. § 22 theorem 2, § 23 example 1). What about the spectrum? Before answering this question we have to develop some tools.

THEOREM 1. *Let A be a linear operator in $\mathfrak{H}$ such that $\overline{\mathfrak{D}}_A = \mathfrak{H}$. Then*

$$(A - \lambda I)^* = A^* - \bar{\lambda} I.$$

PROOF. It makes sense to ask for the adjoint of $(A - \lambda I)$ since $\mathfrak{D}_{A - \lambda I} = \mathfrak{D}_A$

is everywhere dense in $\mathfrak{H}$ (cf. § 17 definition 1). For any $g \in \mathfrak{D}_{A^*} = \mathfrak{D}_{A^* - \bar{\lambda}I}$ we have for all $f \in \mathfrak{D}_A$

$$\langle (A - \lambda I) f, g \rangle = \langle Af, g \rangle - \lambda \langle f, g \rangle$$
$$= \langle f, A^* g \rangle - \langle f, \bar{\lambda} g \rangle$$
$$= \langle f, (A^* - \bar{\lambda}I) g \rangle.$$

This implies $\mathfrak{D}_{A^* - \bar{\lambda}I} \subset \mathfrak{D}_{(A - \lambda I)^*}$ and $(A^* - \bar{\lambda}I) \subset (A - \lambda I)^*$. On the other hand, for any $g \in \mathfrak{D}_{(A - \lambda I)^*}$ and $g^* = (A - \lambda I)^* g$ we get for all $f \in \mathfrak{D}_A$

$$\langle (A - \lambda I) f, g \rangle = \langle f, g^* \rangle,$$
$$\langle Af, g \rangle = \langle f, g^* \rangle + \lambda \langle f, g \rangle$$
$$= \langle f, g^* + \bar{\lambda} g \rangle.$$

This implies $\mathfrak{D}_{(A - \lambda I)^*} \subset \mathfrak{D}_{A^*} = \mathfrak{D}_{A^* - \bar{\lambda}I}$ and therefore $A^* - \bar{\lambda}I = (A - \lambda I)^*$ (cf. § 17 definition 2, § 12 definition 2). $\square$

COROLLARY 1.1. *If the linear operator A in $\mathfrak{H}$ is selfadjoint and if λ is real, then $(A - \lambda I)$ is selfadjoint.*

COROLLARY 1.2. *Let A be a linear operator in $\mathfrak{H}$ such that $\overline{\mathfrak{D}}_A = \overline{\mathfrak{D}}_{A^*} = \mathfrak{H}$ and $A = A^{**}$. Then $\lambda \in \Sigma(A)$ iff $\bar{\lambda} \in \Sigma(A^*)$; equivalently,*

$$\Sigma(A^*) = \{\bar{\lambda} : \lambda \in \Sigma(A)\}.$$

PROOF. If λ is a regular value for A, then $\overline{\mathfrak{D}}_A = \overline{\mathfrak{D}}_{A - \lambda I} = \mathfrak{D}_{(A - \lambda I)^{-1}} = \mathfrak{H}$. By § 17 theorem 5 also the operator $(A - \lambda I)^* = A^* - \bar{\lambda}I$ is one-to-one and we have $(A^* - \bar{\lambda}I)^{-1} = [(A - \lambda I)^{-1}]^*$. Since the last operator is bounded and defined on all of $\mathfrak{H}$ the number $\bar{\lambda}$ is a regular value for A^*. Reversing this reasoning we see that λ is a regular value for A iff $\bar{\lambda}$ is a regular value for A^*. $\square$

Originally we have defined eigenvalues by the existence of eigenvectors (which by definition have to be non-zero; cf. § 22 definition 1). Later we have seen that eigenvalues may also be characterized by $(A - \lambda I)$ not being one-to-one (cf. § 23 theorem 1). For linear operators A with everywhere dense domain of A and A^* and having the property that $A = A^{**}$ (in fact these are precisely the closed linear operators; cf. appendix A theorem 6) we shall now establish a third characterization of eigenvalues in terms of the size of $(A^* - \bar{\lambda}I) \mathfrak{D}_{A^*}$.

Parallel to these three characterizations of eigenvalues we obtain three characterizations of the eigenspace $\mathfrak{M}_\lambda$ corresponding to the eigenvalue λ. In the first place it was defined as the set of all corresponding eigenvectors

together with o. Secondly we obtained $\mathfrak{M}_\lambda$ as the set of all vectors annihilated by $(A - \lambda I)$. Now we shall meet $\mathfrak{M}_\lambda$ again in the form of the orthogonal complement of $(A^* - \bar{\lambda}I)\,\mathfrak{D}_{A^*}$.

The reader to whom these statements and the following ones sound familiar will feel justified after another look at § 22 theorem 5 or § 23 exercise.

THEOREM 2. *Let A be a linear operator in $\mathfrak{H}$ such that $\overline{\mathfrak{D}}_A = \overline{\mathfrak{D}}_{A^*} = \mathfrak{H}$ and
$A = A^{**}$ and let the complex number λ be given. Then*

$$\{\,f \in \mathfrak{D}_A : Af = \lambda f\,\} = [(A^* - \bar{\lambda}I)\,\mathfrak{D}_{A^*}]^\perp.$$

As a consequence, λ is an eigenvalue of A iff $\overline{(A^* - \bar{\lambda}I)\,\mathfrak{D}_{A^*}} \neq \mathfrak{H}$.
PROOF. For any $f \in \mathfrak{H}$ the following statements are equivalent:

$$Af - \lambda f = o,$$

$$\langle (A - \lambda I)\,f, g \rangle = 0 \quad \text{for all } g \in \mathfrak{D}_{A^*},$$

$$\langle f, (A^* - \bar{\lambda}I)\,g \rangle = 0 \quad \text{for all } g \in \mathfrak{D}_{A^*} \quad (\text{cf. theorem 1}),$$

$$f \perp (A^* - \bar{\lambda}I)\,\mathfrak{D}_{A^*}$$

(note that each of these statements also implies $f \in \mathfrak{D}_A = \mathfrak{D}_{A^{**}}$). $\square$

COROLLARY 2.1. *Let A be a selfadjoint operator in $\mathfrak{H}$. A complex number λ is
an eigenvalue of A iff $\overline{(A - \lambda I)\,\mathfrak{D}_A} \neq \mathfrak{H}$. If λ is an eigenvalue of A, then the
corresponding eigenspace is*

$$\mathfrak{M}_\lambda = [(A - \lambda I)\,\mathfrak{D}_A]^\perp.$$

PROOF. If λ is an eigenvalue of A, then λ is real by § 22 theorem 2. Further-
more, by theorem 2 we have $\overline{(A - \lambda I)\,\mathfrak{D}_A} \neq \mathfrak{H}$ and $\mathfrak{M}_\lambda = [(A - \lambda I)\,\mathfrak{D}_A]^\perp$. If λ
is not an eigenvalue of A, then $\bar{\lambda}$ is not an eigenvalue either and therefore
$\overline{(A - \lambda I)\,\mathfrak{D}_A} = \mathfrak{H}$ by theorem 2 (applied to $\bar{\lambda}$ in place of λ). $\square$

Also § 22 theorem 5 may be considered as a corollary of theorem 2 if we
take into account (as has been done there in the proof) that the eigenspace
of U^{-1} corresponding to $\bar{\lambda}$ coincides with the eigenspace of U corresponding
to λ.

THEOREM 3. *The spectrum of a selfadjoint operator in $\mathfrak{H}$ is real.*
PROOF. Let A be a selfadjoint operator in $\mathfrak{H}$ and let $\lambda = \alpha + \beta i$, $\alpha \in \mathbf{R}$, $\beta \in \mathbf{R}$,
$\beta \neq 0$. Then λ is not an eigenvalue of A (cf. § 22 theorem 2) and the resolvent
$(A - \lambda I)^{-1}$ exists and is a closed operator (cf. § 23 theorem 1, § 23 theorem 6).
We shall show that $(A - \lambda I)^{-1}$ is in fact bounded. Indeed, for any

$g \in \mathfrak{D}_{(A-\lambda I)^{-1}} = (A - \lambda I) \mathfrak{D}_A$ we have $f = (A-\lambda I)^{-1} g \in \mathfrak{D}_A$ and

$$
\begin{aligned}
\|g\|^2 &= \langle (A - \lambda I) f, (A - \lambda I) f \rangle \\
&= \langle (A - \alpha I) f - i\beta f, (A - \alpha I) f - i\beta f \rangle \\
&= \|(A - \alpha I) f\|^2 - i\beta \langle f, (A - \alpha I) f \rangle + i\beta \langle (A - \alpha I) f, f \rangle + |\beta|^2 \|f\|^2 \\
&= \|(A - \alpha I) f\|^2 + |\beta|^2 \|f\|^2 \\
&\geqslant |\beta|^2 \|f\|^2 = |\beta|^2 \|(A - \lambda I)^{-1} g\|^2 .
\end{aligned}
$$

We conclude

$$
\|(A - \lambda I)^{-1} g\| \leqslant \frac{\|g\|}{|\beta|} \quad \text{for all } g \in \mathfrak{D}_{(A-\lambda I)^{-1}},
$$

$$
\|(A - \lambda I)^{-1}\| \leqslant \frac{1}{|\beta|}.
$$

Since $(A - \lambda I)^{-1}$ is closed and bounded its domain $(A - \lambda I) \mathfrak{D}_A$ must be closed (cf. § 20 theorem 1). Since λ is not an eigenvalue this domain is everywhere dense in $\mathfrak{H}$ and therefore necessarily coincides with $\mathfrak{H}$ (cf. corollary 2.1). As a consequence, λ is a regular value for A (cf. § 23 definition 1). This proves the assertion. $\square$

THEOREM 4. *The spectrum of a selfadjoint operator in $\mathfrak{H}$ consists of generalized eigenvalues.*

PROOF. Let A be a selfadjoint operator in $\mathfrak{H}$ and suppose $\lambda \in \Sigma(A)$ is not an eigenvalue. Then $(A - \lambda I) \mathfrak{D}_A$ is everywhere dense in $\mathfrak{H}$ by corollary 2.1 and $(A - \lambda I)^{-1}$ is closed by § 23 theorem 6 (and in fact even selfadjoint). If $(A - \lambda I)^{-1}$ were bounded, then its domain $(A - \lambda I) \mathfrak{D}_A$ would be closed by § 20 theorem 1 and therefore coincide with $\mathfrak{H}$. As a consequence, λ would be regular in contradiction to our assumption $\lambda \in \Sigma(A)$. We conclude that $(A - \lambda I)^{-1}$ is unbounded and that λ is a generalized eigenvalue by § 23 theorem 2. $\square$

In case of a selfadjoint operator A in $\mathfrak{H}$ there is even a way to locate the spectrum on the real axis without looking at the resolvent. Observe that corollary 2.1 states that λ is not an eigenvalue iff $\overline{(A - \lambda I) \mathfrak{D}_A} = \mathfrak{H}$. We continue by showing that λ does not belong to the spectrum of A iff $(A - \lambda I) \mathfrak{D}_A = \mathfrak{H}$.

THEOREM 5. *Let A be a selfadjoint operator in $\mathfrak{H}$. A complex number λ is a regular value for A iff $(A - \lambda I) \mathfrak{D}_A = \mathfrak{H}$.*

PROOF. If λ is regular, then $(A - \lambda I) \mathfrak{D}_A = \mathfrak{D}_{(A-\lambda I)^{-1}} = \mathfrak{H}$ by § 23 definition 1.

Conversely, suppose $(A-\lambda I)\,\mathfrak{D}_A=\mathfrak{H}$. If $\lambda\notin\mathbf{R}$, then λ is regular by theorem 3. If $\lambda\in\mathbf{R}$, then λ is not an eigenvalue by corollary 2.1, $(A-\lambda I)$ is selfadjoint by corollary 1.1, and so is $(A-\lambda I)^{-1}$ by § 17 theorem 6. Being defined on all of $\mathfrak{H}$ the resolvent $(A-\lambda I)^{-1}$ is bounded by § 17 corollary 8.2. Therefore λ is regular also in this case. $\square$

In fact the same method can be applied to locate the spectrum of a unitary operator U on the complex unit circle. Recall that $\lambda\in\mathbf{C}$ is not an eigenvalue of U iff $\overline{(U-\lambda I)\,\mathfrak{H}}=\mathfrak{H}$ (cf. § 22 theorem 5). Again it can be shown that λ does not belong to the spectrum of U iff $(U-\lambda I)\,\mathfrak{H}=\mathfrak{H}$. The essential step in the proof is simple if one has at one's disposal the closed graph theorem (cf. appendix A theorem 7): if $(U-\lambda I)\,\mathfrak{H}=\mathfrak{H}$, then λ is not an eigenvalue and $(U-\lambda I)^{-1}$ is a closed linear operator defined on all of $\mathfrak{H}$. As a consequence $(U-\lambda I)^{-1}$ is bounded and λ is regular. If the closed graph theorem is not available, then it is still possible to arrive at the same conclusion with the help of some other theorems stated earlier, as suggested in the exercise of the preceding section. As shown there, the above mentioned statements may even be extended to bounded normal operators on $\mathfrak{H}$ in general.

The facts mentioned so far suggest that we may distinguish further between the points of the spectrum of a unitary or selfadjoint operator A in $\mathfrak{H}$ according to the size of the linear manifold $(A-\lambda I)\,\mathfrak{D}_A$ (the range of the operator $A-\lambda I$). Although each of the following statements could analogously be formulated for a unitary operator, we shall restrict ourselves to the case of a selfadjoint operator where we can refer to a theorem which has explicitly been stated and proved.

Suppose therefore that A is a selfadjoint operator in $\mathfrak{H}$. We distinguish the following four cases.

(a) $(A-\lambda I)\,\mathfrak{D}_A=\overline{(A-\lambda I)\,\mathfrak{D}_A}=\mathfrak{H}$.

In this case (which in particular occurs for $\lambda\notin\mathbf{R}$) λ is a regular value for A by theorem 5. The resolvent is a bounded linear operator defined everywhere on $\mathfrak{H}$.

(b) $(A-\lambda I)\,\mathfrak{D}_A\neq\overline{(A-\lambda I)\,\mathfrak{D}_A}=\mathfrak{H}$.

In this case λ is still not an eigenvalue but a generalized eigenvalue of A by corollary 2.1 and theorem 4. The resolvent is an unbounded selfadjoint operator in $\mathfrak{H}$ (note that λ is real, just as in the cases (c) and (d) considered below).

(c) $(A - \lambda I)\, \mathfrak{D}_A = \overline{(A - \lambda I)\, \mathfrak{D}_A} \neq \mathfrak{H}$.

In this case λ is an eigenvalue by corollary 2.1 and there is no resolvent. The eigenspace $\mathfrak{M}_\lambda$ is annihilated by $(A - \lambda I)$ and reduces A, moreover the restriction A_1 of A to $\mathfrak{M}_\lambda^\perp \cap \mathfrak{D}_A$ is a selfadjoint operator in $\mathfrak{M}_\lambda^\perp$ (cf. § 22 theorem 6, § 21 theorem 1). From $\mathfrak{M}_\lambda \subset \mathfrak{D}_A$ and

$$\mathfrak{D}_A = \mathfrak{M}_\lambda + (\mathfrak{M}_\lambda^\perp \cap \mathfrak{D}_A)$$

we deduce

$$
\begin{aligned}
(A - \lambda I)\, \mathfrak{D}_A &= (A - \lambda I)\, \mathfrak{M}_\lambda + (A - \lambda I)\,(\mathfrak{M}_\lambda^\perp \cap \mathfrak{D}_A) \\
&= (A - \lambda I)\,(\mathfrak{M}_\lambda^\perp \cap \mathfrak{D}_A) \\
&= (A_1 - \lambda I)\, \mathfrak{D}_{A_1}
\end{aligned}
$$

and further

$$(A_1 - \lambda I)\, \mathfrak{D}_{A_1} = (A - \lambda I)\, \mathfrak{D}_A = \overline{(A - \lambda I)\, \mathfrak{D}_A} = \overline{(A_1 - \lambda I)\, \mathfrak{D}_{A_1}}.$$

Since λ is not anymore an eigenvalue of A_1 we have

$$(A_1 - \lambda I)\, \mathfrak{D}_{A_1} = \overline{(A_1 - \lambda I)\, \mathfrak{D}_{A_1}} = \mathfrak{M}_\lambda^\perp$$

and λ is therefore a regular value for A_1. Within $\mathfrak{M}_\lambda^\perp$ the operator $(A_1 - \lambda I)^{-1}$ is a bounded linear (in fact even selfadjoint) operator defined everywhere on $\mathfrak{M}_\lambda^\perp$.

(d) $(A - \lambda I)\, \mathfrak{D}_A \neq \overline{(A - \lambda I)\, \mathfrak{D}_A} \neq \mathfrak{H}$.

In this case again λ is an eigenvalue of A and the same reasoning applies as in (c) up to the point where there has been shown that

$$(A - \lambda I)\, \mathfrak{D}_A = (A_1 - \lambda I)\, \mathfrak{D}_{A_1}.$$

Continuing we now conclude

$$(A_1 - \lambda I)\, \mathfrak{D}_{A_1} = (A - \lambda I)\, \mathfrak{D}_A \neq \overline{(A - \lambda I)\, \mathfrak{D}_A} = \overline{(A_1 - \lambda I)\, \mathfrak{D}_{A_1}}$$

and since λ is not anymore an eigenvalue of A_1

$$(A_1 - \lambda I)\, \mathfrak{D}_{A_1} \neq \overline{(A_1 - \lambda I)\, \mathfrak{D}_{A_1}} = \mathfrak{M}_\lambda^\perp.$$

Thus λ still belongs to the spectrum of A_1 and within $\mathfrak{M}_\lambda^\perp$ the inverse $(A_1 - \lambda I)^{-1}$ is an unbounded selfadjoint operator in $\mathfrak{M}_\lambda^\perp$ as in (b).

DEFINITION 1. Let A be a selfadjoint operator in $\mathfrak{H}$. The set

$$\{\lambda \in \mathbf{R} : \overline{(A - \lambda I)\, \mathfrak{D}_A} \neq \mathfrak{H}\}$$

(which coincides with the set of eigenvalues of A) is called the *discrete spectrum* (or *point spectrum*) of A. The set

$$\{\lambda \in \mathbf{R} : (A - \lambda I)\, \mathfrak{D}_A \neq \overline{(A - \lambda I)\, \mathfrak{D}_A}\}$$

is called the *continuous spectrum* of A.

Comparing this definition with the cases distinguished above we see that the numbers $\lambda \in \mathbf{R}$ as described in (c) and (d) together make up the discrete spectrum of A, while the numbers $\lambda \in \mathbf{R}$ as described in (b) and (d) together make up the continuous spectrum of A. Whereas the spectrum of A is simply the union of the discrete and the continuous spectrum of A, a number $\lambda \in \mathbf{R}$ may well belong at the same time to the discrete and to the continuous spectrum of A (this happens precisely in case (d)).

Example 1. Let $\{e_k\}_{k=1}^{\infty}$ be the standard basis of ℓ_2 and let A be the bounded selfadjoint operator on ℓ_2 defined by

$$Ae_k = \frac{1}{k}\, e_k \quad \text{for } k \geqslant 1$$

(cf. § 17 example 1). Since $1/k$ is an eigenvalue for every $k \geqslant 1$, the spectrum of A as a closed subset of the real axis at least contains the sequence $\{1/k\}_{k=1}^{\infty}$ and its limit point 0. We shall show that in fact the discrete spectrum of A coincides with the sequence $\{1/k\}_{k=1}^{\infty}$ while the continuous spectrum only consists of the number 0.

Using the same reasoning as demonstrated in § 22 example 1 and § 22 example 2 we first see that the numbers $1/k$ $(1 \leqslant k < \infty)$ are the only eigenvalues of A and that the eigenspace $\mathfrak{M}_k$ corresponding to the eigenvalue $1/k$ is spanned by e_k (for technical reasons we have departed from the notation $\mathfrak{M}_\lambda$ as used earlier in this section). Observe that this already implies that the discrete spectrum of A coincides with the sequence $\{1/k\}_{k=1}^{\infty}$ and that 0 belongs to the continuous spectrum of A. Regarded as a linear operator in the orthogonal complement $\mathfrak{M}_k^{\perp} = \vee \{e_l\}_{l=1, l \neq k}^{\infty}$ the operator $(A - I/k)$ is defined by

$$\left(A - \frac{1}{k} I\right) e_l = \frac{k - l}{kl}\, e_l \quad \text{for } 1 \leqslant l < \infty, l \neq k.$$

For any vector

$$f = \sum_{\substack{l=1 \\ l \neq k}}^{\infty} \alpha_l e_l \in \mathfrak{M}_k^{\perp} \quad \left(\sum_{\substack{l=1 \\ l \neq k}}^{\infty} |\alpha_l|^2 < \infty\right)$$

we have

$$g = \sum_{\substack{l=1 \\ l \neq k}}^{\infty} \alpha_l \frac{kl}{k-l} e_l \in \mathfrak{M}_k^{\perp}$$

(since $\lim_{l \to \infty} kl/(k-l) = -k$ the sequence $\{kl/(k-l)\}_{l=1, l \neq k}^{\infty}$ is bounded and consequently $\sum_{l=1, l \neq k}^{\infty} |\alpha_l|^2 \{kl/(k-l)\}^2 < \infty$) and

$$\left(A - \frac{1}{k} I \right) g = f .$$

We conclude

$$\left(A - \frac{1}{k} I \right) \mathfrak{M}_k^{\perp} = \mathfrak{M}_k^{\perp}$$

and therefore $1/k$ does not belong to the continuous spectrum of A.

Similarly, if $\lambda \in \mathbf{R}$ is different from zero and from $1/k$ for all $k \geqslant 1$, then

$$(A - \lambda I) e_k = \frac{1 - \lambda k}{k} e_k \quad \text{for all } k \geqslant 1 .$$

For any vector

$$f = \sum_{k=1}^{\infty} \alpha_k e_k \in \ell_2 \quad \left(\sum_{k=1}^{\infty} |\alpha_k|^2 < \infty \right)$$

we have

$$g = \sum_{k=1}^{\infty} \alpha_k \frac{k}{1 - \lambda k} e_k \in \ell_2$$

since $1 - \lambda k \neq 0$ for all $k \geqslant 1$ and $\sum_{k=1}^{\infty} |\alpha_k|^2 k^2/(1 - \lambda k)^2 < \infty$. We obtain

$$(A - \lambda I) g = f$$

and therefore $(A - \lambda I) \ell_2 = \ell_2$. This shows that λ is a regular value for A (cf. theorem 5).

Finally, for $\lambda = 0$ the resolvent $(A - \lambda I)^{-1} = A^{-1}$ coincides with the unbounded selfadjoint operator $\bar{B}$ considered in § 20 example 1 and § 22 example 1. This agrees with the fact, already mentioned above, that 0 belongs to the continuous spectrum of A. Actually the continuous spectrum here only consists of the single point 0.

Note that

(1)
$$A = \sum_{k=1}^{\infty} \frac{1}{k} P_k$$

where P_k is the projection upon $\mathfrak{M}_k = \vee \{e_k\}$ and where the series converges in the operator norm in $\mathscr{B}$ (cf. § 12 theorem 3 and compare this statement with (2) in § 22 example 1 and with what has been said in § 22 remark 3). In fact for $f = \sum_{k=1}^{\infty} \alpha_k e_k \in \ell_2$ we have

$$\|(A - \sum_{k=1}^{n-1} \frac{1}{k} P_k) f \|^2 = \| \sum_{k=n}^{\infty} \frac{1}{k} \alpha_k e_k \|^2 = \sum_{k=n}^{\infty} \frac{|\alpha_k|^2}{k^2}$$

$$\leqslant \frac{1}{n^2} \sum_{k=n}^{\infty} |\alpha_k|^2 \leqslant \frac{1}{n^2} \|f\|^2 ,$$

$$\|A - \sum_{k=1}^{n-1} \frac{1}{k} P_k\| \leqslant \frac{1}{n} .$$

Remark 1. Although it does not quite agree with the terminology introduced in definition 1 it has become customary to say that a unitary or selfadjoint operator A has purely discrete spectrum if there exists a basis of $\mathfrak{H}$ which consists of eigenvectors of A (cf. § 22 theorem 4). In this sense the operator A considered in example 1 has purely discrete spectrum, although its continuous spectrum (containing the zero) is not empty (cf. also § 22 exercise d)).

Similarly as in example 1 it is possible to show that for the selfadjoint operator $\bar{B}$ in ℓ_2 considered in § 20 example 1 and § 22 example 1 the spectrum coincides with the discrete spectrum while the continuous spectrum is empty (in particular all eigenvalues are of the type as described in (c)). For the selfadjoint differentiation and multiplication operators D and E in $\mathfrak{L}_2(-\infty, +\infty)$ the spectrum coincides with the continuous spectrum while the discrete spectrum is empty (cf. § 23 example 3). The same statement is true in general for the multiplication operator E in $\mathfrak{L}_2(\alpha, \beta)$ $(-\infty \leqslant \alpha < \beta \leqslant +\infty$; cf. § 23 example 1).

By similar methods as applied in example 1 (and using the same terminology) it can be shown that for the (unitary) Fourier-Plancherel operator F in $\mathfrak{L}_2(-\infty, +\infty)$ the spectrum coincides with the discrete spectrum, consisting of ± 1 and $\pm i$ (cf. § 22 example 2) while the continuous spectrum is empty.

The decomposition (1) of the operator A in example 1 meets all requirements of the programme laid out in § 21 in a particularly simple way. Its essential features are that the operator A is represented as a converging series of scalar multiples of projections upon mutually orthogonal finite dimensional subspaces. Linear operators on $\mathfrak{H}$ admitting a decomposition of this type may be characterized in a way which at first sight seems not to have anything to do with such a decomposition. We shall study them in detail in the following chapter. The statements about the spectrum of the particular operator A considered in example 1 above will appear to be specializations of general theorems about the spectral properties of operators of the type in question.

Exercise. Let A be a selfadjoint operator in a separable Hilbert space $\mathfrak{H}$ and suppose there exists a basis $\{e_k\}_{k=1}^{\infty}$ consisting of eigenvectors of A. Furthermore, let Δ denote the discrete spectrum of A ($=$ the set of all eigenvalues of A; cf. definition 1).

a) If $f=\sum_{k=1}^{\infty} \alpha_k e_k \in \mathfrak{D}_A$, then $Af=\sum_{k=1}^{\infty} \alpha_k Ae_k$. (Hint: consider the inner product $\langle Af, e_n \rangle$.)

b) If $\lambda \in \Delta$, then there exists a basis vector e_k such that $Ae_k=\lambda e_k$; in other words, Δ consists precisely of the eigenvalues corresponding to the basisvectors $e_k(1 \leqslant k < \infty)$. (Hint: imitate the reasoning used in § 22 example 1.)

c) If $\lambda \notin \bar{\Delta}$, then λ is regular; consequently we have $\Sigma(A)=\bar{\Delta}$. (Hint: use a) and the fact that $\delta = \inf_{\mu \in \Delta} |\lambda - \mu| > 0$ to show that $(A-\lambda I)^{-1}$ is bounded.)

d) If $\lambda \in \Delta$ is an accumulation point of Δ, then λ belongs to the continuous spectrum of A. (Hint: if $\mathfrak{M}_\lambda$ is the corresponding eigenspace, then there exists a subsequence of basisvectors $\{e_{k_n}\}_{n=1}^{\infty} \subset \mathfrak{M}_\lambda^{\perp}$ such that $\lim_{n \to \infty} (A-\lambda I)e_{k_n}=0$; cf. § 23 theorem 2 and case (d) above.)

e) The continuous spectrum coincides with the set of accumulation points of Δ. (Hint: use c) and d).)

CHAPTER V

SPECTRAL ANALYSIS OF COMPACT LINEAR OPERATORS

§ 25. Compact linear operators

Let us have another look at the bounded selfadjoint operator A in ℓ_2 considered in § 24 example 1. If $\mathfrak{S}$ is the closed unit sphere in ℓ_2 (cf. § 3 example 4), then for any square summable sequence $g = \{\beta_k\}_{k=1}^{\infty} \in A\mathfrak{S}$ we have

$$g = Af, \quad f = \{\alpha_k\}_{k=1}^{\infty} \in \mathfrak{S},$$

$$|\beta_k| = \frac{|\alpha_k|}{k} \leqslant \frac{1}{k} \quad \text{for all } k \geqslant 1.$$

We see that A maps $\mathfrak{S}$ into the set

$$\mathfrak{C} = \{\{\gamma_k\}_{k=1}^{\infty} : |\gamma_k| \leqslant \frac{1}{k} \quad \text{for all } k \geqslant 1\}$$

which is again contained in ℓ_2. This set $\mathfrak{C}$ is also called the *Hilbert cube* or the *Hilbert parallelotope*.

In our present context the following property of the Hilbert cube is of special interest: *every sequence* $\{h_n\}_{n=1}^{\infty}$ *of vectors in* $\mathfrak{C}$ *contains a converging subsequence*. The proof is based on the familiar procedure of taking the diagonal sequence in an infinite matrix the rows of which are successively subsequences of the preceding ones (the reader having done the exercise in § 4 is invited to compare his own solution with the one presented here):

Let $\{e_k\}_{k=1}^{\infty}$ be the standard basis of ℓ_2 and observe that $|\langle h_n, e_k \rangle| \leqslant 1/k$ for all $n \geqslant 1$ and all $k \geqslant 1$. We can therefore extract from $\{h_n\}_{n=1}^{\infty}$ a subsequence $\{h_{1,n}\}_{n=1}^{\infty}$ such that the limit

$$\gamma_1 = \lim_{n \to \infty} \langle h_{1,n}, e_1 \rangle, \quad |\gamma_1| \leqslant 1,$$

exists. Let $\{h_{2,n}\}_{n=1}^{\infty}$ be a subsequence of $\{h_{1,n}\}_{n=1}^{\infty}$ such that also the limit

$$\gamma_2 = \lim_{n \to \infty} \langle h_{2,n}, e_2 \rangle, \quad |\gamma_2| \leqslant \tfrac{1}{2},$$

exists. Continuing this process we obtain a descending chain of subsequences of $\{h_n\}_{n=1}^{\infty}$

$$\{h_{1,n}\}_{n=1}^{\infty} \supset \{h_{2,n}\}_{n=1}^{\infty} \supset \cdots \supset \{h_{k,n}\}_{n=1}^{\infty} \supset \cdots$$

which coincide with the rows of the infinite matrix

$$
\begin{pmatrix}
h_{1,1} & h_{1,2} & \cdots & h_{1,n} & \cdots \\
h_{2,1} & h_{2,2} & \cdots & h_{2,n} & \cdots \\
\cdots\cdots\cdots\cdots\cdots\cdots\cdots\cdots\cdots\cdots \\
h_{k,1} & h_{k,2} & \cdots & h_{k,n} & \cdots \\
\cdots\cdots\cdots\cdots\cdots\cdots\cdots\cdots\cdots\cdots
\end{pmatrix}
$$

and which have the property that for all $k \geqslant 1$ the limit

$$
\gamma_k = \lim_{n \to \infty} \langle h_{k,n}, e_k \rangle, \quad |\gamma_k| \leqslant \frac{1}{k},
$$

exists. Since the "diagonal" sequence $\{h_{n,n}\}_{n=1}^{\infty}$ in this matrix is a sub-sequence of every row (disregarding some corresponding initial segment) we also have

$$
\gamma_k = \lim_{n \to \infty} \langle h_{n,n}, e_k \rangle \quad \text{for all } k \geqslant 1.
$$

We assert that the sequence $\{h_{n,n}\}_{n=1}^{\infty}$ in fact converges to the vector

$$
h = \sum_{k=1}^{\infty} \gamma_k e_k \in \mathfrak{C}.
$$

In order to prove this statement, given any $\varepsilon > 0$ we choose first the natural number l so large that

$$
\sum_{k=l+1}^{\infty} \frac{1}{k^2} < \tfrac{1}{8}\varepsilon^2
$$

and then m so large that

$$
|\gamma_k - \langle h_{n,n}, e_k \rangle|^2 < \frac{\varepsilon^2}{2l} \quad \text{for } 1 \leqslant k \leqslant l \text{ and all } n \geqslant m.
$$

Then for $n \geqslant m$ we have

$$
\|h - h_{n,n}\|^2 = \sum_{k=1}^{l} |\gamma_k - \langle h_{n,n}, e_k \rangle|^2 + \sum_{k=l+1}^{\infty} |\gamma_k - \langle h_{n,n}, e_k \rangle|^2
$$

$$
\leqslant l \cdot \frac{\varepsilon^2}{2l} + \sum_{k=l+1}^{\infty} \frac{4}{k^2} < \tfrac{1}{2}\varepsilon^2 + \tfrac{1}{2}\varepsilon^2 = \varepsilon^2.
$$

This proves our assertion.

DEFINITION 1. A subset $\mathfrak{A}$ of $\mathfrak{H}$ is called *relatively compact* if every sequence in $\mathfrak{A}$ contains a fundamental (and therefore converging) subsequence. The subset $\mathfrak{A}$ is called *compact* if every sequence in $\mathfrak{A}$ contains a subsequence converging to some vector in $\mathfrak{A}$.

Formulated in this terminology the statement proved above says: *the Hilbert cube in ℓ_2 is compact.* Before going on let us also formulate for further reference the essential "diagonalization lemma" shown and used in the course of the proof. Looking carefully at what actually has been used there we may even state it in the following general formulation:

THEOREM 1. *Let $\{e_k\}_{k=1}^{\infty}$ be any sequence in a Hilbert space $\mathfrak{H}$ and let $\{h_n\}_{n=1}^{\infty} \subset \mathfrak{H}$ be a sequence such that for every $k \geqslant 1$*

$$|\langle h_n, e_k \rangle| \leqslant \delta_k \quad \text{for all } n \geqslant 1 \quad (\delta_k \geqslant 0).$$

Then there exists a subsequence $\{h_n'\}_{n=1}^{\infty}$ of $\{h_n\}_{n=1}^{\infty}$ such that the limit

$$\gamma_k = \lim_{n \to \infty} \langle h_n', e_k \rangle, \quad |\gamma_k| \leqslant \delta_k,$$

exists for every $k \geqslant 1$.

It is an immediate consequence of definition 1 that a compact set is also relatively compact and that a subset of a relatively compact set is again relatively compact.

For the case of a one-dimensional Hilbert space the well-known theorem of BOLZANO-WEIERSTRASS (which we have used in the proof of theorem 1 without explicit mentioning) asserts that a bounded sequence of complex numbers contains a converging subsequence. Combining this with some simple topological facts one arrives at the statement, again well-known from elementary analysis, that a subset of the complex plane is compact iff it is closed and bounded. Let us have a look at the connections between compactness, closure, and boundedness in case of a subset of a Hilbert space $\mathfrak{H}$ in general. A subset $\mathfrak{A}$ of $\mathfrak{H}$ will be called *bounded* if there exists an $\alpha > 0$ such that $\|f\| \leqslant \alpha$ for all $f \in \mathfrak{A}$.

THEOREM 2. *A subset $\mathfrak{A} \subset \mathfrak{H}$ is relatively compact iff $\overline{\mathfrak{A}}$ is compact.*

PROOF. If $\mathfrak{A}$ is relatively compact and if $\{f_n\}_{n=1}^{\infty}$ is any sequence in $\overline{\mathfrak{A}}$, then for every $n \geqslant 1$ we may choose a vector $g_n \in \mathfrak{A}$ such that

$$(1) \qquad \qquad \|f_n - g_n\| < \frac{1}{n}.$$

If $\{g_{n_k}\}_{k=1}^{\infty}$ is a subsequence of $\{g_n\}_{n=1}^{\infty}$ converging to some vector $f \in \overline{\mathfrak{A}}$, then because of (1) we also have

$$\|f - f_{n_k}\| \leqslant \|f - g_{n_k}\| + \frac{1}{n_k}.$$

The subsequence $\{f_{n_k}\}_{k=1}^{\infty}$ therefore also converges to $f \in \mathfrak{A}$. Thus $\mathfrak{A}$ is compact. The assertion going the other way immediately follows from definition 1. $\square$

COROLLARY 2.1. *A closed relatively compact set is compact.*

THEOREM 3. *A compact subset of $\mathfrak{H}$ is closed and bounded.*
PROOF. Let $\mathfrak{A} \subset \mathfrak{H}$ be compact and let $\{f_n\}_{n=1}^{\infty}$ be a fundamental sequence in $\mathfrak{A}$. The vector $f = \lim_{n \to \infty} f_n$ is also the limit of every subsequence of $\{f_n\}_{n=1}^{\infty}$ and therefore f must belong to $\mathfrak{A}$. As a consequence, $\mathfrak{A}$ is closed. If $\mathfrak{A}$ were not bounded, then $\mathfrak{A}$ would contain a sequence $\{f_n\}_{n=1}^{\infty}$ such that $\lim_{n \to \infty} \| f_n \| = \infty$. Then we would also have $\lim_{k \to \infty} \| f_{n_k} \| = \infty$ for every subsequence $\{f_{n_k}\}_{k=1}^{\infty}$ of $\{f_n\}_{n=1}^{\infty}$ in contradiction to the compactness of $\mathfrak{A}$. $\square$

The following theorem shows that in contrast to what happens in finite-dimensional Euclidean or unitary space, a closed and bounded subset of a Hilbert space need not be compact in general.

THEOREM 4. *The closed unit sphere in $\mathfrak{H}$ is compact iff the dimension of $\mathfrak{H}$ is finite.*
PROOF. If $\mathfrak{H}$ is the m-dimensional unitary space with basis $\{e_k\}_{k=1}^{m}$ and if $\{h_n\}_{n=1}^{\infty}$ is a sequence in the closed unit sphere in $\mathfrak{H}$, then by a trivial modification of theorem 1 ($e_k = o$ for $k > m$) there exists a subsequence $\{h'_n\}_{n=1}^{\infty}$ of $\{h_n\}_{n=1}^{\infty}$ such that the limit

$$\gamma_k = \lim_{n \to \infty} \langle h'_n, e_k \rangle, \quad |\gamma_k| \leqslant 1$$

exists for $1 \leqslant k \leqslant m$. For $h = \sum_{k=1}^{m} \gamma_k e_k$ we obtain

$$\lim_{n \to \infty} \| h - h'_n \|^2 = \sum_{k=1}^{m} \lim_{n \to \infty} |\gamma_k - \langle h'_n, e_k \rangle|^2 = 0.$$

From $\| h'_n \| \leqslant 1$ for $1 \leqslant n < \infty$ we finally also conclude $\| h \| = \lim_{n \to \infty} \| h'_n \| \leqslant 1$.

If $\mathfrak{H}$ is infinite-dimensional, let $\{e_k\}_{k=1}^{\infty}$ be any orthonormal sequence in $\mathfrak{H}$. This sequence does not contain any converging subsequence since the distance between any two of its terms is $\sqrt{2}$. The closed unit sphere in $\mathfrak{H}$ is therefore not compact. (The proof in fact shows that the closed unit sphere is not even relatively compact in this case; this, however, does not supply any additional information in view of corollary 2.1.) $\square$

COROLLARY 4.1. *A bounded subset of a finite-dimensional subspace of $\mathfrak{H}$ is relatively compact.*
PROOF. Suppose $\mathfrak{A}$ is a bounded subset of a finite-dimensional subspace $\mathfrak{M}$ of $\mathfrak{H}$ and $\| f \| \leqslant \alpha$ ($\alpha > 0$) for all $f \in \mathfrak{A}$. If $\{f_n\}_{n=1}^{\infty}$ is any sequence in $\mathfrak{A}$, then

$\{f_n/\alpha\}_{n=1}^\infty$ is a sequence in the closed unit sphere in $\mathfrak{M}$ which is compact by theorem 4. Therefore $\{f_n/\alpha\}_{n=1}^\infty$ contains a converging subsequence and so does $\{f_n\}_{n=1}^\infty$. $\square$

DEFINITION 2. A linear operator A on $\mathfrak{H}$ is called *compact* (or *completely continuous*) if it maps the closed unit sphere into a relatively compact set.

As we have seen in the first paragraph of this section the bounded self-adjoint operator A on ℓ_2 considered in § 24 example 1 maps the closed unit sphere into the Hilbert cube which we have shown to be compact. Thus A is a compact operator. In fact this statement appears to be a special case of a more general one supplied in the following example.

Example 1. Let $\{e_k\}_{k=1}^\infty$ be a basis of a separable Hilbert space $\mathfrak{H}$ and let the bounded linear operator A on $\mathfrak{H}$ be given (with respect to this basis) by an infinite matrix

$$A' = (\alpha_{k,l})_{1 \leqslant k, l < \infty}$$

(cf. § 12 (6)) with the additional property that

$$\sum_{k=1}^\infty \sum_{l=1}^\infty |\alpha_{k,l}|^2 < \infty$$

(cf. § 12 example 3). We assert that A is compact. It suffices to show that if $\{f_n\}_{n=1}^\infty$ is any sequence in the closed unit sphere in $\mathfrak{H}$ then $\{Af_n\}_{n=1}^\infty$ contains a converging subsequence. Note that

(2) $$|\langle Af_n, e_k \rangle| = |\sum_{l=1}^\infty \alpha_{k,l} \langle f_n, e_l \rangle| \leqslant \left(\sum_{l=1}^\infty |\alpha_{k,l}|^2 \right)^{\frac{1}{2}}$$

for all $n \geqslant 1$ and all $k \geqslant 1$. By theorem 1 there exists a subsequence $\{g_n\}_{n=1}^\infty$ of $\{Af_n\}_{n=1}^\infty$ having the property that the limit

$$\gamma_k = \lim_{n \to \infty} \langle g_n, e_k \rangle$$

exists for every $k \geqslant 1$. From (2) we deduce

$$|\gamma_k| \leqslant \left(\sum_{l=1}^\infty |\alpha_{k,l}|^2 \right)^{\frac{1}{2}},$$

$$\sum_{k=1}^\infty |\gamma_k|^2 \leqslant \sum_{k=1}^\infty \sum_{l=1}^\infty |\alpha_{k,l}|^2 < \infty,$$

$$g = \sum_{k=1}^\infty \gamma_k e_k \in \mathfrak{H}.$$

It remains to show that $g = \lim_{n \to \infty} g_n$. To this end for any given $\varepsilon > 0$ we choose the natural number m such that

$$\sum_{k=m+1}^{\infty} \sum_{l=1}^{\infty} |\alpha_{k,l}|^2 < \tfrac{1}{8}\varepsilon^2$$

and further the natural number $n(\varepsilon)$ such that

$$|\gamma_k - \langle g_n, e_k \rangle|^2 < \frac{\varepsilon^2}{2m} \quad \text{for } 1 \leqslant k \leqslant m \text{ and all } n \geqslant n(\varepsilon).$$

Then for $n \geqslant n(\varepsilon)$ we have

$$\|g - g_n\|^2 = \sum_{k=1}^{m} |\gamma_k - \langle g_n, e_k \rangle|^2 + \sum_{k=m+1}^{\infty} |\gamma_k - \langle g_n, e_k \rangle|^2$$
$$< m \cdot \frac{\varepsilon^2}{2m} + \sum_{k=m+1}^{\infty} 4 \sum_{l=1}^{\infty} |\alpha_{k,l}|^2$$
$$< \tfrac{1}{2}\varepsilon^2 + \tfrac{1}{2}\varepsilon^2 = \varepsilon^2.$$

This completes the proof.

Example 2. In § 12 example 4 it has been shown that a Hilbert-Schmidt operator on $\mathfrak{L}_2(\beta, \gamma)\,(-\infty \leqslant \beta < \gamma \leqslant +\infty)$ is given (with respect to any basis of $\mathfrak{L}_2(\beta, \gamma)$) by an infinite matrix $A' = (\alpha_{k,l})_{1 \leqslant k, l < \infty}$ having the property that $\sum_{k=1}^{\infty} \sum_{l=1}^{\infty} |\alpha_{k,l}|^2 < \infty$. From what has been shown in example 1 it follows that a Hilbert-Schmidt operator on $\mathfrak{L}_2(\beta, \gamma)$ is compact.

Example 3. Let U be a unitary operator in an infinite dimensional Hilbert space $\mathfrak{H}$. The operator U maps the closed unit sphere $\mathfrak{S}$ onto itself. Since $\mathfrak{S}$ is not compact (cf. theorem 4) U is not a compact operator. In particular this applies to the identity operator I. As a further consequence no operator of the type described in example 1, in particular no Hilbert-Schmidt operator on $\mathfrak{L}_2(\beta, \gamma)\,(-\infty \leqslant \beta < \gamma \leqslant +\infty)$, can be unitary (cf. § 14 example 4).

The following two theorems to some extent motivate the alternative terminology "completely continuous" for a compact linear operator.

THEOREM 5. *A compact linear operator is bounded (and therefore continuous).*
PROOF. If A is a compact linear operator on $\mathfrak{H}$, then A maps the closed unit sphere into a relatively compact and therefore bounded set (cf. theorem 2 and theorem 3). As a consequence we have $\|A\| = \sup_{\|f\|=1} \|Af\| < \infty$. $\square$

THEOREM 6. *A compact linear operator maps every bounded set into a relatively compact set.*

PROOF. Let A be a compact linear operator on $\mathfrak{H}$. If $\mathfrak{A}$ is a bounded subset of $\mathfrak{H}$, say

$$\|f\| \leqslant \alpha \quad \text{for all } f \in \mathfrak{A} \, (\alpha > 0),$$

and if $\{f_n\}_{n=1}^\infty$ is any sequence in $\mathfrak{A}$, then $\{f_n/\alpha\}_{n=1}^\infty$ is a subsequence of the closed unit sphere and the sequence $\{Af_n/\alpha\}_{n=1}^\infty$ has a fundamental subsequence. As a consequence also $\{Af_n\}_{n=1}^\infty$ has a fundamental subsequence. This proves the assertion. $\square$

We finally exhibit a special class of compact linear operators which will turn out later to reflect quite typically the structure of compact linear operators in general.

DEFINITION 3. A bounded linear operator A on $\mathfrak{H}$ is said to be of *finite rank* if the linear manifold $A\mathfrak{H}$ is of *finite rank*, i.e. if $A\mathfrak{H}$ consists of all complex linear combinations of some finite set of vectors in $A\mathfrak{H}$.

Remark 1. Note that a linear manifold $\mathfrak{M}$ of finite rank (such as $A\mathfrak{H}$ above) is even closed and therefore a finite-dimensional subspace of $\mathfrak{H}$. While this statement is true for any normed space (cf. § 3 exercise), in Hilbert space a proof may be conducted as follows:

Suppose every vector of $\mathfrak{M}$ is a linear combination of $g_1, ..., g_n$ (without loss of generality we may assume that $g_1, ..., g_n$ are linearly independent). Applying Gram-Schmidt orthonormalization to the finite sequence $\{g_k\}_{k=1}^n$ we obtain an orthonormal family $\{e_k\}_{k=1}^n$ such that every vector in $\mathfrak{M}$ is a linear combination of $e_1, ..., e_n$ (cf. § 8 theorem 1, § 8 corollary 1.1 c)). For every fixed $k(1 \leqslant k \leqslant n)$ the set $\mathfrak{M}_k = \{\lambda e_k : \lambda \in \mathbf{C}\}$ is a subspace since for any converging sequence $\{f_n\}_{n=1}^\infty = \{\lambda_n e_k\}_{n=1}^\infty \subset \mathfrak{M}_k$ we have

$$\lim_{m,\,n \to \infty} |\lambda_m - \lambda_n| = \lim_{m,\,n \to \infty} \|f_m - f_n\| = 0 \, ;$$

therefore also the sequence $\{\lambda_n\}_{n=1}^\infty$ converges to some number $\lambda \in \mathbf{C}$ and

$$\lim_{n \to \infty} f_n = \lim_{n \to \infty} \lambda_n e_k = \lambda e_k \in \mathfrak{M}_k \, .$$

We obtain

$$\mathfrak{M} = \sum_{k=1}^n \mathfrak{M}_k$$

(cf. § 6 definition 3) and by § 7 theorem 2 (or better by its extension to finitely many summands) $\mathfrak{M}$ is a subspace.

THEOREM 7. *A bounded linear operator on $\mathfrak{H}$ of finite rank is compact.*

PROOF. If A is a bounded linear operator of finite rank and if $\mathfrak{S}$ is the closed unit sphere in $\mathfrak{H}$, then $A\mathfrak{S}$ is a bounded subset of the finite-dimensional subspace $A\mathfrak{H}$ and therefore relatively compact by corollary 4.1. $\square$

THEOREM 8. *A linear operator A on $\mathfrak{H}$ is bounded and of finite rank iff there exist vectors $g_1, ..., g_n, h_1, ..., h_n$ such that*

(3) $$Af = \sum_{k=1}^{n} \langle f, g_k \rangle h_k \quad \text{for all } f \in \mathfrak{H}.$$

PROOF. If A is bounded and of finite rank let $e_1, ..., e_n$ be a basis of the subspace $A\mathfrak{H}$. Then we have

$$Af = \sum_{k=1}^{n} \langle Af, e_k \rangle e_k = \sum_{k=1}^{n} \langle f, A^* e_k \rangle e_k.$$

Conversely, from (3) we deduce that $A\mathfrak{H}$ is a linear manifold of finite rank (spanned by $h_1, ..., h_n$) and

$$\|Af\| \leqslant \sum_{k=1}^{n} |\langle f, g_k \rangle| \, \|h_k\| \leqslant \|f\| \left(\sum_{k=1}^{n} \|g_k\| \, \|h_k\| \right),$$

$$\|A\| \leqslant \sum_{k=1}^{n} \|g_k\| \, \|h_k\| . \quad \square$$

Exercise. Let A, B, C be bounded linear operators on $\mathfrak{H}$ and suppose A and B are of finite rank. Then so are the operators $A+B$, $\lambda A (\lambda \in \mathbf{C})$, AC, CA, and A^*. (Hint: in order to prove that A^* is of finite rank show that $A^* g = o$ for all $g \in (A\mathfrak{H})^{\perp}$; cf. also § 24 theorem 2.)

§ 26. Weakly convergent sequences

We shall now characterize compact linear operators in another way. This new characterization will not only reveal some further interesting features of the individual compact linear operators but it will also help us to gain some insight in the algebraic and topological structure of the set of all compact linear operators on a given Hilbert space $\mathfrak{H}$. For this purpose we have to introduce the concept of a *weakly convergent sequence* of vectors.

DEFINITION 1. A sequence $\{f_n\}_{n=1}^{\infty} \subset \mathfrak{H}$ is said to *converge weakly* (to *converge weakly to some vector f*) if for every $g \in \mathfrak{H}$ the sequence $\{\langle f_n, g \rangle\}_{n=1}^{\infty}$ converges (converges to $\langle f, g \rangle$).

In what follows we shall sometimes also use the term "*strong convergence*" for convergence in the usual (norm) sense as opposed to weak convergence.

Example 1. Let $\{e_k\}_{k=1}^{\infty}$ be any orthonormal sequence in $\mathfrak{H}$. For any $g \in \mathfrak{H}$ we have

$$\|g\|^2 \geqslant \sum_{k=1}^{\infty} |\langle g, e_k \rangle|^2$$

by BESSEL's inequality (§ 2 theorem 9). We conclude

$$\lim_{k \to \infty} \langle e_k, g \rangle = 0 = \langle o, g \rangle.$$

Thus the sequence $\{e_k\}_{k=1}^{\infty}$ converges weakly to o while it does not converge strongly at all.

THEOREM 1. *If the sequence* $\{f_n\}_{n=1}^{\infty}$ *converges strongly to* f, *then it also converges weakly to* f.

PROOF. From $f = \lim_{n \to \infty} f_n$ we conclude

$$\langle f, g \rangle = \lim_{n \to \infty} \langle f_n, g \rangle \quad \text{for every } g \in \mathfrak{H}. \ \square$$

THEOREM 2. *A weakly convergent sequence is bounded.*

PROOF. If $\{f_n\}_{n=1}^{\infty}$ converges weakly, then for every $g \in \mathfrak{H}$ the sequence $\{\langle f_n, g \rangle\}_{n=1}^{\infty}$ is bounded. The sequence $\{\|f_n\|\}_{n=1}^{\infty}$ is then bounded by the Banach-Steinhaus theorem (§ 17 theorem 7). $\square$

THEOREM 3. *Let* $\{f_n\}_{n=1}^{\infty}$ *be a weakly convergent sequence. Then there exists a unique vector* f, *called the weak limit of* $\{f_n\}_{n=1}^{\infty}$, *such that* $\{f_n\}_{n=1}^{\infty}$ *converges weakly to* f.

PROOF. By theorem 2 for some constant $\gamma > 0$ we have

$$\|f_n\| \leqslant \gamma \quad \text{for all } n \geqslant 1.$$

The linear functional Φ defined on $\mathfrak{H}$ by

$$\Phi g = \lim_{n \to \infty} \overline{\langle f_n, g \rangle}$$

is bounded since

$$|\Phi g| = \lim_{n \to \infty} |\langle f_n, g \rangle| \leqslant \gamma \|g\|.$$

By the Riesz representation theorem (§ 11 theorem 4) there exists a unique vector f such that

$$\langle f, g \rangle = \overline{\Phi g} = \lim_{n \to \infty} \langle f_n, g \rangle \quad \text{for all } g \in \mathfrak{H}. \ \square$$

The following theorem ought to be compared with § 25 theorem 4. Using an analogous terminology we may formulate it in short: *the closed unit sphere in* $\mathfrak{H}$ *is weakly compact.*

THEOREM 4. *Let $\mathfrak{S}$ be the closed unit sphere in $\mathfrak{H}$. Every sequence in $\mathfrak{S}$ contains a subsequence which weakly converges to some vector in $\mathfrak{S}$.*

PROOF. Let $\{f_n\}_{n=1}^{\infty} \subset \mathfrak{S}$ be given and let $\mathfrak{M} = \vee \{f_n\}_{n=1}^{\infty}$. If $\mathfrak{M}$ is a finite-dimensional subspace, then $\{f_n\}_{n=1}^{\infty}$, being contained in the unit sphere in $\mathfrak{M}$, by § 25 theorem 4 contains a strongly converging subsequence which is also weakly convergent by theorem 1. Otherwise, let $\{e_k\}_{k=1}^{\infty}$ be a basis of $\mathfrak{M}$ (obtained for instance by successively omitting from $\{f_n\}_{n=1}^{\infty}$ all terms that linearly depend on the preceding ones and applying Gram-Schmidt orthonormalization to the remaining sequence). Because of $\{f_n\}_{n=1}^{\infty} \subset \mathfrak{S}$ we have

$$|\langle f_n, e_k \rangle| \leqslant 1 \quad \text{for all } n \geqslant 1 \text{ and all } k \geqslant 1.$$

By § 25 theorem 1 there exists a subsequence $\{f_n'\}_{n=1}^{\infty}$ of $\{f_n\}_{n=1}^{\infty}$ such that the limit

$$\lim_{n \to \infty} \langle f_n', e_k \rangle$$

exists for every $k \geqslant 1$. Then also the limit

$$\lim_{n \to \infty} \langle f_n', \sum_{k=1}^{l} \alpha_k e_k \rangle = \sum_{k=1}^{l} \bar{\alpha}_k \lim_{n \to \infty} \langle f_n', e_k \rangle$$

exists for every finite linear combination $\sum_{k=1}^{l} \alpha_k e_k$. We assert that even for every $g \in \mathfrak{H}$ the limit

$$\lim_{n \to \infty} \langle f_n', g \rangle$$

exists. Indeed, suppose

$$g = g_1 + g_2, \quad g_1 \in \mathfrak{M}, \ g_2 \in \mathfrak{M}^{\perp}$$

and let for any given $\varepsilon > 0$ the index $l \geqslant 1$ and thereby the vector $g_1' = \sum_{k=1}^{l} \langle g, e_k \rangle e_k \in \mathfrak{M}$ be chosen such that

$$\|g_1 - g_1'\| < \tfrac{1}{4}\varepsilon.$$

Then for sufficiently large $n(\varepsilon)$ and $n \geqslant n(\varepsilon)$, $m \geqslant n(\varepsilon)$ we have

$$\begin{aligned}
|\langle f_m', g \rangle - \langle f_n', g \rangle| = |\langle f_m' - f_n', g \rangle| &= |\langle f_m' - f_n', g_1 \rangle| \\
&\leqslant |\langle f_m' - f_n', g_1' \rangle| + |\langle f_m' - f_n', g_1 - g_1' \rangle| \\
&\leqslant \tfrac{1}{2}\varepsilon + \|f_m' - f_n'\| \, \|g_1 - g_1'\| \leqslant \tfrac{1}{2}\varepsilon + 2 \cdot \tfrac{1}{4}\varepsilon = \varepsilon.
\end{aligned}$$

This proves that $\{f_n'\}_{n=1}^{\infty}$ converges weakly. By theorem 3 there exists a weak limit f such that for every $g \in \mathfrak{H}$ we have

$$\langle f, g \rangle = \lim_{n \to \infty} \langle f_n', g \rangle,$$

$$|\langle f, g \rangle| = \lim_{n \to \infty} |\langle f_n', g \rangle| \leqslant \|g\|.$$

We conclude

$$\|f\| = \sup_{\|g\|=1} |\langle f, g \rangle| \leqslant 1$$

(cf. § 2 corollary 4.2). This shows $f \in \mathfrak{S}$ and completes the proof. □

COROLLARY 4.1. *Every bounded sequence in $\mathfrak{H}$ contains a weakly converging subsequence.*

PROOF. Suppose $\|f_n\| \leqslant \alpha (\alpha > 0)$ for $1 \leqslant n < \infty$. The sequence $\{f_n/\alpha\}_{n=1}^\infty$ is contained in the closed unit sphere in $\mathfrak{H}$ and therefore contains a weakly converging subsequence $\{f_n'/\alpha\}_{n=1}^\infty$. Then also $\{f_n'\}_{n=1}^\infty$ which is a subsequence of $\{f_n\}_{n=1}^\infty$ converges weakly. □

What happens if we apply a bounded linear operator on $\mathfrak{H}$ to a weakly converging sequence? The result is pleasing but hardly interesting.

THEOREM 5. *Let A be a bounded linear operator on $\mathfrak{H}$ and let the sequence $\{f_n\}_{n=1}^\infty$ converge weakly to f. Then the sequence $\{Af_n\}_{n=1}^\infty$ converges weakly to Af.*

PROOF. For every $g \in \mathfrak{H}$ we have

$$\lim_{n \to \infty} \langle Af_n, g \rangle = \lim_{n \to \infty} \langle f_n, A^*g \rangle = \langle f, A^*g \rangle = \langle Af, g \rangle. \quad \square$$

What happens if we apply a compact linear operator to a weakly converging sequence? Something unexpected which even serves to characterize compactness of the operator in question.

THEOREM 6. *A bounded linear operator on $\mathfrak{H}$ is compact iff it maps every weakly converging sequence into a strongly converging sequence.*

PROOF. a) Let A be a compact linear operator on $\mathfrak{H}$ and let $\{f_n\}_{n=1}^\infty$ be a weakly converging sequence. Then $\{f_n\}_{n=1}^\infty$ is bounded by theorem 2, and the set $\mathfrak{A} = \{Af_n : 1 \leqslant n < \infty\}$ is relatively compact (cf. § 25 theorem 6). If the sequence $\{Af_n\}_{n=1}^\infty$ did not converge, then by the relative compactness of $\mathfrak{A}$ it would have to contain at least two subsequences $\{Af_n'\}_{n=1}^\infty$ and $\{Af_n''\}_{n=1}^\infty$ converging (strongly) to different limit vectors g' and g'' respectively. We obtain

$$\langle g', h \rangle = \lim_{n \to \infty} \langle Af_n', h \rangle = \lim_{n \to \infty} \langle f_n', A^*h \rangle$$

$$= \lim_{n \to \infty} \langle f_n, A^*h \rangle$$

$$= \lim_{n \to \infty} \langle f_n'', A^*h \rangle = \lim_{n \to \infty} \langle Af_n'', h \rangle$$

$$= \langle g'', h \rangle \quad \text{for all } h \in \mathfrak{H}$$

which implies $g' = g''$, a contradiction. We conclude that the sequence $\{Af_n\}_{n=1}^{\infty}$ must converge.

b) Suppose A is a bounded linear operator on $\mathfrak{H}$ mapping every weakly converging sequence into a strongly converging one. Let $\mathfrak{S}$ be the closed unit sphere in $\mathfrak{H}$ and let the sequence $\{g_n\}_{n=1}^{\infty} \subset A\mathfrak{S}$ be given. Without loss of generality we may suppose $g_n = Af_n$, $\|f_n\| \leqslant 1$ for all $n \geqslant 1$. By theorem 4 there exists a weakly converging subsequence $\{f_n'\}_{n=1}^{\infty}$ of $\{f_n\}_{n=1}^{\infty}$ which by hypothesis is mapped by A into the strongly converging sequence $\{Af_n'\}_{n=1}^{\infty}$. Since this is a subsequence of $\{g_n\}_{n=1}^{\infty}$ we see that $A\mathfrak{S}$ is relatively compact and that A is a compact linear operator. $\square$

COROLLARY 6.1. *Let A be a compact linear operator on $\mathfrak{H}$ and let the sequence $\{f_n\}_{n=1}^{\infty}$ converge weakly to f. Then the sequence $\{Af_n\}_{n=1}^{\infty}$ converges strongly to Af.*

PROOF. By theorem 6 the sequence $\{Af_n\}_{n=1}^{\infty}$ converges strongly to some vector g. Then it also converges weakly to g by theorem 1. On the other hand, by theorem 5 the sequence $\{Af_n\}_{n=1}^{\infty}$ converges weakly to Af. Since there is only one weak limit of the sequence $\{Af_n\}_{n=1}^{\infty}$ by theorem 3 we conclude $g = Af$. $\square$

Let $\mathscr{B}$ be the Banach algebra of bounded linear operators on $\mathfrak{H}$ (cf. § 12 theorem 3). What can be said about the subset of compact operators?

THEOREM 7. *Let A and B be compact linear operators on $\mathfrak{H}$ and let C be a bounded linear operator on $\mathfrak{H}$. Then $A + B$, λA $(\lambda \in \mathbf{C})$, AC and CA are compact linear operators.*

PROOF. Let $\{f_n\}_{n=1}^{\infty}$ be a weakly converging sequence. By theorem 6 the sequences $\{Af_n\}_{n=1}^{\infty}$ and $\{Bf_n\}_{n=1}^{\infty}$ converge strongly and therefore so do the sequences

$$\{(A + B)f_n\}_{n=1}^{\infty} = \{Af_n + Bf_n\}_{n=1}^{\infty},$$
$$\{(\lambda A)f_n\}_{n=1}^{\infty} = \{\lambda(Af_n)\}_{n=1}^{\infty},$$
$$\{(CA)f_n\}_{n=1}^{\infty} = \{C(Af_n)\}_{n=1}^{\infty}.$$

Moreover the sequence $\{Cf_n\}_{n=1}^{\infty}$ is weakly convergent by theorem 5. Therefore again the sequence

$$\{(AC)f_n\}_{n=1}^{\infty} = \{A(Cf_n)\}_{n=1}^{\infty}$$

converges strongly. Applying theorem 6 in the other direction we see that all operators in question are compact. $\square$

THEOREM 8. *If A is a compact linear operator on $\mathfrak{H}$ then so is A^*.*

PROOF. If the sequence $\{f_n\}_{n=1}^{\infty}$ is weakly convergent, then it is bounded by theorem 2, say $\|f_n\| \leqslant \alpha$ $(\alpha > 0)$ for all $n \geqslant 1$. We obtain

$$\begin{aligned}
\|A^*f_m - A^*f_n\|^2 &= \langle A^*(f_m - f_n), A^*(f_m - f_n) \rangle \\
&= \langle AA^*(f_m - f_n), f_m - f_n \rangle \\
&\leqslant \|AA^*f_m - AA^*f_n\| \, \|f_m - f_n\| \\
&\leqslant 2\alpha \, \|AA^*f_m - AA^*f_n\| \, .
\end{aligned}$$

The operator AA^* is compact by theorem 7. As a consequence, the sequence $\{AA^*f_n\}_{n=1}^{\infty}$ converges strongly and

$$\lim_{m,\,n \to \infty} \|A^*f_m - A^*f_n\| \leqslant 2\alpha \lim_{m,\,n \to \infty} \|AA^*f_m - AA^*f_n\| = 0\,.$$

The sequence $\{A^*f_n\}_{n=1}^{\infty}$ therefore converges strongly and A^* is compact by theorem 6. $\square$

THEOREM 9. *If $\{A_k\}_{k=1}^{\infty}$ is a fundamental sequence of compact linear operators on $\mathfrak{H}$ (with respect to the operator norm in $\mathcal{B}$), then the operator $A = \lim_{k \to \infty} A_k$ is compact.*

PROOF. Let $\{f_n\}_{n=1}^{\infty}$ be a weakly converging sequence and suppose $\|f_n\| \leqslant \alpha$ $(\alpha > 0)$ for all $n \geqslant 1$ (cf. theorem 2). Then for every $k \geqslant 1$ we have

$$\begin{aligned}
(1) \qquad \|Af_m - Af_n\| &\leqslant \|(A - A_k)f_m\| + \|(A - A_k)f_n\| + \|A_kf_m - A_kf_n\| \\
&\leqslant 2\alpha \|A - A_k\| + \|A_kf_m - A_kf_n\|\,.
\end{aligned}$$

Given any $\varepsilon > 0$ we choose k so large that $\|A - A_k\| < \varepsilon/4\alpha$. Keeping k fixed we then choose $n(\varepsilon)$ so large that

$$\|A_kf_m - A_kf_n\| < \tfrac{1}{2}\varepsilon \quad \text{for all } m \geqslant n(\varepsilon) \text{ and all } n \geqslant n(\varepsilon)$$

(the sequence $\{A_kf_n\}_{n=1}^{\infty}$ converges strongly by theorem 6). From (1) we conclude

$$\|Af_m - Af_n\| < 2\alpha\,\frac{\varepsilon}{4\alpha} + \tfrac{1}{2}\varepsilon = \varepsilon \quad \text{for all } m \geqslant n(\varepsilon) \text{ and all } n \geqslant n(\varepsilon)\,.$$

The sequence $\{Af_n\}_{n=1}^{\infty}$ is therefore fundamental and converges strongly. By theorem 6 the operator A is compact. $\square$

Theorem 7 states that the set $\mathscr{C}$ of compact linear operators on $\mathfrak{H}$ is closed not only under addition and scalar multiplication but also under left and right multiplication by arbitrary bounded linear operators on $\mathfrak{H}$ (cf. also § 25 exercise). This is formulated in short by saying that $\mathscr{C}$ is a *two-sided ideal* in $\mathcal{B}$.

Theorem 8 states that $\mathscr{C}$ is also closed under passing to adjoints. This is formulated in short by saying that the ideal $\mathscr{C}$ is *symmetric*.

Theorem 9 finally states that $\mathscr{C}$ is topologically closed (under passing to the limit with respect to the operator norm in $\mathscr{B}$).

All these statements are combined in the following condensed statement: *the set of compact linear operators on $\mathfrak{H}$ is a closed symmetric two-sided ideal in $\mathscr{B}$.*

Exercise. Let A be a compact linear operator on an infinite-dimensional Hilbert space $\mathfrak{H}$.

a) If A is one-to-one, then A^{-1} is unbounded. (Hint: consider the images of an orthonormal sequence in $\mathfrak{H}$ under A.)

b) If A is not of finite rank (cf. § 25 definition 3), then there exists a sequence $\{f_k\}_{k=1}^\infty$ such that $\{Af_k\}_{k=1}^\infty$ is an orthonormal sequence. (Hint: choose the sequence $\{g_k\}_{k=1}^\infty$ such that the sequence $\{Ag_k\}_{k=1}^\infty$ is a linearly independent family and apply Gram-Schmidt orthonormalization to the sequence $\{Ag_k\}_{k=1}^\infty$. Note that the compactness of A is not used in this part.)

c) If $\{Af_k\}_{k=1}^\infty$ is an orthonormal sequence, then $\lim_{k\to\infty} \|f_k\| = \infty$. (Hint: assume a subsequence of $\{f_k\}_{k=1}^\infty$ were bounded and choose this subsequence to be weakly convergent.)

d) $A^*\mathfrak{H} \neq \mathfrak{H}$. (Hint: if $A^*\mathfrak{H} = \mathfrak{H}$ observe that A is not of finite rank either (cf. § 25 exercise); then use the Banach-Steinhaus theorem (§ 17 theorem 7) to show that a sequence $\{f_k\}_{k=1}^\infty$ as in b) and c) would have to be bounded.)

e) $A\mathfrak{H} \neq \mathfrak{H}$. (Hint: apply d) to A^* in place of A.)

§ 27. The spectrum of a compact linear operator

As it is to be expected, the special features displayed by compact operators in general also have some effect upon the spectrum of such an operator: apart from the point 0 it only consists of eigenvalues; moreover, if there are infinitely many eigenvalues, then they may be arranged to a sequence converging to 0.

Having betrayed the main result of this section we shall now attack the problem more systematically. The first theorem to follow rather has the character of a lemma.

THEOREM 1. *If $\{e_n\}_{n=1}^\infty$ is an orthonormal sequence and if A is a compact linear operator on $\mathfrak{H}$, then*

$$\lim_{n\to\infty} \langle Ae_n, e_n \rangle = 0.$$

PROOF. The sequence $\{e_n\}_{n=1}^\infty$ converges weakly to zero (cf. § 26 example 1). The sequence $\{Ae_n\}_{n=1}^\infty$ then converges strongly to zero by § 26 corollary 6.1. By CAUCHY's inequality we obtain

$$\lim_{n\to\infty} |\langle Ae_n, e_n\rangle| \leqslant \lim_{n\to\infty} \|Ae_n\| = 0. \quad \square$$

THEOREM 2. *Let A be a compact linear operator on $\mathfrak{H}$ and let $\varrho > 0$ be given. Every family of linearly independent eigenvectors of A corresponding to eigenvalues with absolute values not smaller than ϱ is finite.*

PROOF. Assume the contrary and let $\{f_n\}_{n=1}^\infty$ be an infinite sequence of linearly independent eigenvectors of A such that for the corresponding eigenvalues λ_n we have $|\lambda_n| \geqslant \varrho$ for all $n \geqslant 1$. Let the orthonormal sequence $\{e_n\}_{n=1}^\infty$ be obtained from $\{f_n\}_{n=1}^\infty$ by Gram-Schmidt orthonormalization (cf. § 8 theorem 1, § 8 corollary 1.1 c)). We obtain

$$(A - \lambda_n I)\, e_n = (A - \lambda_n I) \sum_{k=1}^{n} \alpha_{n,k} f_k = \sum_{k=1}^{n-1} \alpha_{n,k}(\lambda_k - \lambda_n)\, f_k$$

$$= g_n \in \bigvee \{f_k\}_{k=1}^{n-1}$$

and therefore $g_n \perp e_n$. We conclude

$$\langle Ae_n, e_n\rangle = \langle g_n + \lambda_n e_n, e_n\rangle = \langle \lambda_n e_n, e_n\rangle = \lambda_n$$

and by theorem 1

$$\lim_{n\to\infty} \lambda_n = \lim_{n\to\infty} \langle Ae_n, e_n\rangle = 0$$

in contradiction to our assumption $|\lambda_n| \geqslant \varrho > 0$ for $1 \leqslant n < \infty$. $\quad \square$

COROLLARY 2.1. *Let A be a compact linear operator on $\mathfrak{H}$. If $\lambda \neq 0$ is an eigenvalue of A, then the corresponding eigenspace is a finite-dimensional subspace.*

COROLLARY 2.2. *Let A be a compact linear operator on $\mathfrak{H}$. The only possible accumulation point of the eigenvalues of A in the complex plane is 0.*

PROOF. If $\{\lambda_n\}_{n=1}^\infty$ were a sequence of different eigenvalues converging to some point $\lambda \neq 0$, then a sequence of corresponding eigenvectors would be linearly independent by § 22 theorem 1 and therefore would violate the conclusion of theorem 2. $\quad \square$

COROLLARY 2.3. *Let A be a compact linear operator on $\mathfrak{H}$. There exist at most countably many different eigenvalues of A. If A has infinitely many eigenvalues $\lambda_n (1 \leqslant n < \infty)$, then*

$$\lim_{n\to\infty} \lambda_n = 0.$$

So far nothing has been said about the rest of the spectrum of A which, as we know, for bounded linear operators in general need not consist only of eigenvalues. The statement which we are looking for in the compact case (the rest of the spectrum consists at most of the point 0) is quite satisfactory, but the theorem laying the ground for this statement seems somewhat technical-confusing rather than satisfactory.

THEOREM 3. *Let A be a compact linear operator on $\mathfrak{H}$ and let the complex number $\lambda \neq 0$ be given. Then there exists a constant $\gamma_\lambda > 0$ with the following property:*

For every $g \in (A - \lambda I) \mathfrak{H}$ there exists some vector $f(g)$ (depending on g) such that

(1)
$$\begin{cases} (A - \lambda I) f(g) = g, \\ \|f(g)\| \leqslant \gamma_\lambda \|g\|. \end{cases}$$

Before entering upon the proof, let us try to clarify the meaning of this theorem. Asking whether the point λ belongs to the spectrum of A we are interested in the invertibility of $(A - \lambda I)$. An inverse of $(A - \lambda I)$ should assign to every $g \in (A - \lambda I) \mathfrak{H}$ a unique vector f. Moreover, as a bounded operator, it should do this in such a way that $\|f\| \leqslant \gamma_\lambda \|g\|$ for some constant $\gamma_\lambda > 0$ not depending on g (f being the image of g). Theorem 3 (1) now states that we can always reverse the action of $(A - \lambda I)$ in a bounded way: never mind uniqueness, for every given $g \in (A - \lambda I) \mathfrak{H}$ there is always a candidate f associated with g by some sort of bounded inverse of $(A - \lambda I)$, the bound being γ_λ.

PROOF. Let P_λ be the projection upon the subspace

$$\mathfrak{M}_\lambda = \{f \in \mathfrak{H}, (A - \lambda I) f = o\}$$

(if λ is an eigenvalue of A, then $\mathfrak{M}_\lambda$ is the corresponding eigenspace; if λ is not an eigenvalue, then $\mathfrak{M}_\lambda = \mathfrak{O}$). Given any $g \in (A - \lambda I) \mathfrak{H}$ and any f such that

$$(A - \lambda I) f = g$$

we observe that

$$(A - \lambda I) (f - h) = g$$

iff $h \in \mathfrak{M}_\lambda$. For $h = P_\lambda f \in \mathfrak{M}_\lambda$ and $f(g) = f - P_\lambda f$ we obtain

(2)
$$(A - \lambda I) f(g) = g,$$
$$\|f(g)\| = \|f - P_\lambda f\| = \min \{\|f - h'\| : h' \in \mathfrak{M}_\lambda\}$$

(cf. § 7 theorem 6) and therefore (putting $f - h' = f'$)

(3)
$$\|f(g)\| = \min \{\|f'\| : (A - \lambda I) f' = g\}.$$

Note that in this way we have associated with every $g \in (A - \lambda I) \mathfrak{H}$ a unique vector $f(g)$ such that (2) holds. We assert that there exists a constant $\gamma_\lambda > 0$ such that

$$\|f(g)\| \leqslant \gamma_\lambda \|g\| \quad \text{for all } g \in (A - \lambda I) \mathfrak{H}.$$

Assuming the contrary we have

$$\sup \left\{ \frac{\|f(g)\|}{\|g\|} : g \neq o, g \in (A - \lambda I) \mathfrak{H} \right\} = \infty.$$

We can therefore choose a sequence $\{g_n\}_{n=1}^\infty \subset (A - \lambda I) \mathfrak{H}$ such that $g_n \neq o$, $f(g_n) \neq o$ for all $n \geqslant 1$ and

$$\lim_{n \to \infty} \frac{\|f(g_n)\|}{\|g_n\|} = \infty.$$

For $f_n' = f(g_n)/\|f(g_n)\|$, $g_n' = g_n/\|f(g_n)\|$ we obtain from (2) and (3)

(4) $$(A - \lambda I) f_n' = g_n',$$

(5) $$\|f_n'\| = \min \{\|f'\| : (A - \lambda I) f' = g_n'\} = 1,$$
            $$\lim_{n \to \infty} g_n' = o.$$

Let $\{f_{n_k}'\}_{k=1}^\infty$ be a weakly converging subsequence of $\{f_n'\}_{n=1}^\infty$ (cf. § 26 corollary 4.1). Then $\{A f_{n_k}'\}_{k=1}^\infty$ converges strongly to some vector h and, as a consequence,

$$\lim_{k \to \infty} \lambda f_{n_k}' = \lim_{k \to \infty} (A f_{n_k}' - g_{n_k}') = h.$$

Because of $\lambda \neq 0$ also the sequence $\{f_{n_k}'\}_{k=1}^\infty$ converges strongly and its limit is the vector $h' = h/\lambda$. From

$$(A - \lambda I) h' = \lim_{k \to \infty} (A - \lambda I) f_{n_k}' = \lim_{k \to \infty} g_{n_k}' = o$$

and (4) we conclude

$$(A - \lambda I)(f_n' - h') = g_n'$$

while

$$\lim_{k \to \infty} \|f_{n_k}' - h'\| = 0$$

and therefore $\|f_n' - h'\| < 1$ for infinitely many $n \geqslant 1$. This, however, contradicts (5). $\square$

Recall that in § 24 the spectrum of a selfadjoint operator in $\mathfrak{H}$ has been classified in terms of the size of the linear manifold $(A - \lambda I) \mathfrak{H}$. In particular, the points of the continuous spectrum have been characterized by the inequality $(A-\lambda I)\,\mathfrak{H}\neq\overline{(A-\lambda I)\,\mathfrak{H}}$. It turns out that this method is quite suitable for the spectral analysis even of compact linear operators on $\mathfrak{H}$ in general. The following theorem already indicates that there will hardly be anything like a continuous spectrum in this case.

THEOREM 4. *If A is a compact linear operator on $\mathfrak{H}$ and if $\lambda\neq0$, then $(A-\lambda I)\,\mathfrak{H}$ is closed and therefore a subspace of $\mathfrak{H}$.*

PROOF. Suppose the sequence $\{g_n\}_{n=1}^{\infty}\subset(A-\lambda I)\,\mathfrak{H}$ converges to some vector $g\in\overline{(A-\lambda I)\,\mathfrak{H}}$. The sequence $\{g_n\}_{n=1}^{\infty}$ is bounded and therefore so is the sequence $\{f(g_n)\}_{n=1}^{\infty}$ corresponding to it by theorem 3 (cf. (1)). Let $\{f(g_{n_k})\}_{k=1}^{\infty}$ be a weakly converging subsequence (cf. § 26 corollary 4.1). Then the sequence $\{Af(g_{n_k})\}_{k=1}^{\infty}$ converges strongly and so does the sequence $\{f(g_{n_k})\}_{k=1}^{\infty}$ since

$$f(g_{n_k})=\frac{1}{\lambda}\left[Af(g_{n_k})-g_{n_k}\right]$$

(cf. (1)). For the vector

$$f=\lim_{k\to\infty}f(g_{n_k})$$

we obtain

$$(A-\lambda I)f=\lim_{k\to\infty}(A-\lambda I)f(g_{n_k})$$
$$=\lim_{k\to\infty}g_{n_k}=g.$$

We conclude $g\in(A-\lambda I)\,\mathfrak{H}$. This proves the assertion. $\square$

THEOREM 5. *Let A be a compact linear operator on $\mathfrak{H}$. A complex number $\lambda\neq0$ is a regular value for A iff $(A-\lambda I)\,\mathfrak{H}=\mathfrak{H}$.*

PROOF. If λ is a regular value for A, then $(A-\lambda I)\,\mathfrak{H}=\mathfrak{H}$ by definition (cf. § 23 definition 1). In the converse direction it suffices to show the following: if $(A-\lambda I)\,\mathfrak{H}=\mathfrak{H}$, then λ is not an eigenvalue of A. In fact, the resolvent $(A-\lambda I)^{-1}$ is then defined on all of $\mathfrak{H}$ (cf. § 23 theorem 1, § 23 definition 3) and bounded by theorem 3 (in this case $f(g)$ is simply the unique solution f of the equation $(A-\lambda I)f=g$). Therefore λ is a regular value.

Suppose therefore that λ were an eigenvalue of A and let f_1 be a corresponding eigenvector. Because of our assumption

$$(A-\lambda I)\,\mathfrak{H}=\mathfrak{H}$$

we can inductively construct a sequence $\{f_n\}_{n=1}^{\infty}$ such that

$$(A-\lambda I)f_n=f_{n-1}\quad\text{for all }n\geqslant1\ (f_0=o).$$

Again by induction we show that the vectors $f_n (1 \leqslant n < \infty)$ must be linearly independent. In fact we have $f_1 \neq o$ by definition of an eigenvector. Suppose $f_1, ..., f_{n-1}$ are linearly independent and

(6)
$$\sum_{k=1}^{n} \alpha_k f_k = o .$$

We obtain

$$o = (A - \lambda I) \sum_{k=1}^{n} \alpha_k f_k = \sum_{k=1}^{n} \alpha_k (A - \lambda I) f_k$$
$$= \sum_{k=1}^{n} \alpha_k f_{k-1} \quad (f_0 = o).$$

We conclude $\alpha_2 = \alpha_3 = \cdots = \alpha_n = 0$ and by (6), since $f_1 \neq o$, also $\alpha_1 = 0$. Therefore also $f_1, ..., f_n$ are linearly independent. Let the orthonormal sequence $\{e_n\}_{n=1}^{\infty}$ be obtained from $\{f_n\}_{n=1}^{\infty}$ by Gram-Schmidt orthonormalization (cf. § 8 theorem 1, § 8 corollary 1.1 c)). As in the proof of theorem 2 we conclude

$$(A - \lambda I) e_n = (A - \lambda I) \sum_{k=1}^{n} \alpha_{n, k} f_k = \sum_{k=1}^{n} \alpha_{n, k} (A - \lambda I) f_k$$
$$= \sum_{k=1}^{n} \alpha_{n, k} f_{k-1} = g_n \in \bigvee \{f_n\}_{k=1}^{n-1}$$

and therefore $g_n \perp e_n$. We obtain

$$\langle Ae_n, e_n \rangle = \langle g_n + \lambda e_n, e_n \rangle = \lambda \langle e_n, e_n \rangle = \lambda \neq 0 \quad \text{for all } n \geqslant 1$$

which contradicts the conclusion of theorem 1. $\square$

THEOREM 6. *Let A be a compact linear operator on $\mathfrak{H}$. A complex number $\lambda \neq 0$ is an eigenvalue of A iff $\bar{\lambda}$ is an eigenvalue of A^*.*
PROOF. Suppose $\bar{\lambda}$ is an eigenvalue of A^* which again is a compact operator by § 26 theorem 8. Then $(A^* - \bar{\lambda} I) \mathfrak{H}$ is a proper subspace of $\mathfrak{H}$ by theorem 4 and theorem 5. Applying § 24 theorem 2 we see that λ is an eigenvalue of A with corresponding eigenspace

(7) $\mathfrak{M}_\lambda = [(A^* - \bar{\lambda} I) \mathfrak{H}]^\perp .$

A symmetric reasoning in the other direction completes the proof. $\square$

THEOREM 7. *Let A be a compact linear operator on $\mathfrak{H}$. A complex number $\lambda \neq 0$ is either a regular value for A or an eigenvalue of A.*
PROOF. If λ is not a regular value then $(A - \lambda I) \mathfrak{H}$ is a proper subspace of $\mathfrak{H}$ by theorem 4 and theorem 5. By § 24 theorem 2 again we conclude that $\bar{\lambda}$ is an eigenvalue of A^*. Then λ is an eigenvalue of A by theorem 6. $\square$

Note that theorem 6 cannot be extended to bounded linear operators on $\mathfrak{H}$ in general. If $\{e_k\}_{k=1}^\infty$ is the standard basis in ℓ_2, then 0 is an eigenvalue and e_1 is a corresponding eigenvector of the left shift operator A in ℓ_2 (cf. § 14 example 2). However, its adjoint A^*, the right shift operator, does not have any eigenvalues (cf. § 22 example 3). Consequently 1 is an eigenvalue of $A + I$ but not of $(A+I)^* = A^* + I$. Still, an assertion analogous to theorem 6 holds for any bounded normal operator on $\mathfrak{H}$ (cf. § 23 exercise b)) and, even for any linear operator A in $\mathfrak{H}$ such that $\overline{\mathfrak{D}}_A = \overline{\mathfrak{D}}_{A^*} = \mathfrak{H}$ and $A = A^{**}$, if the spectrum is considered in place of the set of eigenvalues (cf. § 24 corollary 1.2 and appendix A theorem 6).

The theorems stated so far in this section are illustrated by § 24 example 1. This is also the case with the following theorem which clears up the role of the exceptional number 0 (cf. also § 26 exercise). Note that in case of a finite-dimensional unitary space every linear operator on $\mathbf{C}^n$ is not only bounded (cf. § 12 example 1) but also compact (cf. § 25 theorem 7); the spectrum only consists of finitely many eigenvalues (cf. § 22 remark 2) and 0 clearly need not be one of them.

THEOREM 8. *Let A be a compact linear operator on an infinite-dimensional Hilbert space $\mathfrak{H}$. Then 0 is a generalized eigenvalue and therefore belongs to the spectrum of A.*

PROOF. Let $\{e_n\}_{n=1}^\infty$ be any orthonormal sequence in $\mathfrak{H}$. The sequence $\{e_n\}_{n=1}^\infty$ converges weakly to zero (cf. § 26 example 1) and therefore $\{Ae_n\}_{n=1}^\infty$ converges strongly to zero (cf. § 26 corollary 6.1). This already proves the assertion (cf. § 23 definition 2). $\square$

Our last theorem in this section is, properly speaking, just a summary of previous results. We shall discuss some of its applications to integral equations in § 29.

THEOREM 9 (FREDHOLM ALTERNATIVE). *Let A be a compact linear operator and let the complex number $\lambda \neq 0$ be given. Either the inhomogeneous equations*

$$(8) \qquad (A - \lambda I) f = g, \qquad (8^*) \qquad (A^* - \bar{\lambda} I) f^* = g^*$$

have solutions f and f^* for every given g and g^*, or the homogeneous equations

$$(9) \qquad (A - \lambda I) f = o, \qquad (9^*) \qquad (A^* - \bar{\lambda} I) f^* = o$$

have non-zero solutions f and f^*.

In the first case the solutions f and f^* of (8) and (8*) are unique and depend continuously on g and g^* respectively. In the second case (8) has a solution f iff g is orthogonal to all solutions of (9*), and (8*) has a solution f^* iff g^* is orthogonal to all solutions of (9).

Proof. Either λ is a regular value for A and $\bar{\lambda}$ is a regular value for A^* or λ is an eigenvalue of A and $\bar{\lambda}$ is an eigenvalue of A^* (cf. theorem 6 and theorem 7). In the first case we have

$$f = (A - \lambda I)^{-1}g, \quad f^* = (A^* - \bar{\lambda}I)^{-1}g^*$$

where $(A - \lambda I)^{-1}$ and $(A^* - \bar{\lambda}I)^{-1}$ are bounded and therefore continuous linear operators on $\mathfrak{H}$. In the second case, if $\mathfrak{M}_\lambda$ is the eigenspace of A corresponding to λ, then we have

$$(A^* - \bar{\lambda}I)\,\mathfrak{H} = \mathfrak{M}_\lambda^\perp$$

by (7) (we use the fact that $(A^* - \bar{\lambda}I)\,\mathfrak{H}$ is a subspace; cf. theorem 4 and § 7 theorem 9). Equation (8*) has a solution f^* iff $g^* \in (A - \bar{\lambda}I)\,\mathfrak{H}$ or, in other words, iff g^* is orthogonal to the eigenspace $\mathfrak{M}_\lambda$ which in turn consists of all solutions of (9). Similarly, if $\mathfrak{M}_{\bar{\lambda}}^*$ is the eigenspace of A^* corresponding to $\bar{\lambda}$, then we have

$$(A - \lambda I)\,\mathfrak{H} = \mathfrak{M}_{\bar{\lambda}}^{*\perp}$$

from which we conclude the remaining assertion. $\square$

The following exercise shows that a non-zero compact linear operator on $\mathfrak{H}$ may not have any eigenvalues at all (the spectrum then reduces to its absolute minimum, consisting of one single point only; cf. theorem 8 and § 23 remark 2). At the same time it demonstrates that the conclusion of theorem 6 breaks down for $\lambda = 0$.

Exercise. Let the linear operator on ℓ_2 be defined for all sequences $\{\alpha_k\}_{k=1}^\infty \in \ell_2$ by

$$A\{\alpha_k\}_{k=1}^\infty = \{\alpha_{k-1}/k\}_{k=1}^\infty \quad (\alpha_0 = 0),$$

i.e.

$$A\{\alpha_1, \alpha_2, \alpha_3, \ldots\} = \{0, \tfrac{1}{2}\alpha_1, \tfrac{1}{3}\alpha_2, \ldots\}.$$

a) The operator A is compact. (Hint: cf. § 25 example 1.)

b) The operator A does not have any eigenvalues. (Hint: cf. § 22 example 3.)

c) The only eigenvalue of A^* is 0. (Hint: cf. § 14 example 2.)

d) The operator A is not normal.

§ 28. The spectral decomposition of a compact selfadjoint operator

With all information that we have so far accumulated about the spectrum of a compact operator on $\mathfrak{H}$ we shall now try to recapture the operator by means of its spectral characteristics, namely its eigenvalues and the projections upon the corresponding eigenspaces, thus ultimately obtaining the desired decomposition of the operator into simpler parts. This can conveniently be done if the operator in question is also selfadjoint, a fact which supplies us with a considerable amount of additional information. Another good reason for turning our attention mainly to selfadjoint compact operators on $\mathfrak{H}$ lies in the fact that every compact linear operator on $\mathfrak{H}$ can be written as a linear combination of two of these (cf. § 14 theorem 7 (6) (7)). Although this will not give us a decomposition as looked for of a compact linear operator in general, our insight in the structure of such an operator will still profit by the decomposition obtained for selfadjoint compact operators on $\mathfrak{H}$.

If the compact linear operator A on $\mathfrak{H}$ is selfadjoint, then all its eigenvalues are real (cf. § 22 theorem 2). Moreover, since there are at most countably many different eigenvalues $\lambda_k \neq 0$ and since, in case there are infinitely many eigenvalues, we have $\lim_{k \to \infty} \lambda_k = 0$, we may without loss of generality assume

$$|\lambda_1| \geqslant |\lambda_2| \geqslant \cdots \geqslant |\lambda_k| \geqslant \cdots,$$
$$\lambda_k \neq \lambda_l \quad \text{for } k \neq l.$$

Furthermore we know $|\lambda_1| \leqslant \|A\|$ by § 23 corollary 3.2 if an eigenvalue λ_1 exists at all (cf. § 27 exercise b)). We start with showing that this is indeed the case and with locating λ_1 more precisely on the real axis.

THEOREM 1. *Let A be a compact selfadjoint operator on $\mathfrak{H}$. Then there exists an eigenvalue λ of A with $|\lambda| = \|A\|$.*

PROOF. The assertion is trivial for $\|A\| = 0$. We may therefore assume $\|A\| > 0$. Since A is selfadjoint we have $\langle Af, f \rangle \in \mathbf{R}$ for all $f \in \mathfrak{H}$ and

$$\|A\| = \sup_{\|f\|=1} |\langle Af, f \rangle|$$

(cf. § 14 theorem 5 and § 14 corollary 5.1). We can therefore select a sequence $\{f_n\}_{n=1}^{\infty}$ of unit vectors having the property that

$$\|A\| = \lim_{n \to \infty} |\langle Af_n, f_n \rangle|.$$

Without loss of generality we may assume that the sequence $\{f_n\}_{n=1}^{\infty}$ converges weakly (cf. § 26 corollary 4.1) and that the sequence $\{\langle Af_n, f_n \rangle\}_{n=1}^{\infty}$ converges to some limit λ which then has to coincide with either $+\|A\|$ or $-\|A\|$ (if necessary we simply replace $\{f_n\}_{n=1}^{\infty}$ by a suitably chosen subsequence). Since A is compact the sequence $\{Af_n\}_{n=1}^{\infty}$ converges strongly to some vector g. We obtain

$$\lambda = \lim_{n \to \infty} \langle Af_n, f_n \rangle,$$

$$g = \lim_{n \to \infty} Af_n,$$

$$\|g\| = \lim_{n \to \infty} \|Af_n\| \leqslant \|A\| \quad (\text{since } \|f_n\| = 1),$$

$$\|Af_n - \lambda f_n\|^2 = \langle Af_n - \lambda f_n, Af_n - \lambda f_n \rangle$$
$$= \|Af_n\|^2 - 2\lambda \langle Af_n, f_n \rangle + \lambda^2,$$

$$0 \leqslant \lim_{n \to \infty} \|Af_n - \lambda f_n\|^2 = \|g\|^2 - 2\lambda^2 + \lambda^2 = \|g\|^2 - \lambda^2$$
$$= \|g\|^2 - \|A\|^2 \leqslant 0.$$

We conclude

$$\|g\| = \|A\| \neq 0,$$

$$\lim_{n \to \infty} \lambda f_n = \lim_{n \to \infty} Af_n = g \neq o,$$

$$\lim_{n \to \infty} f_n = \frac{g}{\lambda} \neq o,$$

$$(A - \lambda I)\frac{g}{\lambda} = \lim_{n \to \infty} (A - \lambda I) f_n = o.$$

Thus λ is an eigenvalue of A with corresponding eigenvector g/λ. $\square$

Remark 1. Note that in the proof of theorem 1 actually the following statement has been shown: if $\{f_n\}_{n=1}^{\infty}$ is a weakly converging sequence of unit vectors such that

$$\lambda = \lim_{n \to \infty} \langle Af_n, f_n \rangle = \pm \|A\|,$$

then $\{f_n\}_{n=1}^{\infty}$ even converges strongly to an eigenvector corresponding to the eigenvalue λ.

In § 30 theorem 7 and § 30 theorem 8 it will be shown that for any bounded selfadjoint operator A on $\mathfrak{H}$ the numbers $\alpha_1 = \inf_{\|f\|=1} \langle Af, f \rangle$ and $\alpha_2 = \sup_{\|f\|=1} \langle Af, f \rangle$ belong to the spectrum of A and that moreover the spectrum of A is contained in the interval $[\alpha_1, \alpha_2]$. As a consequence, if A is a compact selfadjoint operator on $\mathfrak{H}$, then those of the two numbers α_1 and α_2 which are non-zero are also eigenvalues and all eigenvalues of A (as well as the point 0) lie in the interval $[\alpha_1, \alpha_2]$.

Let P_k be the projection upon the eigenspace $\mathfrak{M}_k$ corresponding to the eigenvalue λ_k $(k \geqslant 1)$. By § 22 theorem 4 we have $\mathfrak{M}_k \perp \mathfrak{M}_l$ (or, equivalently, $P_k \perp P_l$) for $k \neq l$ (cf. § 22 remark 4). We now distinguish two cases: either there are only finitely many different non-zero eigenvalues

$$|\lambda_1| \geqslant |\lambda_2| \geqslant \cdots \geqslant |\lambda_\varkappa| > 0, \quad \varkappa < \infty,$$

or there is a sequence $\{\lambda_k\}_{k=1}^\infty$ of different non-zero eigenvalues and

(1) $$\begin{cases} |\lambda_1| \geqslant |\lambda_2| \geqslant \cdots \geqslant |\lambda_k| \geqslant \cdots, \\ \lim_{k \to \infty} \lambda_k = 0. \end{cases}$$

In the first case let

$$\mathfrak{M} = \sum_{k=1}^\varkappa \mathfrak{M}_k.$$

Since $\mathfrak{M}$ is invariant under A so is $\mathfrak{M}_0 = \mathfrak{M}^\perp$ (cf. § 21 corollary 4.1). The restriction of A to $\mathfrak{M}_0$ is selfadjoint and still maps every weakly convergent sequence into a strongly convergent one (cf. § 21 theorem 1). Therefore A induces in $\mathfrak{M}_0$ a compact selfadjoint operator A_0. Every eigenvector of A_0 in $\mathfrak{M}_0$ is also an eigenvector of A. Since such an eigenvector does not belong to any of the eigenspaces $\mathfrak{M}_k$ ($1 \leqslant k \leqslant \varkappa$) the corresponding eigenvalue must be 0. Thus the only eigenvalue of A_0 is 0 and by theorem 1 we have $\|A_0\| = 0$, in other words $Af = o$ for all $f \in \mathfrak{M}_0$. If P_0 is the projection upon $\mathfrak{M}_0$, then for every vector

$$f = \sum_{k=0}^\varkappa P_k f$$

we have $P_k f \in \mathfrak{M}_k$ for $0 \leqslant k \leqslant \varkappa$ and therefore

$$Af = \sum_{k=1}^\varkappa \lambda_k P_k f \quad \text{for all } f \in \mathfrak{H},$$

(2) $$A = \sum_{k=1}^\varkappa \lambda_k P_k.$$

Since every eigenspace $\mathfrak{M}_k$ ($1 \leqslant k \leqslant \varkappa$) is finite-dimensional so is $\sum_{k=1}^\varkappa \mathfrak{M}_k$ and the operator A is of finite rank (cf. § 25 definition 3).

In the second case let

$$\mathfrak{M} = \sum_{k=1}^\infty \mathfrak{M}_k$$

(cf. § 7 theorem 4). Again $\mathfrak{M}$ is invariant under A and therefore so is $\mathfrak{M}_0 = \mathfrak{M}^\perp$. The same reasoning as in case one applies to show that $Af = o$ for all $f \in \mathfrak{M}_0$. Let P_0 again be the projection upon $\mathfrak{M}_0$. From the orthogonal decomposition

$$\mathfrak{H} = \mathfrak{M}_0 + \sum_{k=1}^\infty \mathfrak{M}_k$$

we deduce

$$f = \sum_{k=0}^{\infty} P_k f \quad \text{for all } f \in \mathfrak{H}$$

(cf. § 7 theorem 5), and further

(3) $$Af = \sum_{k=0}^{\infty} AP_k f = \sum_{k=1}^{\infty} \lambda_k P_k f \quad \text{for all } f \in \mathfrak{H}.$$

Note, however, that in contrast to the case of a finite sum (2) this does not yet authorize us to write

(4) $$A = \sum_{k=1}^{\infty} \lambda_k P_k.$$

In fact, in the Banach algebra $\mathscr{B}$ of bounded linear operators on $\mathfrak{H}$, (4) is not defined by (3) but by

(5) $$\lim_{n \to \infty} \| A - \sum_{k=1}^{n} \lambda_k P_k \| = 0.$$

We shall now justify (4) by proving (5). For every $f \in \mathfrak{H}$ we have

$$(A - \sum_{k=1}^{n} \lambda_k P_k) f = \sum_{k=1}^{\infty} \lambda_k P_k f - \sum_{k=1}^{n} \lambda_k P_k f$$

$$= \sum_{k=n+1}^{\infty} \lambda_k P_k f,$$

$$\| (A - \sum_{k=1}^{n} \lambda_k P_k) f \|^2 = \sum_{k=n+1}^{\infty} \lambda_k^2 \| P_k f \|^2 \leqslant \lambda_{n+1}^2 \sum_{k=n+1}^{\infty} \| P_k f \|^2$$

$$\leqslant \lambda_{n+1}^2 \sum_{k=0}^{\infty} \| P_k f \|^2 = \lambda_{n+1}^2 \| f \|^2,$$

(6) $$\| A - \sum_{k=1}^{n} \lambda_k P_k \| \leqslant |\lambda_{n+1}|.$$

By (1) this implies (5). Moreover, the inequality (6) may actually be sharpened to an equality. Indeed, for an eigenvector $f \in \mathfrak{M}_{n+1}$ we get

$$(A - \sum_{k=1}^{n} \lambda_k P_k) f = \lambda_{n+1} f,$$

$$\| (A - \sum_{k=1}^{n} \lambda_k P_k) f \| = |\lambda_{n+1}| \| f \|,$$

which together with (6) implies

$$\| A - \sum_{k=1}^{n} \lambda_k P_k \| = |\lambda_{n+1}|.$$

The result of this discussion is collected in the following theorem. In order to treat both cases at once we use the letter κ to denote either a natural number (case one) or the symbol ∞ (case two).

THEOREM 2 (SPECTRAL THEOREM FOR COMPACT SELFADJOINT OPERATORS). *Let A be a compact selfadjoint operator on $\mathfrak{H}$. Let $\{\lambda_k\}_{k=1}^{\kappa}$ ($\kappa < \infty$ or $\kappa = \infty$) be the sequence of different non-zero eigenvalues of A, arranged in such a way that*

$$|\lambda_1| \geqslant |\lambda_2| \geqslant \cdots \geqslant |\lambda_k| \geqslant \cdots,$$
$$\lambda_k \neq \lambda_l \quad \text{for } k \neq l,$$

and let P_k be the projection upon the eigenspace $\mathfrak{M}_k$ corresponding to the eigenvalue λ_k ($1 \leqslant k_{(\leqq)} \kappa$). The projections P_k are of finite rank and mutually orthogonal and

$$(7) \qquad\qquad A = \sum_{k=1}^{\kappa} \lambda_k P_k.$$

If $\kappa < \infty$, then A is of finite rank. If $\kappa = \infty$, then the series (7) converges in the operator norm in $\mathscr{B}$ and

$$\left\| A - \sum_{k=1}^{n} \lambda_k P_k \right\| = |\lambda_{n+1}|.$$

COROLLARY 2.1. *Let A be a compact selfadjoint operator on $\mathfrak{H}$ and let $\mathfrak{M} = \overline{A\mathfrak{H}}$. The subspace $\mathfrak{M}$ is separable and reduces A. Its orthogonal complement $\mathfrak{M}^{\perp}$ is the set of all vectors annihilated by A (= the eigenspace corresponding to the eigenvalue 0 if $\mathfrak{M}^{\perp} \neq \mathfrak{D}$). Moreover $\mathfrak{M}$ admits a basis $\{e_k\}_{k=1}^{\kappa'}$ ($\kappa' < \infty$ or $\kappa' = \infty$) such that e_k is an eigenvector corresponding to the eigenvalue $\lambda'_k \neq 0$ of A for $1 \leqslant k_{(\leqq)} \kappa'$ and*

$$Af = \sum_{k=1}^{\kappa'} \lambda'_k \langle f, e_k \rangle e_k \quad \text{for all } f \in \mathfrak{H}.$$

PROOF. From (2) and (4) we deduce

$$\mathfrak{M}_l \subset A\mathfrak{H} \subset \sum_{k=1}^{\kappa} \mathfrak{M}_k \quad \text{for } 1 \leqslant l_{(\leqq)} \kappa$$

and therefore by § 7 theorem 4

$$\sum_{l=1}^{\kappa} \mathfrak{M}_l \subset \overline{A\mathfrak{H}} \subset \sum_{k=1}^{\kappa} \mathfrak{M}_k,$$
$$\mathfrak{M} = \overline{A\mathfrak{H}} = \sum_{k=1}^{\kappa} \mathfrak{M}_k,$$
$$\mathfrak{M}^{\perp} = \mathfrak{M}_0.$$

Choosing in each of the mutually orthogonal finite-dimensional eigenspaces $\mathfrak{M}_k (1 \leqslant k_{(} \leqslant_{)} \kappa)$ a basis we obtain a basis $\{e_k\}_{k=1}^{\kappa'}$ of $\mathfrak{M}$ with the desired properties. Since this basis of $\mathfrak{M}$ is countable, the subspace $\mathfrak{M}$ is separable (cf. § 8 theorem 2). Note that in contrast to the sequence $\{\lambda_k\}_{k=1}^{\kappa}$ where we have $\lambda_k \neq \lambda_l$ for $k \neq l$, every non-zero eigenvalue appears as many times in the sequence $\{\lambda_k'\}_{k=1}^{\kappa'}$ as indicated by its *multiplicity*, i.e. by the dimension of the corresponding eigenspace. $\square$

Theorem 2 shows that a compact selfadjoint operator A on $\mathfrak{H}$ is completely determined by its non-zero eigenvalues and the corresponding eigenspaces. The expansion (7) is called the *spectral decomposition* of A. An example for this spectral decomposition is again furnished by § 24 example 1 (1). The next theorem also illustrates the close relationship between the operator A and the projections turning up in its spectral decomposition.

THEOREM 3. *Let A be a compact selfadjoint operator on $\mathfrak{H}$ and let*

$$A = \sum_{k=1}^{\kappa} \lambda_k P_k \quad (\kappa < \infty \text{ or } \kappa = \infty)$$

be its spectral decomposition. If B is a bounded linear operator on $\mathfrak{H}$, then $AB = BA$ iff

(8) $$P_k B = B P_k \quad for \ 1 \leqslant k_{(} \leqslant_{)} \kappa.$$

PROOF. From (8) we deduce

$$AB = \left(\sum_{k=1}^{\kappa} \lambda_k P_k\right) B = \sum_{k=1}^{\kappa} \lambda_k (P_k B) = \sum_{k=1}^{\kappa} \lambda_k (B P_k) = B \left(\sum_{k=1}^{\kappa} \lambda_k P_k\right) = BA.$$

Conversely, suppose $AB = BA$ and therefore also $B^* A = AB^*$. Let $\mathfrak{M}_k$ be the eigenspace of A corresponding to λ_k. For $f \in \mathfrak{M}_k$ we obtain

$$(A - \lambda_k I) Bf = B (A - \lambda_k I) f = o,$$
$$(A - \lambda_k I) B^* f = B^* (A - \lambda_k I) f = o.$$

We conclude $Bf \in \mathfrak{M}_k$ and $B^* f \in \mathfrak{M}_k$. Thus $\mathfrak{M}_k$ is invariant under B and B^* and therefore reduces B (cf. § 21 theorem 4). As a consequence we have $P_k B = B P_k$ (cf. § 21 theorem 3). $\square$

Remark 2. If P_0 is the projection upon $\mathfrak{M}_0 = (\sum_{k=1}^{\kappa} \mathfrak{M}_k)^{\perp} = \{f \in \mathfrak{H} : Af = o\}$, then the argument used above also shows that $AB = BA$ at the same time implies $P_0 B = B P_0$.

We shall now apply this theorem in order to get an analogous spectral decomposition for a compact normal operator A (cf. § 14 definition 1). Recall that we have

$$A = B + iC$$

where the operators

$$B = \tfrac{1}{2}(A + A^*),$$

$$C = \frac{1}{2i}(A - A^*)$$

are selfadjoint (cf. § 14 theorem 7 (6) (7)) and

(9) $BC = CB$.

Moreover, as linear combinations of compact linear operators also B and C are compact (cf. § 26 theorem 7 and § 26 theorem 8). Let

$$B = \sum_{k=1}^{\infty} \xi_k P_k,$$

$$C = \sum_{l=1}^{\infty} \eta_l Q_l$$

be the spectral decompositions of B and C (the cases that some of these series reduce to finite sums are dealt with by minor modifications; we omit these in order to avoid unnecessary complications). Let $\mathfrak{M}_k$ and $\mathfrak{N}_l$ be the eigenspaces of B and C corresponding to the eigenvalues ξ_k and η_l respectively ($1 \leqslant k < \infty$, $1 \leqslant l < \infty$). Let P_0 and Q_0 be the projections upon the eigenspaces $\mathfrak{M}_0$ and $\mathfrak{N}_0$ of B and C respectively corresponding to the eigenvalue $\xi_0 = \eta_0 = 0$ (if 0 is an eigenvalue at all; otherwise these projections reduce to the zero operator). We assert that

(10) $P_k Q_l = Q_l P_k$ for $0 \leqslant k < \infty$, $0 \leqslant l < \infty$,

which is equivalent with the statement that $P_k Q_l$ is the projection upon the subspace $\mathfrak{M}_k \cap \mathfrak{N}_l$ (cf. § 15 theorem 4 and § 15 corollary 4.1). Indeed (9) implies $P_k C = CP_k$ for $0 \leqslant k < \infty$ by theorem 3 and remark 2. Applying theorem 3 and remark 2 once more we obtain the assertion (10). Since $P_k \perp P_{k'}$ for $k \neq k'$ and $Q_l \perp Q_{l'}$ for $l \neq l'$ we further get

$$P_k Q_l \perp P_{k'} Q_{l'}   \text{ for } k \neq k' \text{ or } l \neq l'.$$

Consider now an arbitrary vector $f \in \mathfrak{H}$. From

$$\mathfrak{H} = \sum_{l=0}^{\infty} \mathfrak{N}_l$$

we obtain

$$f = \sum_{l=0}^{\infty} Q_l f \quad \text{(cf. § 7 theorem 5)},$$

$$Bf = \sum_{k=0}^{\infty} \xi_k P_k \left(\sum_{l=0}^{\infty} Q_l f \right)$$

$$= \sum_{k=0}^{\infty} \left(\sum_{l=0}^{\infty} \xi_k P_k Q_l f \right).$$

Similarly we obtain

$$Cf = \sum_{l=0}^{\infty} \left(\sum_{k=0}^{\infty} \eta_l Q_l P_k f \right) = \sum_{l=0}^{\infty} \left(\sum_{k=0}^{\infty} \eta_l P_k Q_l f \right).$$

Fortunately we may interchange the order of summation, as may be seen as follows: from

$$P_{k'} Cf = \sum_{l=0}^{\infty} P_{k'} \left(\sum_{k=0}^{\infty} \eta_l P_k Q_l f \right) = \sum_{l=0}^{\infty} \eta_l P_{k'} Q_l f$$

we obtain

$$Cf = \sum_{k=0}^{\infty} P_k Cf = \sum_{k=0}^{\infty} \left(\sum_{l=0}^{\infty} \eta_l P_k Q_l f \right),$$

(11)
$$Af = (B + iC) f = \sum_{k=0}^{\infty} \left[\sum_{l=0}^{\infty} (\xi_k + i\eta_l) P_k Q_l f \right].$$

Note that some of the projections $P_k Q_l$ may well be the zero operator and that $\xi_k + i\eta_l \neq \xi_{k'} + i\eta_{l'}$ if $k \neq k'$ or $l \neq l'$. Those numbers $\xi_k + i\eta_l$ for which the space $\mathfrak{M}_k \cap \mathfrak{N}_l$ is non-zero are eigenvalues of A (every non-zero vector $f \in \mathfrak{M}_k \cap \mathfrak{N}_l$ is a corresponding eigenvector of A). Omitting $\xi_0 + i\eta_0 = 0$ we can therefore arrange them to a sequence $\{\lambda_n\}_{n=1}^{\infty}$ such that

$$|\lambda_1| \geqslant |\lambda_2| \geqslant \cdots \geqslant |\lambda_n| \geqslant \cdots,$$

$$\lim_{n \to \infty} \lambda_n = 0.$$

Let $\{R_n\}_{n=1}^{\infty}$ be the sequence of projections upon the corresponding subspaces $\mathfrak{R}_n (1 \leqslant n < \infty)$ (we have $R_n = P_k Q_l$ and $\mathfrak{R}_n = \mathfrak{M}_k \cap \mathfrak{N}_l$ for suitable indices k and l) and let R_0 be the projection upon the subspace $\mathfrak{R}_0 = (\sum_{n=1}^{\infty} \mathfrak{R}_n)^{\perp}$.

From (11) we conclude

$$Af \in \sum_{n=1}^{\infty} \Re_n,$$

$$R_n Af = \lambda_n R_n f \quad \text{for } 1 \leqslant n < \infty,$$

$$Af = \sum_{n=1}^{\infty} R_n Af = \sum_{n=1}^{\infty} \lambda_n R_n f,$$

$$\left\| \left(A - \sum_{n=1}^{m} \lambda_n R_n \right) f \right\|^2 = \left\| \sum_{n=m+1}^{\infty} \lambda_n R_n f \right\|^2 = \sum_{n=m+1}^{\infty} |\lambda_n|^2 \|R_n f\|^2$$

$$\leqslant |\lambda_{m+1}|^2 \sum_{n=0}^{\infty} \|R_n f\|^2 = |\lambda_{m+1}|^2 \|f\|^2.$$

Choosing in particular an eigenvector $f \in \Re_{m+1}$ we obtain

$$\left\| \left(A - \sum_{k=1}^{m} \lambda_n R_n \right) f \right\|^2 = \|Af\|^2 = |\lambda_{m+1}|^2 \|f\|^2.$$

We conclude

$$\left\| A - \sum_{n=1}^{m} \lambda_n R_n \right\| = |\lambda_{m+1}|,$$

$$A = \sum_{n=1}^{\infty} \lambda_n R_n.$$

Finally we show that $\{\lambda_n\}_{n=1}^{\infty}$ is the sequence of all non-zero eigenvalues of A and that $\{\Re_n\}_{n=1}^{\infty}$ is the sequence of corresponding eigenspaces. If f is any eigenvector corresponding to any eigenvalue λ of A, then we have

$$f = \sum_{n=0}^{\infty} R_n f \neq o,$$

$$(A - \lambda I) f = \sum_{n=0}^{\infty} (\lambda_n - \lambda) R_n f = o.$$

Since not all projections $R_n f$ can vanish there must be an index m such that

$$\lambda = \lambda_m,$$
$$R_n f = o \quad \text{for all } n \neq m,$$
$$f = R_m f \in \Re_m.$$

This coincides with our assertion.

Summarizing the result in a theorem we include the trivial modifications omitted in the above discussion and switch back to the notation used in theorem 2 (now the eigenvalues λ_k need not anymore be real numbers).

Except for the hypothesis this theorem then word for word repeats the statement of theorem 2.

THEOREM 4 (SPECTRAL THEOREM FOR COMPACT NORMAL OPERATORS). *Let A be a compact normal operator on $\mathfrak{H}$. Let $\{\lambda_k\}_{k=1}^{\kappa}$ ($\kappa < \infty$ or $\kappa = \infty$) be the sequence of different non-zero eigenvalues of A, arranged in such a way that*

$$|\lambda_1| \geqslant |\lambda_2| \geqslant \cdots \geqslant |\lambda_n| \geqslant \cdots,$$
$$\lambda_k \neq \lambda_l \quad \textit{for } k \neq l,$$

and let P_k be the projection upon the eigenspace $\mathfrak{M}_k$ corresponding to the eigenvalue λ_k ($1 \leqslant k \leqslant \kappa$). The projections P_k are of finite rank and mutually orthogonal and

$$A = \sum_{k=1}^{\kappa} \lambda_k P_k.$$

If $\kappa < \infty$, then A is of finite rank. If $\kappa = \infty$, then the series converges in the operator norm in $\mathcal{B}$ and

$$\left\| A - \sum_{k=1}^{n} \lambda_k P_k \right\| = |\lambda_{n+1}|.$$

We omit a repetition of corollary 2.1 for the case of a compact normal operator.

If A is any compact linear operator on $\mathfrak{H}$ we still have

$$A = B + \mathrm{i}C, \quad B = \tfrac{1}{2}(A + A^*), \quad C = \frac{1}{2\mathrm{i}}(A - A^*),$$

where B and C are compact selfadjoint operators but in general $BC \neq CB$. The reasoning above therefore does not apply in the general case. The best we can get using the spectral decomposition of B and C is

$$B = \sum_{k=1}^{\infty} \xi_k P_k,$$

$$C = \sum_{l=1}^{\infty} \eta_l Q_l,$$

(12)
$$A = \sum_{k=1}^{\infty} \xi_k P_k + \mathrm{i} \sum_{l=1}^{\infty} \eta_l Q_l.$$

Still we can draw a useful conclusion which serves to give yet another characterization of compact linear operators.

THEOREM 5. *A bounded linear operator A on $\mathfrak{H}$ is compact iff for every $\varepsilon > 0$ there exists a bounded linear operator A_ε on $\mathfrak{H}$ of finite rank such that*

$$\|A - A_\varepsilon\| < \varepsilon.$$

PROOF. If the stated condition is satisfied and if we let ε run through the sequence $\{1/n\}_{n=1}^\infty$, then we obtain

$$A = \lim_{n \to \infty} A_{1/n}$$

where $A_{1/n}$ is a compact linear operator for $n \geqslant 1$ since it is of finite rank (cf. § 25 theorem 7). Then also A is compact by § 26 theorem 9.

Conversely, if A is compact and if $\varepsilon > 0$ is given, then by (12) there exist indices n and m such that

$$\left\| A - \sum_{k=1}^n \xi_k P_k - \mathrm{i} \sum_{l=1}^m \eta_l Q_l \right\| < \varepsilon.$$

Since the operator $A_\varepsilon = \sum_{k=1}^n \xi_k P_k + \mathrm{i} \sum_{l=1}^m \eta_l Q_l$ is of finite rank the assertion is proved. $\square$

In an alternative formulation theorem 5 asserts that a bounded linear operator A on $\mathfrak{H}$ is compact iff it can be approximated (in norm) by bounded linear operators of finite rank.

The following exercise among other things implies that a spectral decomposition as described in theorem 2 and theorem 4 is indeed only possible for compact normal operators.

Exercise. Let $\{P_k\}_{k=1}^\infty$ be a sequence of mutually orthogonal non-zero projections of finite rank and let $\{\lambda_k\}_{k=1}^\infty$ be a sequence of different non-zero complex numbers such that

$$|\lambda_1| \geqslant |\lambda_2| \geqslant \cdots \geqslant |\lambda_k| \geqslant \cdots,$$

$$\lim_{k \to \infty} \lambda_k = 0.$$

Then the series $\sum_{k=1}^\infty \lambda_k P_k$ converges (in norm) to a compact normal operator A with the spectral decomposition

$$A = \sum_{k=1}^\infty \lambda_k P_k.$$

The operator A is selfadjoint iff $\{\lambda_k\}_{k=1}^\infty \subset \mathbf{R}$.

§ 29. Fredholm integral equations

Let A be a Hilbert-Schmidt operator on $\mathfrak{L}_2(\beta, \gamma)$ $(-\infty \leqslant \beta < \gamma \leqslant +\infty)$

with kernel $a \in \mathfrak{L}_2(X)$ where $X =]\beta, \gamma[\times]\beta, \gamma[$ (cf. § 12 example 4, § 25 example 2). In the present section we shall simply collect and interpret for this case the results on compact linear operators discussed in a more general context in the preceding sections.

Recall that also A^* is a Hilbert-Schmidt operator and that for $f \in \mathfrak{L}_2(\beta, \gamma)$ we have

$$\left. \begin{aligned} Af(\xi) &= \int_\beta^\gamma a(\xi, \eta) f(\eta) \, d\eta \\ A^*f(\xi) &= \int_\beta^\gamma \overline{a(\eta, \xi)} f(\eta) \, d\eta \end{aligned} \right\} \quad \text{for almost all (a.a.) } \xi \in]\beta, \gamma[$$

(cf. § 14 example 3). Moreover we have

$$\|A\| = \|A^*\| \leqslant \alpha = \left[\int_\beta^\gamma \int_\beta^\gamma |a(\xi, \eta)|^2 \, d\xi \, d\eta \right]^{\frac{1}{2}}.$$

An equation of the form

$$(A - \lambda I) f = g$$

or, more explicitely,

$$(1) \qquad \int_\beta^\gamma a(\xi, \eta) f(\eta) \, d\eta - \lambda f(\xi) = g(\xi) \quad \text{for a.a. } \xi \in]\beta, \gamma[$$

where $0 \neq \lambda \in \mathbb{C}$ and $g \in \mathfrak{L}_2(\beta, \gamma)$ are given is called a *Fredholm integral equation*. The problem is to find a function $f \in \mathfrak{L}_2(\beta, \gamma)$ (if such a function exists at all) which satisfies (1). The Fredholm alternative (§ 27 theorem 9) describes the two situations that can arise when we attempt to solve (1), depending on whether λ is an eigenvalue of A or not.

Let us first consider the case that λ is not an eigenvalue of A. Then (1) admits a solution $f \in \mathfrak{L}_2(\beta, \gamma)$ for every given $g \in \mathfrak{L}_2(\beta, \gamma)$. This solution is given by

$$f = (A - \lambda I)^{-1} g$$

and $(A - \lambda I)^{-1}$ is a bounded linear operator on $\mathfrak{L}_2(\beta, \gamma)$.

The situation becomes particularly pleasant if $|\lambda| > \|A\|$ (this is certainly the case if $|\lambda| > \alpha$). Then we have

$$(2) \qquad (A - \lambda I)^{-1} = - \sum_{k=0}^\infty \frac{1}{\lambda^{k+1}} A^k$$

(cf. § 23 corollary 3.2). We shall now show that also A^k (which we already know to be compact) is actually a Hilbert-Schmidt operator for all $k \geqslant 1$. If $|\lambda| > \alpha$, then even the operator $\sum_{k=1}^{\infty} A^k/\lambda^{k+1}$ will turn out to be a Hilbert-Schmidt operator.

THEOREM 1. *Let A be a Hilbert-Schmidt operator on $\mathfrak{L}_2(\beta, \gamma)$ $(-\infty \leqslant \beta < \gamma \leqslant \leqslant +\infty)$ with kernel $a \in \mathfrak{L}_2(X)$ $(X =]\beta, \gamma[\times]\beta, \gamma[)$ and let*

$$\alpha^2 = \int_{\beta}^{\gamma} \int_{\beta}^{\gamma} |a(\xi, \eta)|^2 \, d\xi \, d\eta > 0.$$

Then for all $k \geqslant 2$ the operator A^k is a Hilbert-Schmidt operator with the "iterated kernel" $a^{(k)} \in \mathfrak{L}_2(X)$ given by

$$(3) \qquad \left. \begin{aligned} a^{(k)}(\xi, \eta) &= \int_{\beta}^{\gamma} a(\xi, \zeta) \, a^{(k-1)}(\zeta, \eta) \, d\zeta \\ &= \int_{\beta}^{\gamma} a^{(k-1)}(\xi, \zeta) \, a(\zeta, \eta) \, d\zeta \\ a^{(1)}(\xi, \eta) &= a(\xi, \eta) \end{aligned} \right\} \quad \textit{for a.a. } (\xi, \eta) \in X,$$

and

$$(4) \qquad \int_{\beta}^{\gamma} \int_{\beta}^{\gamma} |a^{(k)}(\xi, \eta)|^2 \, d\xi \, d\eta \leqslant \alpha^{2k}.$$

PROOF. Using induction on k it suffices to show: if A^{k-1} is a Hilbert-Schmidt operator with kernel $a^{(k-1)} \in \mathfrak{L}_2(X)$, then A^k is a Hilbert-Schmidt operator with kernel $a^{(k)} \in \mathfrak{L}_2(X)$ given by (3) and satisfying (4).

First observe that for almost all $\xi \in]\beta, \gamma[$ and almost all $\eta \in]\beta, \gamma[$ we have

$$a(\xi, z) \in \mathfrak{L}_2(\beta, \gamma),$$
$$a^{(k-1)}(z, \eta) \in \mathfrak{L}_2(\beta, \gamma),$$

by FUBINI's theorem (cf. appendix B8). Therefore for these ξ and η in $]\beta, \gamma[$ the function $a(\xi, z) \, a^{(k-1)}(z, \eta)$ is integrable on $]\beta, \gamma[$ (cf. § 5 remark 1) and, again by FUBINI's theorem (applied to the function $a(x, z) \, a^{(k-1)}(z, y)$ on the product of $]\beta, \gamma[$ and bounded subsquares of X), the function $a^{(k)}$ defined a.e. on X by (3) is measurable on X. Furthermore we have by CAUCHY's inequality

$$\int\limits_{\beta}^{\gamma} \int\limits_{\beta}^{\gamma} |a^{(k)}(\xi, \eta)|^2 \, d\xi \, d\eta = \int\limits_{\beta}^{\gamma} \int\limits_{\beta}^{\gamma} \left| \int\limits_{\beta}^{\gamma} a(\xi, \zeta) \, a^{(k-1)}(\zeta, \eta) \, d\zeta \right|^2 \, d\xi \, d\eta$$

$$\leqslant \int\limits_{\beta}^{\gamma} \int\limits_{\beta}^{\gamma} \left[\int\limits_{\beta}^{\gamma} |a(\xi, \zeta)|^2 \, d\zeta \int\limits_{\beta}^{\gamma} |a^{(k-1)}(\zeta, \eta)|^2 \, d\zeta \right] d\xi \, d\eta$$

$$= \int\limits_{\beta}^{\gamma} \int\limits_{\beta}^{\gamma} |a(\xi, \zeta)|^2 \, d\xi \, d\zeta \cdot \int\limits_{\beta}^{\gamma} \int\limits_{\beta}^{\gamma} |a^{(k-1)}(\zeta, \eta)|^2 \, d\zeta \, d\eta$$

$$\leqslant \alpha^2 \cdot \alpha^{2\,(k-1)} = \alpha^{2k} < \infty \,.$$

This proves (4) and implies $a^{(k)} \in \mathfrak{L}_2(X)$.

For $f \in \mathfrak{L}_2(\beta, \gamma)$ we get

$$A^k f(\xi) = A(A^{k-1}f)(\xi)$$

$$= \int\limits_{\beta}^{\gamma} a(\xi, \zeta) \left[\int\limits_{\beta}^{\gamma} a^{(k-1)}(\zeta, \eta) \, f(\eta) \, d\eta \right] d\zeta$$

$$= \int\limits_{\beta}^{\gamma} \left[\int\limits_{\beta}^{\gamma} a(\xi, \zeta) \, a^{(k-1)}(\zeta, \eta) \, f(\eta) \, d\eta \right] d\zeta \quad \text{for a.a. } \xi \in]\beta, \gamma[\,.$$

Again for almost all $\xi \in]\beta, \gamma[$ we have $a(\xi, z) \in \mathfrak{L}_2(\beta, \gamma)$ and therefore

$$a(\xi, z) \, f(y) \in \mathfrak{L}_2(X),$$
$$a^{(k-1)}(z, y) \in \mathfrak{L}_2(X).$$

As a consequence, for these $\xi \in]\beta, \gamma[$ the function $a(\xi, z)f(y) \, a^{(k-1)}(z, y)$ is integrable on X and we can apply FUBINI's theorem, obtaining

$$A^k f(\xi) = \int\limits_{\beta}^{\gamma} \left[\int\limits_{\beta}^{\gamma} a(\xi, \zeta) \, a^{(k-1)}(\zeta, \eta) \, d\zeta \right] f(\eta) \, d\eta$$

$$= \int\limits_{\beta}^{\gamma} a^{(k)}(\xi, \eta) \, f(\eta) \, d\eta \,.$$

The second formula in (3) is obtained similarly and must give the same iterated kernel, since disregarding a set of Lebesgue measure zero in X the kernel is uniquely determined by the operator A^k (cf. § 14 remark 1). $\square$

THEOREM 2. *Let A be a Hilbert-Schmidt operator on $\mathfrak{L}_2(\beta, \gamma)$ as in theorem 1. If λ is any complex number such that $|\lambda| > \alpha$, then the series $\sum_{k=1}^{\infty} a^{(k)}/\lambda^{k+1}$ converges in $\mathfrak{L}_2(X)$ and almost everywhere on X to a function $b \in \mathfrak{L}_2(X)$, and the operator*

$$B = \sum_{k=1}^{\infty} \frac{1}{\lambda^{k+1}} A^k$$

is a Hilbert-Schmidt operator on $\mathfrak{L}_2(\beta, \gamma)$ with kernel b.

PROOF. Convergence of the series $\sum_{k=1}^{\infty} a^{(k)}/\lambda^{k+1}$ in $\mathfrak{L}_2(X)$ to some function $b \in \mathfrak{L}_2(X)$ follows from

$$\lim_{m, n \to \infty} \| \sum_{k=m}^{n} \frac{1}{\lambda^{k+1}} a^{(k)} \| \leqslant \lim_{m, n \to \infty} \sum_{k=m}^{n} \frac{1}{|\lambda|^{k+1}} \alpha^k = 0.$$

In order to verify convergence a.e. on X we note that for almost all $(\xi, \eta) \in X$ we have

$$a(\xi, z) \in \mathfrak{L}_2(\beta, \gamma),$$
$$a(w, \eta) \in \mathfrak{L}_2(\beta, \gamma)$$

and therefore

$$a(\xi, z) \, a(w, \eta) \in \mathfrak{L}_2(X).$$

For these $(\xi, \eta) \in X$ and for $k \geqslant 3$ we obtain by (3)

$$|a^{(k)}(\xi, \eta)| = \left| \int_{\beta}^{\gamma} \int_{\beta}^{\gamma} a(\xi, \zeta) \, a^{(k-2)}(\zeta, \omega) \, a(\omega, \eta) \, \mathrm{d}\zeta \, \mathrm{d}\omega \right|$$
$$\leqslant \| a(\xi, z) \, a(w, \eta) \| \, \| a^{(k-2)} \| \leqslant \gamma_{\xi, \eta} \alpha^{k-2}$$

where $\gamma_{\xi, \eta}$ is a non-negative constant depending on ξ and η but not on k. The assertion now follows from

$$\sum_{k=3}^{\infty} \frac{1}{|\lambda|^{k+1}} |a^{(k)}(\xi, \eta)| \leqslant \gamma_{\xi, \eta} \cdot \frac{1}{|\lambda| \, \alpha^2} \sum_{k=3}^{\infty} |\alpha/\lambda|^k < \infty.$$

Since a subsequence of the partial sums of the series $\sum_{k=1}^{\infty} a^{(k)}/\lambda^{k+1}$ must converge a.e. on X to b (cf. § 5 remark 3) we conclude that the series itself converges to b a.e. on X.

Finally, observe that for every $n \geqslant 1$ the operator

$$\sum_{k=1}^{n} \frac{1}{\lambda^{k+1}} A^k$$

is the Hilbert-Schmidt operator with kernel

$$\sum_{k=1}^{n} \frac{1}{\lambda^{k+1}} a^{(k)}.$$

If B' is the Hilbert-Schmidt operator with kernel b, then we have

$$\lim_{n \to \infty} \|B' - \sum_{k=1}^{n} \frac{1}{\lambda^{k+1}} A^k\| \leqslant \lim_{n \to \infty} \|b - \sum_{k=1}^{n} \frac{1}{\lambda^{k+1}} a^{(k)}\| = 0$$

and therefore

$$B' = \sum_{k=1}^{\infty} \frac{1}{\lambda^{k+1}} A^k = B. \quad \square$$

Using theorem 1 we deduce from (2) that for $|\lambda| > \|A\|$ the solution f of (1) is given, for every $g \in \mathfrak{L}_2(\beta, \gamma)$ by

$$(5) \qquad f = -\frac{g}{\lambda} - \sum_{k=1}^{\infty} \frac{1}{\lambda^{k+1}} \int_{\beta}^{\gamma} a^{(k)}(x, \eta) \, g(\eta) \, d\eta$$

where the series converges in $\mathfrak{L}_2(\beta, \gamma)$ but not necessarily almost everywhere on $]\beta, \gamma[$. For $|\lambda| > \alpha$ we can sharpen (5) to

$$(6) \qquad f(\xi) = -\frac{g(\xi)}{\lambda} - \int_{\beta}^{\gamma} \left[\sum_{k=1}^{\infty} \frac{a^{(k)}(\xi, \eta)}{\lambda^{k+1}} \right] g(\eta) \, d\eta \quad \text{for a.a. } \xi \in]\beta, \gamma[$$

where the series in brackets converges in $\mathfrak{L}_2(X)$ and almost everywhere on X to a kernel $b \in \mathfrak{L}_2(X)$. Thus, in this case we have

$$f = -\left(B + \frac{1}{\lambda} I\right) g$$

where B is again a Hilbert-Schmidt operator (with kernel b). This fact may also be formulated in terms of the resolvent of the operator A as follows.

COROLLARY 2.1. *Let A and α be as in theorem 1. If λ is any complex number such that $|\lambda| > \alpha$, then*

$$(A - \lambda I)^{-1} = -\left(B + \frac{1}{\lambda} I\right)$$

where B is a Hilbert-Schmidt operator as given in theorem 2.

At the same time an analogous statement holds for the "*adjoint equation*"

$$(1^*) \qquad \int_{\beta}^{\gamma} \overline{a(\eta, \xi)} f^*(\eta) \, d\eta - \bar{\lambda} f^*(\xi) = g^*(\xi) \quad \text{for a.a. } \xi \in]\beta, \gamma[$$

(cf. § 27 theorem 9).

If λ is an eigenvalue of A, then $\bar{\lambda}$ is an eigenvalue of A^* (cf. § 27 theorem 6) and there exist non-zero functions $f^* \in \mathfrak{L}_2(\beta, \gamma)$ such that

(7) $$\int_\beta^\gamma \overline{a(\eta, \xi)}\, f^*(\eta)\, \mathrm{d}\eta = \bar{\lambda} f^*(\xi) \quad \text{for a.a. } \xi \in \,]\beta, \gamma[\,.$$

The Fredholm alternative states that in this case a solution f of (1) exists iff

$$\int_\beta^\gamma f^*(\xi)\, \overline{g(\xi)}\, \mathrm{d}\xi = 0$$

for all solutions f^* of (7). Again an analogous statement applies for the adjoint equation (1*).

As we should expect, the situation developes some special features if

$$\overline{a(\eta, \xi)} = a(\xi, \eta) \quad \text{for a.a. } (\xi, \eta) \in X\,,$$

i.e. if the Hilbert-Schmidt operator A is selfadjoint (cf. § 14 example 4). According to § 28 corollary 2.1 there now exists a countable orthonormal family $\{e_k\}_{k=1}^{\kappa'}$ ($\kappa' < \infty$ or $\kappa' = \infty$) of eigenfunctions $e_k \in \mathfrak{L}_2(\beta, \gamma)$ corresponding to non-zero eigenvalues $\lambda_k' \in \mathbf{R}$ such that

(8) $$\begin{cases} \|A\| = |\lambda_1'| \geqslant |\lambda_2'| \geqslant \cdots \geqslant |\lambda_k'| \geqslant \cdots, \qquad |\lambda_k'| \neq 0 \\[2mm] \langle e_k, e_l \rangle = \int_\beta^\gamma e_k(\eta)\, \overline{e_l(\eta)}\, \mathrm{d}\eta = \delta_{k,l}, \\[2mm] A e_k = \int_\beta^\gamma a(x, \eta)\, e_k(\eta)\, \mathrm{d}\eta = \lambda_k' e_k, \\[2mm] A f = \int_\beta^\gamma a(x, \eta)\, f(\eta)\, \mathrm{d}\eta = \sum_{k=1}^{\kappa'} \lambda_k' \langle f, e_k \rangle\, e_k \quad \text{for all } f \in \mathfrak{L}_2(\beta, \gamma). \end{cases}$$

The family $\mathfrak{E} = \{e_k\}_{k=1}^{\kappa'}$ is a basis of the subspace $\mathfrak{M} = \overline{A\mathfrak{H}}$ of $\mathfrak{L}_2(\beta, \gamma)$. Its orthogonal complement $\mathfrak{M}_0 = \mathfrak{M}^\perp$ is a separable Hilbert space in its own right and therefore admits a countable basis $\mathfrak{E}' = \{e_k'\}_{k=1}^{\kappa''}$ (cf. § 8 theorem 2; if $\mathfrak{E}'$ is not empty, then it consists of eigenfunctions corresponding to the eigenvalue 0). Let the union of $\mathfrak{E}$ and $\mathfrak{E}'$, which then is a basis of $\mathfrak{L}_2(\beta, \gamma)$, be arranged to a sequence $\{f_k\}_{k=1}^\infty$. The functions

$$f_{k,l}(x, y) = f_k(x)\, \overline{f_l(y)} \quad (1 \leqslant k < \infty, 1 \leqslant l < \infty)$$

form a basis of $\mathfrak{L}_2(X)$ (cf. § 9 theorem 10). Expanding the kernel $a = a(x, y)$ into a Fourier series with respect to this basis (cf. § 8 remark 1) we get

$$a = \sum_{k=1}^{\infty} \sum_{l=1}^{\infty} \langle a, f_{k,l} \rangle f_{k,l}$$

where

$$\langle a, f_{k,l} \rangle = \int_{\beta}^{\gamma} \int_{\beta}^{\gamma} a(\xi, \eta) \, \overline{f_k(\xi)} \, f_l(\eta) \, d\xi \, d\eta$$

$$= \int_{\beta}^{\gamma} \left(\int_{\beta}^{\gamma} a(\xi, \eta) f_l(\eta) \, d\eta \right) \overline{f_k(\xi)} \, d\xi$$

$$= \begin{cases} \displaystyle\int_{\beta}^{\gamma} \lambda_n' e_n(\xi) \, \overline{e_m(\xi)} \, d\xi = \lambda_n' \delta_{m,n} & \text{if } f_k = e_m \in \mathfrak{E} \text{ and } f_l = e_n \in \mathfrak{E}, \\ 0 & \text{otherwise}. \end{cases}$$

As a consequence we get the following expansion formula (9) for a self-adjoint Hilbert-Schmidt kernel $a \in \mathfrak{L}_2(X)$.

THEOREM 3. *Let A be a selfadjoint Hilbert-Schmidt operator on $\mathfrak{L}_2(\beta, \gamma)$ with kernel $a \in \mathfrak{L}_2(X)$ $(X =]\beta, \gamma[\times]\beta, \gamma[)$. Let $\{e_k\}_{k=1}^{\kappa'}$ ($\kappa' < \infty$ or $\kappa' = \infty$) be an orthonormal family of eigenfunctions of A and let $(\lambda_k')_{k=1}^{\kappa'}$ be the sequence of corresponding eigenvalues satisfying (8). Then*

$$(9) \qquad\qquad a(x, y) = \sum_{k=1}^{\kappa'} \lambda_k' e_k(x) \, \overline{e_k(y)}.$$

In case of a selfadjoint Hilbert-Schmidt kernel $a \in \mathfrak{L}_2(X)$ there is a nice direct way to solve the Fredholm integral equation (1) if the eigenfunctions $\{e_k\}_{k=1}^{\kappa'}$ and the eigenvalues $\{\lambda_k'\}_{k=1}^{\kappa'}$ are known. Suppose f and g are two elements of $\mathfrak{L}_2(\beta, \gamma)$ connected by the equation

$$(1') \qquad\qquad (A - \lambda I) f = g.$$

Using as above the bases $\mathfrak{E} = \{e_k\}_{k=1}^{\kappa'}$ of $\mathfrak{M} = \overline{A\mathfrak{H}}$ and $\mathfrak{E}' = \{e_k'\}_{k=1}^{\kappa''}$ of $\mathfrak{M}_0 = \mathfrak{M}^{\perp}$ we get

$$(10) \qquad \begin{cases} \langle g, e_k \rangle = \langle Af, e_k \rangle - \lambda \langle f, e_k \rangle \\ \qquad\quad = \lambda_k' \langle f, e_k \rangle - \lambda \langle f, e_k \rangle \\ \qquad\quad = (\lambda_k' - \lambda) \langle f, e_k \rangle \quad \text{for all } e_k \in \mathfrak{E}, \\ \langle g, e_k' \rangle = \langle Af, e_k' \rangle - \lambda \langle f, e_k' \rangle \\ \qquad\quad = -\lambda \langle f, e_k' \rangle \qquad\quad \text{for all } e_k' \in \mathfrak{E}'. \end{cases}$$

Suppose first that $\lambda \neq 0$ is not an eigenvalue of A, i.e. $\lambda'_k - \lambda \neq 0$ for $1 \leqslant k_{(\leqq)} \kappa'$. From the Fredholm alternative (§ 27 theorem 9) we already know that then for every $g \in \mathfrak{L}_2(\beta, \gamma)$ a solution $f \in \mathfrak{L}_2(\beta, \gamma)$ of (1') must exist. For its Fourier coefficients with respect to the basis $\mathfrak{E} \cup \mathfrak{E}'$ of $\mathfrak{L}_2(\beta, \gamma)$ we get

$$\langle f, e_k \rangle = \frac{1}{\lambda'_k - \lambda} \langle g, e_k \rangle \quad \text{for all } e_k \in \mathfrak{E},$$

$$\langle f, e'_k \rangle = -\frac{1}{\lambda} \langle g, e'_k \rangle \quad \text{for all } e'_k \in \mathfrak{E}'.$$

We conclude

$$f = \sum_{k=1}^{\kappa'} \langle f, e_k \rangle e_k + \sum_{k=1}^{\kappa''} \langle f, e'_k \rangle e'_k$$

$$= \sum_{k=1}^{\kappa'} \frac{1}{\lambda'_k - \lambda} \langle g, e_k \rangle e_k - \frac{1}{\lambda} \sum_{k=1}^{\kappa''} \langle g, e'_k \rangle e'_k.$$

This may be rewritten in two ways. On the one hand we can complete the second term to the Fourier series of g/λ thus obtaining

$$f = -\frac{g}{\lambda} + \sum_{k=1}^{\kappa'} \frac{1}{\lambda} \langle g, e_k \rangle e_k + \sum_{k=1}^{\kappa'} \frac{1}{\lambda'_k - \lambda} \langle g, e_k \rangle e_k$$

$$= -\frac{g}{\lambda} + \sum_{k=1}^{\kappa'} \left(\frac{1}{\lambda} + \frac{1}{\lambda'_k - \lambda} \right) \langle g, e_k \rangle e_k$$

(11) $$= -\frac{g}{\lambda} + \frac{1}{\lambda} \sum_{k=1}^{\kappa'} \frac{\lambda'_k}{\lambda'_k - \lambda} \langle g, e_k \rangle e_k.$$

This essentially coincides with formula (5) which, however, was valid only for $|\lambda| > \|A\|$. On the other hand, let us denote by $\{\lambda_k\}_{k=1}^{\kappa}$ the sequence of *different* non-zero eigenvalues of A as in § 28 theorem 2, and by P_k the projection upon the eigenspace spanned by all functions in $\mathfrak{E}$ corresponding to the same eigenvalue λ_k. Furthermore, let P_0 be the projection upon $\mathfrak{M}_0$ and define $\lambda_0 = 0$. Then we have

$$\sum_{k=1}^{\kappa''} \langle g, e'_k \rangle e'_k = P_0 g,$$

$$\sum_{k=1}^{\kappa'} \frac{1}{\lambda'_k - \lambda} \langle g, e_k \rangle e_k = \sum_{k=1}^{\kappa} \frac{1}{\lambda_k - \lambda} P_k g,$$

(12) $$f = \sum_{k=0}^{\kappa} \frac{1}{\lambda_k - \lambda} P_k g.$$

This formula can also be obtained directly from the spectral decomposition of A (cf. § 28 theorem 2).

Suppose next that $\lambda \neq 0$ is an eigenvalue of A and therefore coincides with finitely many of the numbers λ'_k (cf. § 27 corollary 2.1). Let $\mathfrak{M}_\lambda$ be the corresponding finite-dimensional eigenspace and suppose $g \perp \mathfrak{M}_\lambda$ according to the Fredholm alternative. It is also seen from (10) that this is a necessary condition for the existence of a solution f of (1′). Moreover, the Fourier coefficients $\langle f, e_k \rangle$ $(e_k \in \mathfrak{E})$ are now determined by (10) only for $e_k \notin \mathfrak{M}_\lambda$. Assigning to the remaining finitely many Fourier coefficients arbitrary values we get by the same reasoning as above, corresponding to formula (11)

$$f = -\frac{g}{\lambda} + \frac{1}{\lambda} \sum_{\lambda'_k \neq \lambda} \frac{\lambda'_k}{\lambda'_k - \lambda} \langle g, e_k \rangle e_k + h, \quad h \in \mathfrak{M}_\lambda$$

where the series is taken over all $k (1 \leqslant k_{(} \leqslant_{)} \kappa')$ such that $\lambda'_k \neq \lambda$ (or, equivalently, $e_k \in \mathfrak{M}_\lambda^\perp$) and where $h \in \mathfrak{M}_\lambda$ is arbitrary. Corresponding to formula (12) we can also rewrite this in the form

$$f = \sum_{\lambda_k \neq \lambda} \frac{1}{\lambda_k - \lambda} P_k g + h, \quad h \in \mathfrak{M}_\lambda.$$

There remains the question how to find the eigenvalues λ'_k and the corresponding eigenfunctions e_k. According to § 28 remark 1 this may theoretically be accomplished in the following way. If $\{f_n\}_{n=1}^\infty$ is a sequence of unit vectors in $\mathfrak{L}_2(\beta, \gamma)$ such that

$$\lim_{n \to \infty} \langle Af_n, f_n \rangle = \pm \|A\| = \pm \sup_{\|f\|=1} |\langle Af, f \rangle|,$$

then we have
$$\lambda'_1 = \lim_{n \to \infty} \langle Af_n, f_n \rangle.$$

The bounded sequence $\{f_n\}_{n=1}^\infty$ certainly contains weakly converging subsequences and any such subsequence necessarily even converges strongly to an eigenfunction corresponding to the eigenvalue λ'_1. Choosing therefore any converging subsequence $\{f_{n_k}\}_{k=1}^\infty$ of $\{f_n\}_{n=1}^\infty$ we may put

$$e_1 = \lim_{k \to \infty} f_{n_k}.$$

If the eigenvalues $\lambda'_1, \ldots, \lambda'_k$ and corresponding eigenfunctions $e_1, \ldots, e_k$ have already been found then the subspace $\mathfrak{N}_k = \vee \{e_l\}_{l=1}^k$ reduces A (cf. § 21 corollary 4.1) and in $\mathfrak{N}_k^\perp$ the operator A induces a compact selfadjoint operator A_{k+1} with norm $\|A_{k+1}\| = |\lambda'_{k+1}|$ (cf. § 21 theorem 1 and § 28 corollary 2.1). The above procedure may then be repeated with the operator A_{k+1}

(given by the same Hilbert-Schmidt kernel a within $\mathfrak{N}_k^{\perp}$) to find λ'_{k+1} and a corresponding eigenfunction e_{k+1}.

The following exercise exhibits a special type of Hilbert-Schmidt operators which do not have any non-zero eigenvalues. At the same time it demonstrates that the series $\sum_{k=1}^{\infty} a^{(k)}/\lambda^{k+1}$ and $\sum_{k=1}^{\infty} A^k/\lambda^{k+1}$ may very well converge also for $|\lambda| \leqslant \alpha$, even for all $\lambda \neq 0$ (cf. theorem 2).

Exercise. Let $X = [\beta, \gamma] \times [\beta, \lambda]$ $(-\infty < \beta < \gamma < +\infty)$ and let A be a Hilbert-Schmidt operator on $\mathfrak{L}_2(\beta, \gamma)$ with a kernel $a = a(x, y) \in \mathfrak{L}_2(X)$ of *Volterra type*, i.e. $a(x, y)$ is continuous on the closed triangle in X below the diagonal $y = x$ and $a(x, y)$ vanishes on the complementary triangle in X above this diagonal.

a) The function Af is continuous on $[\beta, \gamma]$ for every $f \in \mathfrak{L}_2(\beta, \gamma)$. (Hint: observe that $\mathfrak{L}_2(\beta, \gamma) \subset \mathfrak{L}_1(\beta, \gamma)$ (cf. § 5 corollary 2.1) and use the fact that a continuous function a on a closed bounded subset of $\mathbf{R}^2$ is uniformly continuous, i.e. $|a(\xi_1, \eta_1) - a(\xi_2, \eta_2)| < \varepsilon$ if $(\xi_1 - \xi_2)^2 + (\eta_1 - \eta_2)^2 < \delta^2$.)

b) For every $k \geqslant 1$ the iterated kernel $a^{(k)}$ is again of Volterra type. (Hint: observe that a is bounded and use induction on k; cf. theorem 1.)

c) If $|a(\xi, \eta)| \leqslant \theta$ for all $(\xi, \eta) \in X$, then

$$|a^{(k)}(\xi, \eta)| \leqslant \theta^k \frac{(\xi - \eta)^{k-1}}{(k-1)!} \quad \text{for } \beta \leqslant \eta \leqslant \xi \leqslant \gamma$$

and for all $k \geqslant 1$. In particular, for $k \geqslant 2$ the iterated kernels $a^{(k)}$ are continuous on X. (Hint: use induction on k.)

d) For every fixed $\lambda \neq 0$ the series $\sum_{k=1}^{\infty} a^{(k)}/\lambda^{k+1}$ converges uniformly on X and therefore also in $\mathfrak{L}_2(X)$ to a Volterra kernel b. (Hint: use c).)

e) Every complex number $\lambda \neq 0$ is a regular value for A. (Hint: show that the series $\sum_{k=1}^{\infty} A^k/\lambda^{k+1}$ converges to the Hilbert-Schmidt operator B with the kernel b.)

CHAPTER VI

SPECTRAL ANALYSIS OF BOUNDED LINEAR OPERATORS

§ 30. The order relation for bounded selfadjoint operators

In order to prepare the way for the desired spectral decomposition of bounded selfadjoint operators in general we shall first discuss two auxiliary concepts. In the present section we shall fix our attention upon the fact that, just as for real numbers, it sometimes makes sense for two bounded self-adjoint operators on $\mathfrak{H}$, to say that one is less or greater than the other. In continuation thereof we shall see in the next sections that it also makes sense to use a bounded selfadjoint operator A on $\mathfrak{H}$ as an argument for a complex-valued function φ originally defined on $\mathbf{R}$: while $\varphi(\xi)$ is a complex number for $\xi \in \mathbf{R}$, $\varphi(A)$ will be a bounded normal operator on $\mathfrak{H}$.

Recall that a bounded linear operator A on $\mathfrak{H}$ is selfadjoint iff

$$\langle Af, f \rangle \in \mathbf{R} \quad \text{for all } f \in \mathfrak{H}$$

(cf. § 14 theorem 5) and that if A is selfadjoint, then

$$\|A\| = \sup_{\|f\|=1} |\langle Af, f \rangle| = \sup_{\|f\|=\|g\|=1} |\langle Af, g \rangle|$$

(cf. § 13 corollary 5.2, § 14 corollary 5.1).

DEFINITION 1. Let A and B be bounded selfadjoint operators on $\mathfrak{H}$. Then $A \leqslant B$ (in words: *A less than or equal to B*) or, equivalently, $B \geqslant A$ (in words: *B greater than or equal to A*) if

$$\langle Af, f \rangle \leqslant \langle Bf, f \rangle \quad \text{for all } f \in \mathfrak{H}.$$

In particular, A is called *positive* if $A \geqslant O$, i.e. if

$$\langle Af, f \rangle \geqslant 0 \quad \text{for all } f \in \mathfrak{H}.$$

Remark 1. Note that $A \leqslant B$ iff $O \leqslant B - A$. Indeed, we have $\langle Af, f \rangle \leqslant \langle Bf, f \rangle$ for all $f \in \mathfrak{H}$ iff $0 \leqslant \langle (B-A)f, f \rangle$ for all $f \in \mathfrak{H}$.

Example 1. Let P and Q be the projections upon the subspaces $\mathfrak{M}$ and $\mathfrak{N}$ respectively. Then we have $O \leqslant P \leqslant I$ and $O \leqslant Q \leqslant I$ since $0 \leqslant \langle Pf, f \rangle =$

$= \|Pf\|^2 \leqslant \langle f, f \rangle = \|f\|^2$ (similarly for Q). Furthermore we have $P \leqslant Q$ iff $\mathfrak{M} \subset \mathfrak{N}$ by § 15 theorem 6 a), e). At the same time this implies that if $\mathfrak{M} \not\subset \mathfrak{N}$ and $\mathfrak{N} \not\subset \mathfrak{M}$, then we have neither $P \leqslant Q$ nor $Q \leqslant P$. In other words it may very well happen that two bounded selfadjoint operators on $\mathfrak{H}$ are not comparable with each other by means of the *order relation* "$\leqslant$".

Example 2. If A is any bounded linear operator on $\mathfrak{H}$, then A^*A and AA^* are positive selfadjoint operators on $\mathfrak{H}$. Indeed, we have $(A^*A)^* = A^*A^{**} = = A^*A$, $(AA^*)^* = AA^*$, and

$$\left. \begin{array}{l} \langle A^*Af, f \rangle = \langle Af, Af \rangle \geqslant 0 \\ \langle AA^*f, f \rangle = \langle A^*f, A^*f \rangle \geqslant 0 \end{array} \right\} \quad \text{for all } f \in \mathfrak{H}.$$

However, the two operators A^*A and AA^* will not be comparable with each other in general.

The following theorem justifies the use of the familiar notation "less than or equal to" for the relation between bounded selfadjoint operators on $\mathfrak{H}$ as defined above.

THEOREM 1. *Let A, B, C be bounded selfadjoint operators on $\mathfrak{H}$ and let α, β be real numbers.*

a) $A \leqslant A$.
b) *If $A \leqslant B$ and $B \leqslant C$, then $A \leqslant C$.*
c) *If $A \leqslant B$ and $B \leqslant A$, then $A = B$.*
d) *If $A \leqslant B$ and $\alpha \geqslant 0$, then*

$$A + C \leqslant B + C,$$
$$\alpha A \leqslant \alpha B,$$
$$-A \geqslant -B.$$

e) *If $\alpha \leqslant \beta$, then $\alpha A \leqslant \beta A$.*

PROOF. The assertions a), b), d), and e) follow immediately from definition 1.
  c) Define the bilinear forms φ and ψ on $\mathfrak{H}$ by

$$\varphi(f, g) = \langle Af, g \rangle,$$
$$\psi(f, g) = \langle Bf, g \rangle$$

(cf. § 13 corollary 5.1 a)). The hypothesis implies for the associated quadratic forms $\hat{\varphi}$ and $\hat{\psi}$ that

$$\hat{\varphi}(f) = \hat{\psi}(f) \quad \text{for all } f \in \mathfrak{H}.$$

Using the polar identity (cf. § 13 corollary 1.1) we conclude

$$\varphi(f, g) = \psi(f, g) \quad \text{for all } f \in \mathfrak{H} \text{ and all } g \in \mathfrak{H},$$
$$\langle (A - B) f, g \rangle = 0 \quad \text{for all } f \in \mathfrak{H} \text{ and all } g \in \mathfrak{H},$$
$$A - B = O. \; \square$$

Let us see whether the analogy between real numbers and bounded self-adjoint operators may still be extended. It is well known that a bounded monotone sequence of real numbers converges. What about a "bounded" (whatever this may mean) monotone sequence of bounded selfadjoint operators on $\mathfrak{H}$? Before answering this question we state two auxiliary results.

THEOREM 2 (GENERALIZED CAUCHY INEQUALITY). *Let A be a positive bounded selfadjoint operator on $\mathfrak{H}$. Then*

$$|\langle Af, g \rangle|^2 \leqslant \langle Af, f \rangle \langle Ag, g \rangle,$$

PROOF. Define the bilinear form φ on $\mathfrak{H}$ as above by

$$\varphi(f, g) = \langle Af, g \rangle.$$

Since A is selfadjoint φ has all properties required for an inner product on $\mathfrak{H}$ (cf. § 2 definition 4) except that it may well happen that $\varphi(f, f) = 0$ for some $f \neq o$. The assertion then follows from § 2 remark 3. $\square$

THEOREM 3. *Let A be a positive selfadjoint operator on $\mathfrak{H}$. Then*

$$\|Af\|^2 \leqslant \|A\| \langle Af, f \rangle.$$

PROOF. Applying theorem 2 with $g = Af$ we obtain

$$\|Af\|^4 = \langle Af, Af \rangle^2 \leqslant \langle Af, f \rangle \langle A^2 f, Af \rangle$$
$$\leqslant \langle Af, f \rangle \|A^2 f\| \|Af\| \leqslant \langle Af, f \rangle \|A\| \|Af\|^2.$$

The assertion follows after cancellation of $\|Af\|^2$ on both sides (if $Af = o$, then the assertion is trivial). $\square$

THEOREM 4. *Let $A_n (1 \leqslant n < \infty)$ and B be bounded selfadjoint operators on $\mathfrak{H}$ such that*

$$A_1 \leqslant A_2 \leqslant \cdots \leqslant A_n \leqslant \cdots \leqslant B.$$

Then there exists a bounded selfadjoint operator A on $\mathfrak{H}$ such that

$$A_n \leqslant A \leqslant B \quad \text{for all } n \geqslant 1,$$
$$\lim_{n \to \infty} A_n f = Af \quad \text{for all } f \in \mathfrak{H}.$$

PROOF. Note that $B - A_1 \geqslant A_m - A_n \geqslant O$ for $m \geqslant n \geqslant 1$ and

$$\|A_m - A_n\| = \sup_{\|f\|=1} \langle (A_m - A_n) f, f \rangle$$
$$\leqslant \sup_{\|f\|=1} \langle (B - A_1) f, f \rangle = \|B - A_1\|.$$

For any fixed $f \in \mathfrak{H}$ we have

$$\langle A_1 f, f \rangle \leqslant \langle A_2 f, f \rangle \leqslant \cdots \leqslant \langle A_n f, f \rangle \leqslant \cdots \leqslant \langle B f, f \rangle,$$

$$\lim_{m, n \to \infty} \langle (A_m - A_n) f, f \rangle = 0,$$

$$\|(A_m - A_n) f\|^2 \leqslant \|A_m - A_n\| \langle (A_m - A_n) f, f \rangle \quad \text{(by theorem 3)}$$
$$\leqslant \|B - A_1\| \langle (A_m - A_n) f, f \rangle \quad \text{for } m \geqslant n \geqslant 1,$$

$$\lim_{m, n \to \infty} \|A_m f - A_n f\| = 0.$$

Therefore the sequence $\{A_n f\}_{n=1}^{\infty}$ converges to some vector which, in anticipation of the result, we shall denote by Af:

$$\lim_{n \to \infty} A_n f = Af.$$

Since

$$A(\alpha f + \beta g) = \lim_{n \to \infty} A_n(\alpha f + \beta g) = \alpha \lim_{n \to \infty} A_n f + \beta \lim_{n \to \infty} A_n g$$
$$= \alpha Af + \beta Ag,$$

A is a linear operator on $\mathfrak{H}$. Since

$$\langle Af, g \rangle = \lim_{n \to \infty} \langle A_n, g \rangle = \lim_{n \to \infty} \langle f, A_n g \rangle = \langle f, Ag \rangle,$$

A is selfadjoint and consequently bounded by § 17 corollary 8.2. The remaining assertion follows from

$$\langle A_n f, f \rangle \leqslant \lim_{m \to \infty} \langle A_m f, f \rangle = \langle Af, f \rangle \leqslant \langle B f, f \rangle \quad \text{for all } f \in \mathfrak{H}. \;\; \square$$

Remark 2. An analogous assertion holds if the sequence $\{A_n\}_{n=1}^{\infty}$ is decreasing and bounded below by B. In fact, this case may be reduced to the case treated in theorem 4 by switching to the negative sequence $\{-A_n\}_{n=1}^{\infty}$.

Example 3. Let $\{e_k\}_{k=1}^{\infty}$ be a basis of a separable Hilbert space $\mathfrak{H}$ and let P_n be the projection upon the subspace $\mathfrak{M}_n = \bigvee \{e_k\}_{k=1}^{n}$. Then we have

$$\mathfrak{O} \subset \mathfrak{M}_1 \subset \mathfrak{M}_2 \subset \cdots \subset \mathfrak{M}_n \subset \cdots \subset \mathfrak{H}$$

and therefore, by what has been said in example 1,

$$O \leqslant P_1 \leqslant P_2 \leqslant \cdots \leqslant P_n \leqslant \cdots \leqslant I$$

(recall that O and I are the projections upon $\mathfrak{O}$ and $\mathfrak{H}$ respectively). From

$$P_n f = \sum_{k=1}^{n} \langle f, e_k \rangle e_k$$

we conclude

$$\lim_{n \to \infty} P_n f = \sum_{k=1}^{\infty} \langle f, e_k \rangle e_k = f = If \, .$$

In view of theorem 4 we would be inclined to say that the monotone sequence of projections $\{P_n\}_{n=1}^{\infty}$ in some sense "converges" to I. Note, however, that $I - P_n$ is the projection upon the subspace $\mathfrak{M}_n^{\perp} = \bigvee \{e_k\}_{k=n+1}^{\infty} \neq \mathfrak{O}$ (cf. § 15 theorem 3) and therefore

$$\|I - P_n\| = 1 \quad \text{for all } n \geqslant 1 \, .$$

The sequence $\{P_n\}_{n=1}^{\infty}$ therefore does not converge to I in the sense of the norm in $\mathcal{B}$.

Theorem 4 and example 3 motivate the following definition:

DEFINITION 2. A sequence $\{A_n\}_{n=1}^{\infty} \subset \mathcal{B}$ converges strongly to an operator $A \in \mathcal{B}$, called the strong limit of the sequence $\{A_n\}_{n=1}^{\infty}$ (notation: (s) $\lim_{n \to \infty} A_n = A$) if

$$\lim_{n \to \infty} A_n f = Af \quad \text{for all } f \in \mathfrak{H} \, .$$

Note that the very definition of the strong limit of a sequence $\{A_n\}_{n=1}^{\infty} \subset \mathcal{B}$ already implies that it is uniquely determined by the sequence $\{A_n\}_{n=1}^{\infty}$. We shall reserve the notation

$$\lim_{n \to \infty} A_n = A$$

without any further attribute to denote convergence in norm, i.e.

$$\lim_{n \to \infty} \|A - A_n\| = 0 \, .$$

We shall also call this "uniform convergence" as opposed to strong convergence. From

$$\|A_m f - A_n f\| \leqslant \|A_m - A_n\| \, \|f\|$$

it follows that uniform convergence of the sequence $\{A_n\}_{n=1}^{\infty}$ to A implies strong convergence of $\{A_n\}_{n=1}^{\infty}$ to A. The converse of this assertion is false, as demonstrated in example 3 (there the sequence $\{P_n\}_{n=1}^{\infty}$ converges strongly but not uniformly to I). Theorem 4 and remark 2 assert that in any case a bounded monotone sequence of bounded selfadjoint operators on $\mathfrak{H}$ converges strongly to some bounded selfadjoint operator on $\mathfrak{H}$.

The distinction between uniform and strong convergence (of operators) in $\mathscr{B}$ resembles the distinction between strong and weak convergence (of vectors) in $\mathfrak{H}$. In fact it is possible to introduce yet another concept of "*weak convergence*" in $\mathscr{B}$ by defining

$$\text{(w)} \lim_{n \to \infty} A_n = A$$

if and only if

$$\lim_{n \to \infty} \langle A_n f, g \rangle = \langle Af, g \rangle \quad \text{for all } f \in \mathfrak{H} \text{ and all } g \in \mathfrak{H}.$$

We shall, however, not use this notion of weak convergence in $\mathscr{B}$ and we therefore omit a further discussion of it.

The following theorem generalizes the assertion of example 3.

THEOREM 5. *Let $\{P_n\}_{n=1}^{\infty}$ be an increasing (or decreasing) sequence of projections upon corresponding subspaces $\mathfrak{M}_n$ ($1 \leqslant n < \infty$). The strong limit of the sequence $\{P_n\}_{n=1}^{\infty}$ is the projection upon the subspace $\bigvee_{n=1}^{\infty} \mathfrak{M}_n$ (or $\bigcap_{n=1}^{\infty} \mathfrak{M}_n$ respectively).*

PROOF. Suppose

$$P_1 \leqslant P_2 \leqslant \cdots \leqslant P_n \leqslant \cdots \leqslant I$$

(cf. example 1) and let

$$\text{(s)} \lim_{n \to \infty} P_n = P$$

where P is a bounded selfadjoint operator on $\mathfrak{H}$ (cf. theorem 4). For any $f \in \mathfrak{H}$ we have

$$P_m P_n f = P_n f \quad \text{for } m \geqslant n \geqslant 1$$

(cf. § 15 theorem 6 a), b)) and therefore

$$PP_n f = \lim_{m \to \infty} P_m P_n f = P_n f,$$

$$P^2 f = P \lim_{n \to \infty} P_n f = \lim_{n \to \infty} PP_n f = \lim_{n \to \infty} P_n f$$

$$= Pf.$$

We conclude $P^2 = P$ and therefore P is a projection upon a subspace $\mathfrak{M}$ (cf. § 15 theorem 2).

If Q is the projection upon $\bigvee_{n=1}^{\infty} \mathfrak{M}_n$, then we have $P_n \leqslant Q$ for all $n \geqslant 1$ and therefore by theorem 4

$$P_n \leqslant P \leqslant Q \qquad \text{for all } n \geqslant 1,$$

$$\mathfrak{M}_n \subset \mathfrak{M} \subset \bigvee_{m=1}^{\infty} \mathfrak{M}_m \quad \text{for all } n \geqslant 1.$$

This implies $\mathfrak{M} = \bigvee_{n=1}^{\infty} \mathfrak{M}_n$.

Suppose next

$$P_1 \geqslant P_2 \geqslant \cdots \geqslant P_n \geqslant \cdots \geqslant O$$

(cf. example 1). The sequence $\{P_n^{\perp}\}_{n=1}^{\infty} = \{I - P_n\}_{n=1}^{\infty}$ increases and therefore converges strongly to the projection Q upon the subspace $\bigvee_{n=1}^{\infty} (\mathfrak{M}_n^{\perp})$. The sequence $\{P_n\}_{n=1}^{\infty} = \{I - P_n^{\perp}\}_{n=1}^{\infty}$ then converges strongly to the projection $P = I - Q$ upon the orthogonal complement of $\bigvee_{n=1}^{\infty} (\mathfrak{M}_n^{\perp})$. This complement consists of all vectors which are orthogonal to every subspace $\mathfrak{M}_n^{\perp}$ or, equivalently, contained in every subspace $\mathfrak{M}_n$. This proves the alternative assertion. $\square$

What has the order relation got to do with the spectrum of a bounded selfadjoint operator? A first answer is contained in the following three theorems. The proof of theorem 8 presented below has been communicated to me by F. TAKENS (Amsterdam).

THEOREM 6. *Let A be a bounded selfadjoint operator on $\mathfrak{H}$ and let $\alpha_1 = \inf_{\|f\|=1} \langle Af, f \rangle$, $\alpha_2 = \sup_{\|f\|=1} \langle Af, f \rangle$. Then*

$$\alpha_1 = \inf_{\|f\|=1} \langle Af, f \rangle = \max \{\alpha \in \mathbf{R}: \alpha I \leqslant A\},$$

$$\alpha_2 = \sup_{\|f\|=1} \langle Af, f \rangle = \min \{\alpha \in \mathbf{R}: A \leqslant \alpha I\}$$

and $\|A\| = \max \{|\alpha_1|, |\alpha_2|\}$.

PROOF. The last assertion follows from $\|A\| = \sup_{\|f\|=1} |\langle Af, f \rangle|$. We shall prove the first equality in the theorem, the second one being demonstrated similarly. Suppose $\alpha \in \mathbf{R}$ and $\alpha I \leqslant A$. We conclude

$$\alpha \langle f, f \rangle \leqslant \langle Af, f \rangle \quad \text{for all } f \in \mathfrak{H},$$

$$\alpha \leqslant \inf_{\|f\|=1} \langle Af, f \rangle = \alpha_1,$$

(1) $$\qquad \sup \{\alpha \in \mathbf{R}: \alpha I \leqslant A\} \leqslant \alpha_1.$$

On the other hand, by the definition of α_1 we have

$$\alpha_1 \leqslant \langle Af/\|f\|, f/\|f\| \rangle \quad \text{for all } f \neq o,$$
$$\alpha_1 \langle f, f \rangle \leqslant \langle Af, f \rangle \quad \text{for all } f \in \mathfrak{H},$$
$$\alpha_1 I \leqslant A.$$

Combining this with (1) we get

$$\sup \{\alpha \in \mathbf{R}: \alpha I \leqslant A\} = \alpha_1 = \max \{\alpha \in \mathbf{R}: \alpha I \leqslant A\}. \quad \square$$

THEOREM 7. *Let A, α_1, α_2 be as in theorem 6. Then $\alpha_1 \in \Sigma(A)$ and $\alpha_2 \in \Sigma(A)$.*
PROOF. Since $\alpha_1 = \max\{\alpha \in \mathbf{R}: \alpha I \leqslant A\}$, for every number $\alpha_1 + 1/n > \alpha_1$ $(1 \leqslant n < \infty)$ there must exist a vector $f_n \in \mathfrak{H}$ such that

$$(\alpha_1 + 1/n) \langle f_n, f_n \rangle > \langle Af_n, f_n \rangle.$$

Since f_n is necessarily non-zero we may without loss of generality assume $\|f_n\| = 1$ for all $n \geqslant 1$. We conclude

$$0 \leqslant \langle (A - \alpha_1 I) f_n, f_n \rangle = \langle [A - (\alpha_1 + 1/n) I] f_n, f_n \rangle + \frac{1}{n} \langle f_n, f_n \rangle < \frac{1}{n},$$

$$(2) \qquad \lim_{n \to \infty} \langle (A - \alpha_1 I) f_n, f_n \rangle = 0.$$

Suppose now α_1 were a regular value for A, i.e. $(A - \alpha_1 I)^{-1}$ were a bounded linear operator on $\mathfrak{H}$. Then for every $n \geqslant 1$ we would have

$$1 = \|f_n\|^2 = \|(A - \alpha_1 I)^{-1} (A - \alpha_1 I) f_n\|^2$$
$$\leqslant \|(A - \alpha_1 I)^{-1}\|^2 \|(A - \alpha_1 I) f_n\|^2$$
$$\leqslant \|(A - \alpha_1 I)^{-1}\|^2 \|A - \alpha_1 I\| \langle (A - \alpha_1 I) f_n, f_n \rangle$$

(the last inequality follows from theorem 3 since $(A - \alpha_1 I)$ is a positive self-adjoint operator). This, however, contradicts (2).

By the same token the number

$$- \alpha_2 = \max \{\alpha \in \mathbf{R}: \alpha I \leqslant - A\}$$

belongs to the spectrum of $(-A)$. In other words, the operator $(-A + \alpha_2 I) = -(A - \alpha_2 I)$ is not invertible. As a consequence we have $\alpha_2 \in \Sigma(A)$. $\quad \square$

COROLLARY 7.1. *The spectrum of a bounded selfadjoint operator on $\mathfrak{H}$ is not empty.*

Remark 3. The fact that $\Sigma(A)$ is not empty for any bounded linear operator A on $\mathfrak{H}$ has already been mentioned in § 23 remark 2. For bounded normal operators on $\mathfrak{H}$ and for unbounded selfadjoint operators in $\mathfrak{H}$ this can also be deduced from the corresponding spectral theorems § 37 theorem 1 and § 39 theorem 5 (cf. § 34 theorem 4).

THEOREM 8. *Let A, α_1, α_2 be as in theorem 6. Then $\Sigma(A) \subset [\alpha_1, \alpha_2]$.*
PROOF. Observe that the real number λ belongs to the spectrum of A iff the real number $\lambda' = \lambda - \frac{1}{2}(\alpha_1 + \alpha_2)$ belongs to the spectrum of the operator $A' = A - \frac{1}{2}(\alpha_1 + \alpha_2) I$ since

$$A - \lambda I = \left(A - \tfrac{1}{2}(\alpha_1 + \alpha_2) I\right) - \left(\lambda - \tfrac{1}{2}(\alpha_1 + \alpha_2)\right) I.$$

Using the notation of theorem 6 also for the operator A' we have

$$\alpha_1' = \max\{\alpha \in \mathbf{R}: \alpha I \leqslant A'\} = \max\{\alpha \in \mathbf{R}: \alpha I \leqslant A - \tfrac{1}{2}(\alpha_1 + \alpha_2) I\}$$
$$= \max\{\alpha \in \mathbf{R}: \left(\alpha + \tfrac{1}{2}(\alpha_1 + \alpha_2)\right) I \leqslant A\} = \alpha_1 - \tfrac{1}{2}(\alpha_1 + \alpha_2) = \tfrac{1}{2}(\alpha_1 - \alpha_2),$$
$$\alpha_2' = \min\{\alpha \in \mathbf{R}: A' \leqslant \alpha I\} = \alpha_2 - \tfrac{1}{2}(\alpha_1 + \alpha_2) = \tfrac{1}{2}(\alpha_2 - \alpha_1),$$
$$\|A'\| = \max\{|\alpha_1'|, |\alpha_2'|\} = \tfrac{1}{2}(\alpha_2 - \alpha_1).$$

If λ' belongs to the spectrum of A', then by § 23 corollary 3.2 we have

$$|\lambda'| = |\lambda - \tfrac{1}{2}(\alpha_1 + \alpha_2)| \leqslant \|A'\| = \tfrac{1}{2}(\alpha_2 - \alpha_1)$$

which implies

$$\alpha_1 = \tfrac{1}{2}(\alpha_1 + \alpha_2) - \tfrac{1}{2}(\alpha_2 - \alpha_1) \leqslant \lambda \leqslant \tfrac{1}{2}(\alpha_1 + \alpha_2) + \tfrac{1}{2}(\alpha_2 - \alpha_1) = \alpha_2. \quad \square$$

COROLLARY 8.1. *Let A, α_1, α_2 be as in theorem 6. Then*

$$\alpha_1 = \min\{\lambda: \lambda \in \Sigma(A)\} = \min \Sigma(A),$$
$$\alpha_2 = \max\{\lambda: \lambda \in \Sigma(A)\} = \max \Sigma(A),$$
$$\|A\| = \max\{|\lambda|: \lambda \in \Sigma(A)\}.$$

Exercise. Let A be a bounded selfadjoint operator on $\mathfrak{H}$ and let $\alpha_1 = \min \Sigma(A)$, $\alpha_2 = \max \Sigma(A)$.

a) A is positive iff $\alpha_1 \geqslant 0$.
b) A is positive and invertible iff $\alpha_1 > 0$.
c) If $\alpha_1 > 0$, then A^{-1} is a positive selfadjoint operator and

$$\min \Sigma(A^{-1}) = \alpha_2^{-1},$$
$$\max \Sigma(A^{-1}) = \alpha_1^{-1}.$$

(Hint: for $0 < \alpha < \alpha_1$ we have $A - \alpha I = -\alpha A(A^{-1} - \alpha^{-1} I)$.)

§ 31. Polynomials in a bounded operator

In the course of the following sections we shall have to introduce several classes of complex-valued functions defined on $\mathbf{C}$ or $\mathbf{R}$. If the sum, scalar multiple, and product of functions are defined pointwise as usual, then these classes become commutative algebras with the constant function 1 as a unit (cf. § 12 definition 3). Since the functions in question will not be considered as elements of a Hilbert space we shall use for them the notation $\varphi(x)$, $\psi(x)$. Furthermore we shall adhere to the notation

$$(\varphi + \psi)(x) = \varphi(x) + \psi(x),$$
$$(\lambda\varphi)(x) = \lambda\varphi(x),$$
$$(\varphi\psi)(x) = \varphi(x)\,\psi(x)$$

for sum, scalar multiple, and product of these functions, and we shall denote by $\bar{\varphi}(x)$ the complex conjugate function of $\varphi(x)$ (defined by $\bar{\varphi}(\xi) = \overline{\varphi(\xi)}$ for all ξ in the domain of $\varphi(x)$; cf. § 3 remark 1).

As a first class of functions we consider the algebra $\mathscr{P}$ of all polynomials (in one variable) with complex coefficients:

$$\mathscr{P} = \{\varphi(x) = \sum_{k=0}^{n} \alpha_k x^k : \quad \alpha_k \in \mathbf{C} \text{ for } 0 \leqslant k \leqslant n, \, n \geqslant 0\}.$$

THEOREM 1. *Let A be a bounded linear operator on $\mathfrak{H}$ and for*

$$\varphi(x) = \sum_{k=0}^{n} \alpha_k x^k \in \mathscr{P} \quad (x^0 = 1)$$

define

$$\varphi(A) = \sum_{k=0}^{n} \alpha_k A^k \in \mathscr{B} \quad (A^0 = I),$$

$$\bar{\varphi}(A) = \sum_{k=0}^{n} \bar{\alpha}_k A^{*k} \in \mathscr{B}.$$

The mapping $\varphi(x) \to \varphi(A)$ has the following properties:

a) $(\varphi + \psi)(A) = \varphi(A) + \psi(A)$;
b) $(\lambda\varphi)(A) = \lambda\varphi(A)$;
c) $(\varphi\psi)(A) = \varphi(A)\,\psi(A)$;
d) $[\varphi(A)]^* = \bar{\varphi}(A)$.

PROOF. a), b), c) Trivial.

d) By § 14 theorem 2 we get

$$[\varphi(A)]^* = \left[\sum_{k=0}^{n} \alpha_k A^k\right]^* = \sum_{k=0}^{n} \bar{\alpha}_k A^{*k} = \bar{\varphi}(A). \quad \square$$

Remark 1. Note that $\varphi(A)\psi(A)=\psi(A)\varphi(A)$ and that a constant polynomial $\varphi(x)=\alpha_0=\alpha_0 x^0$ is mapped into the operator $\varphi(A)=\alpha_0 I$.

Since we shall come across mappings with properties analogous to a), b), c) also in the following sections let us introduce here the proper terminology.

DEFINITION 1. A mapping $a \to a'$ of an algebra $\mathfrak{A}$ into an algebra $\mathfrak{A}'$ is called a *homomorphism* if it has the following properties:

  a) $(a+b)'=a'+b'$;
  b) $(\lambda a)'=\lambda a'$;
  c) $(ab)'=a'b'$.

Theorem 1 in this terminology asserts that the mapping which associates with any polynomial $\varphi(x)$ the operator $\varphi(A)$ is a homomorphism of $\mathscr{P}$ into $\mathscr{B}$ enjoying the additional property d). Another important property of this mapping concerns the spectrum of the operator $\varphi(A)$.

THEOREM 2. *If $A \in \mathscr{B}$ and $\varphi(x) \in \mathscr{P}$ are given, then*

$$\Sigma(\varphi(A)) = \{\varphi(\xi) \colon \xi \in \Sigma(A)\}.$$

In other words, the spectrum of $\varphi(A)$ consists precisely of all values which $\varphi(x)$ assumes on the spectrum of A; in short

$$\Sigma(\varphi(A)) = \varphi(\Sigma(A)).$$

PROOF. The assertion is trivial if $\varphi(x)$ is a constant polynomial. Suppose therefore $n \geqslant 1$ and $\alpha_n \neq 0$.

  a) Let $\xi \in \Sigma(A)$ be given. Since ξ is a root of the polynomial $\varphi(x)-\varphi(\xi)$ we have

$$\varphi(x) - \varphi(\xi) = (x - \xi)\psi(x), \quad \psi(x) \in \mathscr{P},$$
$$\varphi(A) - \varphi(\xi)I = (A - \xi I)\psi(A) = \psi(A)(A - \xi I)$$

(cf. theorem 1 and remark 1). If $\varphi(\xi)$ were a regular value for $\varphi(A)$, then from

$$[\varphi(A) - \varphi(\xi)I]\psi(A) = \psi(A)[\varphi(A) - \varphi(\xi)I],$$

we would conclude

$$\psi(A)[\varphi(A) - \varphi(\xi)I]^{-1} = [\varphi(A) - \varphi(\xi)I]^{-1}\psi(A),$$
$$I = (A - \xi I)\psi(A)[\varphi(A) - \varphi(\xi)I]^{-1}$$
$$= [\varphi(A) - \varphi(\xi)I]^{-1}\psi(A)(A - \xi I)$$
$$= \psi(A)[\varphi(A) - \varphi(\xi)I]^{-1}(A - \xi I).$$

This, however, would imply that $(A - \xi I)$ is invertible (cf. § 12 definition 5) in contradiction to our assumption. We conclude $\varphi(\xi) \in \Sigma(\varphi(A))$.

b) Conversely, suppose the number η belongs to the spectrum of $\varphi(A)$, i.e. $\eta \in \Sigma(\varphi(A))$. We now have to show that there exists a number $\xi \in \Sigma(A)$ such that $\eta = \varphi(\xi)$. To this end, let

$$\varphi(x) - \eta = \alpha_n(x - \xi_1) \cdots (x - \xi_n), \quad \alpha_n \neq 0$$

be the decomposition of the polynomial $\varphi(x) - \eta$ into a product of linear factors ($\xi_1, \ldots, \xi_n$ are precisely the roots of this polynomial). By theorem 1 we have

$$\varphi(A) - \eta I = \alpha_n(A - \xi_1 I) \cdots (A - \xi_n I).$$

If all roots $\xi_k (1 \leqslant k \leqslant n)$ were regular values for A, then every factor $(A - \xi_k I)$ would be invertible and the operator

$$(A - \xi_n I)^{-1} \cdots (A - \xi_1 I)^{-1}$$

would be an inverse of the operator $\varphi(A) - \eta I$. This, however, contradicts our assumption $\eta \in \Sigma(\varphi(A))$. For some root $\xi = \xi_k$ we therefore have $\xi \in \Sigma(A)$. Since $\varphi(\xi) - \eta = 0$ this completes the proof. □

COROLLARY 2.1. *Let $A \in \mathscr{B}$ be selfadjoint and suppose $\varphi(x) \in \mathscr{P}$ has real coefficients. Then $\varphi(A)$ is selfadjoint and*

(1) $$\|\varphi(A)\| = \max\{|\varphi(\xi)| : \xi \in \Sigma(A)\}.$$

PROOF. The operator $\varphi(A)$ is selfadjoint since by theorem 1 d) we have

$$[\varphi(A)]^* = \bar{\varphi}(A) = \varphi(A).$$

By theorem 2 the spectrum of $\varphi(A)$ coincides with the set $\{\varphi(\xi) : \xi \in \Sigma(A)\}$. It only remains to apply § 30 corollary 8.1. □

Exercise. Let $A \in \mathscr{B}$ and $\varphi(x) \in \mathscr{P}$ be given.

a) $\varphi(A)$ is invertible iff $\varphi(\xi) \neq 0$ for all $\xi \in \Sigma(A)$.

b) If $\varphi(A)$ is invertible, then

$$\Sigma(\varphi(A)^{-1}) = \{\varphi(\xi)^{-1} : \xi \in \Sigma(A)\}.$$

(Hint: for $\eta \neq 0$ we have $\varphi(A)^{-1} - \eta I = -\eta \varphi(A)^{-1}[\varphi(A) - \eta^{-1} I]$.)

§ 32. Continuous functions of a bounded selfadjoint operator

In the present section and in the following one we shall fix our attention upon a *selfadjoint* operator $A \in \mathscr{B}$. In order to simplify the notation, if $\varphi(x)$

is any function on **R** we define

$$\|\varphi(x)\|_A = \sup\{|\varphi(\xi)| : \xi \in \Sigma(A)\}$$

(cf. § 3 example 1 and fig. 18).

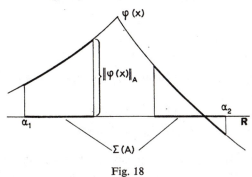

Fig. 18

If $\varphi(x)$ is continuous, in particular if $\varphi(x)$ is a polynomial, then the supremum is actually assumed for some point $\xi \in \Sigma(A)$ since $\Sigma(A)$ is a bounded and closed subset of **R** (cf. § 23 corollary 3.2 and § 23 corollary 4.1). The supremum may then be written as a maximum as already done in § 31 corollary 2.1. In fact, formula (1) in this corollary may now be written in the form

$$\|\varphi(A)\| = \|\varphi(x)\|_A,$$

$\varphi(x)$ being a polynomial with real coefficients.

Note that for bounded functions $\varphi(x)$, restricted to $\Sigma(A)$, $\|\ \|_A$ has all properties of a norm (cf. § 3 definition 1). If $\varphi(x)$ and $\varphi_n(x)$ $(1 \leqslant n < \infty)$ are functions on **R** which are bounded on $\Sigma(A)$ and satisfy

$$\lim_{n \to \infty} \|\varphi(x) - \varphi_n(x)\|_A = 0,$$

then we therefore have

$$\|\varphi(x)\|_A = \lim_{n \to \infty} \|\varphi_n(x)\|_A$$

(cf. § 3 theorem 6 b)).

Considering the polynomials turning up in § 31 as functions on **R** we shall now extend the correspondence between these polynomials $\varphi(x)$ and the operators $\varphi(A)$ to a larger class of functions, namely to the algebra $\mathscr{S}$ of all continuous complex-valued functions on **R**:

$$\mathscr{S} = \{\varphi(x) : \varphi(x) \text{ is a continuous complex-valued function on } \mathbf{R}\}.$$

THEOREM 1. *Let the selfadjoint operator* $A \in \mathscr{B}$ *and the function* $\varphi(x) \in \mathscr{S}$ *be given. Then there exists a unique operator* $\varphi(A) \in \mathscr{B}$ *with the following property: whenever* $\{\varphi_n(x)\}_{n=1}^{\infty}$ *is a sequence (of polynomials) in* $\mathscr{P}$ *such that*

(1)
$$\lim_{n \to \infty} \| \varphi(x) - \varphi_n(x) \|_A = 0,$$

then

$$\varphi(A) = \lim_{n \to \infty} \varphi_n(A).$$

The mapping $\varphi(x) \to \varphi(A)$ *is a homomorphism of the algebra* $\mathscr{S}$ *into* $\mathscr{B}$ *with the additional properties*

$$[\varphi(A)]^* = \bar{\varphi}(A),$$
$$\| \varphi(A) \| \leqslant 2 \| \varphi(x) \|_A.$$

PROOF. Let $\{\varphi_n(x)\}_{n=1}^{\infty}$ be any sequence in $\mathscr{P}$ satisfying (1). (Note that by WEIERSTRASS' approximation theorem in the second formulation given in appendix B1 there certainly exists such a sequence which converges to $\varphi(x)$ uniformly on the interval $[\alpha_1 = \min \Sigma(A), \alpha_2 = \max \Sigma(A)]$ and which therefore satisfies (1).) Let

$$\varphi^{(1)}(x) = \tfrac{1}{2}\{\varphi(x) + \bar{\varphi}(x)\} \in \mathscr{S}, \quad \varphi^{(2)}(x) = \frac{1}{2i}\{\varphi(x) - \bar{\varphi}(x)\} \in \mathscr{S},$$

$$\varphi_n^{(1)}(x) = \tfrac{1}{2}\{\varphi_n(x) + \bar{\varphi}_n(x)\} \in \mathscr{P}, \quad \varphi_n^{(2)}(x) = \frac{1}{2i}\{\varphi_n(x) - \bar{\varphi}_n(x)\} \in \mathscr{P}$$

be the real and imaginary parts of the functions $\varphi(x)$ and $\varphi_n(x)$ respectively. By § 31 corollary 2.1 we have

$$\lim_{m, n \to \infty} \| \varphi_m^{(k)}(A) - \varphi_n^{(k)}(A) \| = \lim_{m, n \to \infty} \| \varphi_m^{(k)}(x) - \varphi_n^{(k)}(x) \|_A$$

$$\leqslant \lim_{m, n \to \infty} \| \varphi_m(x) - \varphi_n(x) \|_A = 0 \quad \text{for } k = 1, 2.$$

Therefore in $\mathscr{B}$ there exist the limits

$$\varphi^{(k)}(A) = \lim_{n \to \infty} \varphi_n^{(k)}(A) \quad \text{for } k = 1, 2$$

and we have

$$\| \varphi^{(k)}(A) \| = \lim_{n \to \infty} \| \varphi_n^{(k)}(A) \| = \lim_{n \to \infty} \| \varphi_n^{(k)}(x) \|_A$$

$$= \| \varphi^{(k)}(x) \|_A \leqslant \| \varphi(x) \|_A.$$

Defining

$$\varphi(A) = \varphi^{(1)}(A) + i\varphi^{(2)}(A)$$

we get

$$\varphi(A) = \lim_{n \to \infty} \left[\varphi_n^{(1)}(A) + i\varphi_n^{(2)}(A) \right] = \lim_{n \to \infty} \varphi_n(A),$$

$$\|\varphi(A)\| \leqslant \|\varphi^{(1)}(A)\| + \|\varphi^{(2)}(A)\| \leqslant 2\|\varphi(x)\|_A.$$

If $\{\psi_n(x)\}_{n=1}^\infty$ is any other sequence in $\mathscr{P}$ satisfying

$$\lim_{n \to \infty} \|\varphi(x) - \psi_n(x)\|_A = 0,$$

then splitting $\psi_n(x)$ again into real and imaginary parts we conclude

$$\lim_{n \to \infty} \|\varphi_n^{(k)}(A) - \psi_n^{(k)}(A)\| = \lim_{n \to \infty} \|\varphi_n^{(k)}(x) - \psi_n^{(k)}(x)\|_A$$

$$\leqslant \lim_{n \to \infty} \left[\|\varphi_n^{(k)}(x) - \varphi^{(k)}(x)\|_A + \|\varphi^{(k)}(x) - \psi_n^{(k)}(x)\|_A \right]$$

$$= 0 \quad \text{for } k = 1, 2,$$

and therefore

$$\lim_{n \to \infty} \|\varphi^{(k)}(A) - \psi^{(k)}(A)\| \leqslant \lim_{n \to \infty} \left[\|\varphi^{(k)}(A) - \varphi_n^{(k)}(A)\| + \|\varphi_n^{(k)}(A) - \psi_n^{(k)}(A)\| \right]$$

$$= 0 \quad \text{for } k = 1, 2,$$

$$\lim_{n \to \infty} \psi_n^{(k)}(A) = \lim_{n \to \infty} \varphi_n^{(k)}(A) = \varphi^{(k)}(A) \quad \text{for } k = 1, 2,$$

$$\lim_{n \to \infty} \psi_n(A) = \varphi(A).$$

This also shows that there is only one operator $\varphi(A)$ having the property stated in theorem 1, and that for $\varphi(x) \in \mathscr{P}$ the operator $\varphi(A)$ coincides with the operator defined in § 31 theorem 1 (let $\varphi_n(x) = \varphi(x)$ for all $n \geqslant 1$).

If $\varphi(x) \in \mathscr{S}$ and $\psi(x) \in \mathscr{S}$ are given and if $\{\varphi_n(x)\}_{n=1}^\infty$ and $\{\psi_n(x)\}_{n=1}^\infty$ are sequences in $\mathscr{P}$ such that

$$\lim_{n \to \infty} \|\varphi(x) - \varphi_n(x)\|_A = 0,$$

$$\lim_{n \to \infty} \|\psi(x) - \psi_n(x)\|_A = 0,$$

then we get

$$\lim_{n \to \infty} \|(\varphi + \psi)(x) - (\varphi_n + \psi_n)(x)\|_A = 0,$$

$$\lim_{n \to \infty} \|(\lambda\varphi)(x) - (\lambda\varphi_n)(x)\|_A = 0,$$

$$\lim_{n \to \infty} \|(\varphi\psi)(x) - (\varphi_n\psi_n)(x)\|_A = 0,$$

$$\lim_{n \to \infty} \|\bar{\varphi}(x) - \bar{\varphi}_n(x)\|_A = 0,$$

and therefore

$$(\varphi + \psi)(A) = \lim_{n \to \infty} (\varphi_n + \psi_n)(A) = \lim_{n \to \infty} \varphi_n(A) + \lim_{n \to \infty} \psi_n(A)$$
$$= \varphi(A) + \psi(A),$$
$$(\lambda\varphi)(A) = \lim_{n \to \infty} (\lambda\varphi_n)(A) = \lambda \lim_{n \to \infty} \varphi_n(A)$$
$$= \lambda\varphi(A),$$
$$(\varphi\psi)(A) = \lim_{n \to \infty} (\varphi_n\psi_n)(A) = \lim_{n \to \infty} \varphi_n(A) \cdot \lim_{n \to \infty} \psi_n(A)$$
$$= \varphi(A)\,\psi(A),$$
$$\bar{\varphi}(A) = \lim_{n \to \infty} \bar{\varphi}_n(A) = \lim_{n \to \infty} [\varphi_n(A)]^* = [\lim_{n \to \infty} \varphi_n(A)]^*$$
$$= [\varphi(A)]^* . \ \square$$

COROLLARY 1.1. *Let A and $\varphi(x)$ be as in theorem* 1. *Then $\varphi(A)$ is normal. If $\varphi(x)$ is real-valued, then $\varphi(A)$ is selfadjoint.*
PROOF. $\varphi(A)\,[\varphi(A)]^* = \varphi(A)\,\bar{\varphi}(A) = (\varphi\bar{\varphi})(A) = (\bar{\varphi}\varphi)(A) = \bar{\varphi}(A)\,\varphi(A)$
$$= [\varphi(A)]^*\,\varphi(A). \ \square$$

Example 1. Let $A \in \mathscr{B}$ be selfadjoint and let $\varphi(x) = e^{ix}$. Then we have

$$e^{iA} = \sum_{k=0}^{\infty} \frac{(iA)^k}{k!}.$$

Moreover, e^{iA} is a unitary operator and its inverse is the operator

$$(e^{iA})^* = e^{-iA} = \sum_{k=0}^{\infty} \frac{(-iA)^k}{k!}.$$

Indeed, the partial sums

$$\varphi_n(x) = \sum_{k=0}^{n} \frac{(ix)^k}{k!} \quad (0 \leqslant n < \infty)$$

converge to the function $\varphi(x) = e^{ix}$ uniformly on any bounded interval containing $\Sigma(A)$. By theorem 1 we get

$$e^{iA} = \varphi(A) = \lim_{n \to \infty} \varphi_n(A) = \sum_{k=0}^{\infty} \frac{(iA)^k}{k!}.$$

The fact that e^{iA} is unitary follows from $(e^{iA})^* = e^{-iA}$ and from the inequality

$$\|e^{iA} e^{-iA} - I\| = \|e^{-iA} e^{iA} - I\| \leqslant 2 \|e^{ix} e^{-ix} - 1\|_A = 0.$$

Example 2. If $\lambda \notin \mathbf{R}$, then $\varphi(x) = 1/(x - \lambda) \in \mathscr{S}$ and $\varphi(A) = (A - \lambda I)^{-1}$. Indeed, from

$$(x - \lambda) \varphi(x) = \varphi(x)(x - \lambda) = 1$$

we conclude by theorem 1

$$(A - \lambda I) \varphi(A) = \varphi(A)(A - \lambda I) = I.$$

This proves the assertion (cf. § 12 definition 5).

If the selfadjoint operator $A \in \mathscr{B}$ and $\varphi(x) \in \mathscr{S}$, $\psi(x) \in \mathscr{S}$ are given, then applying the homomorphism $\varphi(x) \rightarrow \varphi(A)$ to both members of the equation $\varphi(x) \psi(x) = \psi(x) \varphi(x)$ we obtain $\varphi(A) \psi(A) = \psi(A) \varphi(A)$. This result is contained as a special case in the following theorem.

THEOREM 2. *Let the selfadjoint operator $A \in \mathscr{B}$ and the function $\varphi(x) \in \mathscr{S}$ be given and let the operator $B \in \mathscr{B}$ satisfy $AB = BA$. Then $\varphi(A)B = B\varphi(A)$.*
PROOF. Let $\{\varphi_n(x)\}_{n=1}^{\infty}$ be a sequence in $\mathscr{P}$ such that

$$\varphi(A) = \lim_{n \to \infty} \varphi_n(A).$$

The hypothesis implies $\varphi_n(A) B = B\varphi_n(A)$ for all $n \geqslant 1$. We conclude

$$\varphi(A) B = \left[\lim_{n \to \infty} \varphi_n(A) \right] B = \lim_{n \to \infty} \left[\varphi_n(A) B \right]$$
$$= \lim_{n \to \infty} \left[B\varphi_n(A) \right] = B \left[\lim_{n \to \infty} \varphi_n(A) \right] = B\varphi(A). \ \square$$

The next theorem extends § 31 theorem 2. Note, however, that in § 31 theorem 2 the operator A was an arbitrary bounded operator on $\mathfrak{H}$, while now A in addition has to be selfadjoint.

THEOREM 3. *Let the selfadjoint operator $A \in \mathscr{B}$ and the function $\varphi(x) \in \mathscr{S}$ be given. Then*

$$\Sigma(\varphi(A)) = \{\varphi(\xi) : \xi \in \Sigma(A)\} = \varphi(\Sigma(A)).$$

PROOF. a) Let $\xi \in \Sigma(A)$ be given and assume $\varphi(\xi)$ were a regular value for $\varphi(A)$. Then we could choose a polynomial $\psi(x) \in \mathscr{P}$ such that

$$\|\varphi(x) - \psi(x)\|_A \leqslant \frac{1}{3 \|[\varphi(A) - \varphi(\xi) I]^{-1}\|}.$$

We obtain

$$\|[\varphi(A) - \varphi(\xi)\,I] - [\psi(A) - \psi(\xi)\,I]\| \leqslant \|\varphi(A) - \psi(A)\| + |\varphi(\xi) - \psi(\xi)|$$

$$\leqslant 3\,\|\varphi(x) - \psi(x)\|_A < \frac{1}{\|[\varphi(A) - \varphi(\xi)\,I]^{-1}\|}.$$

By § 23 corollary 3.1 also the operator $\psi(A) - \psi(\xi)\,I$ would be invertible. This, however, contradicts the fact that $\psi(\xi) \in \Sigma(\psi(A))$ by § 31 theorem 2.

b) Let $\eta \in \Sigma(\varphi(A))$ be given. We have to show that there exists a $\xi \in \Sigma(A)$ such that $\eta = \varphi(\xi)$. Assume the contrary

$$\varphi(\xi) - \eta \neq 0 \quad \text{for all } \xi \in \Sigma(A).$$

We shall now construct a continuous function $\psi(x) \in \mathscr{S}$ such that

(2)
$$\psi(\xi) = \frac{1}{\varphi(\xi) - \eta} \quad \text{for all } \xi \in \Sigma(A)$$

(fig. 19; the reader familiar with topology is referred to TIETZE's extension theorem, an application of which replaces all but the last paragraph of this proof; cf. [6] (7.40)).

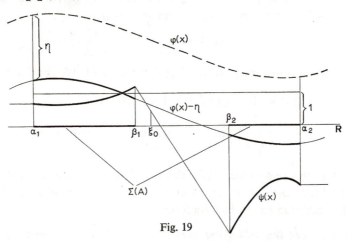

Fig. 19

On $\Sigma(A)$ we define the function $\psi(x)$ by (2). Let $\alpha_1 = \min \Sigma(A)$ and $\alpha_2 = \max \Sigma(A)$. We then define

(3)
$$\psi(\xi) = \begin{cases} \dfrac{1}{\varphi(\alpha_1) - \eta} & \text{for } \xi \leqslant \alpha_1, \\[2ex] \dfrac{1}{\varphi(\alpha_2) - \eta} & \text{for } \xi \geqslant \alpha_2. \end{cases}$$

In order to define $\psi(x)$ for any $\xi_0 \notin \Sigma(A)$, $\alpha_1 < \xi_0 < \alpha_2$, let

$$(4) \qquad \begin{cases} \beta_1 = \max\{\lambda : \lambda < \xi_0; \lambda \in \Sigma(A)\}, \\ \beta_2 = \min\{\lambda : \lambda > \xi_0; \lambda \in \Sigma(A)\}. \end{cases}$$

Since $\Sigma(A)$ is bounded and closed we have $\beta_1 \in \Sigma(A)$, $\beta_2 \in \Sigma(A)$ (this has already been anticipated by using the notation "max" and "min" in place of "sup" and "inf") and therefore $\beta_1 < \xi_0 < \beta_2$ and $\xi \notin \Sigma(A)$ for $\beta_1 < \xi < \beta_2$. We now define $\psi(x)$ linearly in the interval $[\beta_1, \beta_2]$ by

$$(5) \qquad \psi(\xi) = \psi(\beta_1) + \frac{\xi - \beta_1}{\beta_2 - \beta_1} [\psi(\beta_2) - \psi(\beta_1)]$$

$$= \frac{\beta_2 - \xi}{\beta_2 - \beta_1} \psi(\beta_1) + \frac{\xi - \beta_1}{\beta_2 - \beta_1} \psi(\beta_2) \quad \text{for } \beta_1 \leqslant \xi \leqslant \beta_2.$$

Next we shall verify that the function $\psi(x)$ defined on $\mathbf{R}$ by (2), (3) and (5) is continuous on $\mathbf{R}$ (this will allow us to associate with $\psi(x)$ an operator $\psi(A)$ by theorem 1). In fact it suffices to check left continuity, i.e. to show that for any $\xi_1 \in \mathbf{R}$ and for any given $\varepsilon > 0$ there exists a $\delta > 0$ such that

$$(6) \qquad |\psi(\xi_1) - \psi(\xi)| < \varepsilon \quad \text{for all } \xi \in [\xi_1 - \delta, \xi_1].$$

The corresponding inequality in $[\xi_1, \xi_1 + \delta]$, indicating right continuity, is deduced similarly. To this end we distinguish two cases:

α) Suppose there is an interval $[\xi_1 - \delta_1, \xi_1[$ (left closed, right open) which does not contain any points of $\Sigma(A)$. Then $\psi(x)$ is defined on the closed interval $[\xi_1 - \delta_1, \xi_1]$ linearly either by (3) or by (5). As a consequence $\psi(x)$ is continuous on $[\xi_1 - \delta_1, \xi_1]$; in particular (6) is satisfied if $\delta < \delta_1$ is chosen sufficiently small.

β) Suppose for every $\delta > 0$ the interval $[\xi_1 - \delta, \xi_1[$ contains some point of $\Sigma(A)$. Since $\Sigma(A)$ is closed we conclude

$$\xi_1 \in \Sigma(A),$$

$$\psi(\xi_1) = \frac{1}{\varphi(\xi_1) - \eta} \quad \text{(by (2))},$$

and by the continuity of φ we obtain for a sufficiently small chosen δ

$$\left| \psi(\xi_1) - \frac{1}{\varphi(\xi) - \eta} \right| < \varepsilon \quad \text{for all } \xi \in [\xi_1 - \delta, \xi_1].$$

In particular we get

(7) $\qquad |\psi(\xi_1) - \psi(\xi)| < \varepsilon \quad \text{for } \xi \in \Sigma(A),\ \xi \in [\xi_1 - \delta, \xi_1]$.

If $[\xi_1 - \delta, \xi_1] \subset \Sigma(A)$, then we are through. Otherwise, without loss of generality (replacing δ by a smaller constant if necessary), we may assume $\xi_1 - \delta \in \Sigma(A)$. For any $\xi_0 \in [\xi_1 - \delta, \xi_1] \backslash \Sigma(A)$ let $\beta_1 \in \Sigma(A), \beta_2 \in \Sigma(A)$ be defined as in (4). We have then $\beta_k \in [\xi_1 - \delta, \xi_1]$ for $k = 1, 2$ and therefore by (7)

$$|\psi(\xi_1) - \psi(\beta_k)| < \varepsilon \quad \text{for } k = 1, 2.$$

Applying the second formula in (5) we obtain

$$|\psi(\xi_1) - \psi(\xi_0)| \leqslant \frac{\beta_2 - \xi_0}{\beta_2 - \beta_1} |\psi(\xi_1) - \psi(\beta_1)| + \frac{\xi_0 - \beta_1}{\beta_2 - \beta_1} |\psi(\xi_1) - \psi(\beta_2)|$$

$$< \left(\frac{\beta_2 - \xi_0}{\beta_2 - \beta_1} + \frac{\xi_0 - \beta_1}{\beta_2 - \beta_1} \right) \varepsilon = \varepsilon.$$

We conclude

$$|\psi(\xi_1) - \psi(\xi)| < \varepsilon \quad \text{for } \xi \notin \Sigma(A),\ \xi \in [\xi_1 - \delta, \xi_1].$$

Together with (7) this proves the (left) continuity of $\psi(x)$ in ξ_1.

Observe that by the definition of $\psi(x)$ on $\Sigma(A)$ (2) we have

$$\|\psi(x)[\varphi(x) - \eta] - 1\|_A = 0.$$

For the operator $\psi(A)$ associated with $\psi(x)$ by theorem 1 we obtain

$$\|\psi(A)[\varphi(A) - \eta I] - I\| = \|[\varphi(A) - \eta I]\psi(A) - I\|$$
$$\leqslant 2\|\psi(x)[\varphi(x) - \eta] - 1\|_A = 0.$$

As a consequence the operator $\varphi(A) - \eta I$ is invertible and $\psi(A)$ is its inverse. This, however, contradicts our hypothesis $\eta \in \Sigma(\varphi(A))$. The assumption $\eta \notin \varphi(\Sigma(A))$ is therefore disproved. $\square$

COROLLARY 3.1. *Let A and $\varphi(x)$ be as in theorem 3.*

a) *The operator $\varphi(A)$ is selfadjoint iff*

(8) $\qquad \varphi(\xi) \in \mathbf{R} \quad \text{for all } \xi \in \Sigma(A).$

b) *The operator $\varphi(A)$ is unitary iff*

(9) $\qquad |\varphi(\xi)| = 1 \quad \text{for all } \xi \in \Sigma(A).$

c) *The operator $\varphi(A)$ is invertible iff*

$$\varphi(\xi) \neq 0 \quad \text{for all } \xi \in \Sigma(A).$$

PROOF. a) If $\varphi(A)$ is selfadjoint, then $\varphi(\Sigma(A)) = \Sigma(\varphi(A)) \subset \mathbf{R}$ (cf. § 24 theorem 3). Conversely from (8) we conclude

$$\|\varphi(A) - [\varphi(A)]^*\| \leqslant 2\|\varphi(x) - \bar{\varphi}(x)\|_A = 0$$

(cf. theorem 1).

b) If $\varphi(A)$ is unitary and if $\xi \in \Sigma(A)$ is given, then $\varphi(\xi) \in \Sigma(\varphi(A))$ and $|\varphi(\xi)| = 1$ by § 23 theorem 7. Conversely, from (9) we conclude

$$\|\varphi(A)[\varphi(A)]^* - I\| = \|[\varphi(A)]^*\varphi(A) - I\| \leqslant 2\|\varphi(x)\bar{\varphi}(x) - 1\|_A = 0.$$

Therefore $\varphi(A)$ is unitary.

c) The operator $\varphi(A)$ is invertible iff $0 \notin \Sigma(\varphi(A)) = \varphi(\Sigma(A))$. $\square$

COROLLARY 3.2. *Let A and $\varphi(x)$ be as in theorem 3. If $\varphi(A)$ is selfadjoint, then*

$$\|\varphi(A)\| = \|\varphi(x)\|_A.$$

PROOF. By § 30 corollary 8.1 we have

$$\|\varphi(A)\| = \max\{|\lambda| : \lambda \in \Sigma(\varphi(A))\}$$
$$= \max\{|\lambda| : \lambda \in \varphi(\Sigma(A))\} = \|\varphi(x)\|_A.$$

COROLLARY 3.3. *Let the selfadjoint operator $A \in \mathcal{B}$ and the functions $\varphi(x) \in \mathcal{S}$, $\psi(x) \in \mathcal{S}$ be given, then $\varphi(A) = \psi(A)$ iff $\|\varphi(x) - \psi(x)\|_A = 0$.*

PROOF. If $\varphi(A) = \psi(A)$, then the operator $\varphi(A) - \psi(A) = O$ is selfadjoint and

$$\|\varphi(x) - \psi(x)\|_A = \|\varphi(A) - \psi(A)\| = 0.$$

The converse assertion already follows from theorem 1. $\square$

COROLLARY 3.4. *Let A and $\varphi(x)$ be as in theorem 3. If $\varphi(x)$ is real-valued and non-negative on $\Sigma(A)$ (in particular if $\varphi(x) \geqslant 0$, i.e. if $\varphi(\xi) \geqslant 0$ for all $\xi \in \mathbf{R}$), then $\varphi(A) \geqslant O$.*

PROOF. The operator $\varphi(A)$ is selfadjoint by corollary 3.1 a). We conclude

$$\beta_1 = \max\{\beta \in \mathbf{R} : \beta I \leqslant \varphi(A)\}$$
$$= \min \Sigma(\varphi(A)) \qquad \text{(cf. § 30 corollary 8.1)}$$
$$= \min \varphi(\Sigma(A)) \geqslant 0 \quad \text{(cf. theorem 3)}$$

and therefore

$$O \leqslant \beta_1 I \leqslant \varphi(A). \ \square$$

COROLLARY 3.5. *Let $A \in \mathcal{B}$ be selfadjoint. The homomorphism $\varphi(x) \to \varphi(A)$ of $\mathcal{S}$ into $\mathcal{B}$ is order preserving in the following sense: if $\varphi(x) \in \mathcal{S}$ and $\psi(x) \in \mathcal{S}$ are real-valued on $\Sigma(A)$ and*

$$\varphi(\xi) \geqslant \psi(\xi) \quad \text{for all } \xi \in \Sigma(A)$$

(in particular if $\varphi(x) \geqslant \psi(x)$, i.e. if $\varphi(x)$ and $\psi(x)$ are real-valued and $\varphi(x) - \psi(x) \geqslant 0$), then $\varphi(A) \geqslant \psi(A)$.

PROOF. By corollary 3.4 the inequality $\varphi(\xi) - \psi(\xi) \geqslant 0$ for all $\xi \in \Sigma(A)$ implies $\varphi(A) - \psi(A) \geqslant 0$. $\square$

We shall apply these results to derive two interesting facts which, however, will not be needed in the sequel. In the first place, a positive bounded selfadjoint operator on $\mathfrak{H}$ has a unique positive "*square root*". In the second place, analogously to the familiar factorization of a complex number

$$\zeta = |\zeta| \, e^{i \arg \zeta},$$

a bounded normal operator on $\mathfrak{H}$ may be written as a commutative product of a positive selfadjoint operator, representing its "*absolute value*", and a unitary operator, representing the factor "of absolute value one".

THEOREM 4. *If the operator $A \in \mathcal{B}$ is selfadjoint and positive, then there exists a unique positive selfadjoint operator $B = \sqrt{A} \in \mathcal{B}$ such that $B^2 = A$. If A is invertible, then so is B.*

PROOF. By hypothesis we have

$$\alpha_1 = \min \Sigma(A) \geqslant 0$$

(cf. § 30 theorem 6 and § 30 corollary 8.1). Define the function $\varphi(x) \in \mathcal{S}$ by

$$\varphi(\xi) = \begin{cases} 0 & \text{for } \xi \leqslant 0, \\ \sqrt{\xi} & \text{for } \xi \geqslant 0. \end{cases}$$

The operator $B = \varphi(A)$ is selfadjoint and positive by corollary 1.1 and corollary 3.4. Moreover it satisfies

$$\|B^2 - A\| = \|\varphi^2(x) - x\|_A = 0$$

(cf. corollary 3.2).

In order to prove the statement about uniqueness let $C \in \mathcal{B}$ be any positive selfadjoint operator such that $C^2 = A$. Then $\Sigma(A) = \{\xi^2 : \xi \in \Sigma(C)\}$ by theorem 3. If $\{\varphi_n(x)\}_{n=1}^{\infty} \subset \mathcal{S}$ is a sequence of polynomials converging to $\varphi(x)$ (as above) uniformly on $\Sigma(A)$, then $\{\varphi_n(x^2)\}_{n=1}^{\infty}$ is a sequence of polynomials converging to the function $\psi(x) = \varphi(x^2) \in \mathcal{S}$ uniformly on $\Sigma(C)$. Applying theorem 1 to C in place of A we obtain

$$B = \lim_{n \to \infty} \varphi_n(A) = \lim_{n \to \infty} \varphi_n(C^2) = \psi(C)$$

$$\|B - C\| = \|\psi(C) - C\| = \|\psi(x) - x\|_c = \|\varphi(x^2) - x\|_c = 0$$

since $\varphi(\xi^2) = \xi$ for all $\xi \geqslant 0$.

If A is invertible, then $0 \notin \Sigma(A) = \Sigma(B^2)$ and therefore $0 \notin \Sigma(B)$ by theorem 3. As a consequence B is invertible. $\square$

THEOREM 5. *For every operator $A \in \mathscr{B}$ there exists a positive selfadjoint operator $B = |A| \in \mathscr{B}$ and an isometric operator C with domain $\mathfrak{D}_C = \overline{B\mathfrak{H}}$ and range $C\mathfrak{D}_C = \overline{A\mathfrak{H}}$ such that $A = CB$.*

PROOF. The operator A^*A is selfadjoint and positive (cf. § 30 example 2). Define the operator $B = |A| \in \mathscr{B}$ by

$$B = \sqrt{A^*A}$$

(cf. theorem 4; note that if A is selfadjoint then $B = \varphi(A)$ where $\varphi(x) = |x| = \sqrt{x^2}$). Then we have

(10) $\quad \langle Af, Ag \rangle = \langle A^*Af, g \rangle = \langle B^2 f, g \rangle$

$$= \langle Bf, Bg \rangle \quad \text{for all } f \in \mathfrak{H} \text{ and all } g \in \mathfrak{H},$$

in particular

(11) $\qquad\qquad \|Af\| = \|Bf\| \quad \text{for all } f \in \mathfrak{H}.$

From $Af = Ag$ therefore follows

$$0 = \|A(f - g)\| = \|B(f - g)\|,$$
$$Bf = Bg,$$

and conversely. The mapping C defined on $B\mathfrak{H}$ by

(12) $\qquad\qquad C(Bf) = Af \quad \text{for all } f \in \mathfrak{H}$

is therefore one-to-one and bounded (cf. § 11 definition 1). The mapping C is also linear since

$$C(B + Bg) = C[B(f + g)] = A(f + g) = Af + Ag$$
$$= C(Bf) + C(Bg),$$
$$C[\lambda(Bf)] = C[B(\lambda f)] = A(\lambda f) = \lambda Af$$
$$= \lambda C(Bf).$$

By § 11 theorem 3 the mapping C can uniquely be extended to a bounded linear operator with domain $\overline{B\mathfrak{H}}$ which we shall again call C. If f and g are vectors in $\overline{B\mathfrak{H}}$ and

$$f = \lim_{n \to \infty} Bf_n,$$

$$g = \lim_{n \to \infty} Bg_n,$$

then by § 3 theorem 6 c) and (10) we conclude

$$\langle Cf, Cg \rangle = \lim_{n \to \infty} \langle Af_n, Ag_n \rangle = \lim_{n \to \infty} \langle Bf_n, Bg_n \rangle = \langle f, g \rangle.$$

Therefore C is isometric. Furthermore we have

$$Cf = \lim_{n \to \infty} Af_n \in \overline{A\mathfrak{H}}.$$

Conversely, if

$$h = \lim_{n \to \infty} Ah_n \in \overline{A\mathfrak{H}}$$

is given, then we have $Ah_n = CBh_n$ for $n \geq 1$. Since C is isometric, also the sequence $\{Bh_n\}_{n=1}^{\infty}$ converges to some limit

$$h' = \lim_{n \to \infty} Bh_n \in \overline{B\mathfrak{H}}.$$

We conclude

$$h = \lim_{n \to \infty} CBh_n = C \lim_{n \to \infty} Bh_n = Ch'.$$

This shows that C maps $\overline{B\mathfrak{H}}$ onto $\overline{A\mathfrak{H}}$. The equation $A = CB$ already follows from the definition (12) of C. $\square$

COROLLARY 5.1. *If the operator $A \in \mathcal{B}$ is normal, then there exists a positive selfadjoint operator $B = |A| \in \mathcal{B}$ and a unitary operator C such that*

$$A = BC = CB.$$

If A is also invertible, then B and C are uniquely determined by these requirements.

PROOF. If A is normal, then

$$\|Bf\| = \|Af\| = \|A^*f\|$$

by (11) and § 14 theorem 6. We conclude

$$\mathfrak{M} = \{f : Bf = o\} = \{f : Af = o\} = \{f : A^*f = o\},$$
$$(B\mathfrak{H})^{\perp} = \{f : Bf = o\} \qquad \text{(by § 24 corollary 2.1)}$$
$$= \mathfrak{M} = \{f : A^*f = o\} = (A\mathfrak{H})^{\perp} \quad \text{(by § 24 theorem 2)}$$

and therefore

$$\overline{B\mathfrak{H}} = \mathfrak{M}^{\perp} = \overline{A\mathfrak{H}}$$

(cf. § 7 theorem 9). By theorem 5 we already have $C\mathfrak{M}^{\perp} = \mathfrak{M}^{\perp}$. Defining

$$Cf = f \quad \text{for all } f \in \mathfrak{M}$$

we obtain a unitary operator on $\mathfrak{H}$ (which we again denote by C) such that $A = CB$. Moreover, for $f \in \mathfrak{M}$ we have

(13) $$BCf = Bf = o = Af .$$

Since the operator A commutes with A^*A by

$$A(A^*A) = (AA^*) A = (A^*A) A,$$

also the operator $B = \sqrt{A^*A}$ commutes with A

$$AB = BA$$

by theorem 2 (applied to A^*A in place of A). For any vector

$$f = \lim_{n \to \infty} Bf_n \in \mathfrak{M}^{\perp}$$

we therefore have

(14) $$\begin{aligned} BCf &= \lim_{n \to \infty} BCBf_n = \lim_{n \to \infty} BAf_n \\ &= \lim_{n \to \infty} ABf_n = A \lim_{n \to \infty} Bf_n = Af . \end{aligned}$$

From (13) and (14) we conclude $A = BC$.

If A is invertible, then so are the operators A^* (by § 14 theorem 2 e)), A^*A, and $B = \sqrt{A^*A}$ (by theorem 4). In order to prove the statement about uniqueness, suppose

$$A = B'C' = C'B'$$

is another factorization of A with the properties described in corollary 5.1. We conclude

$$A^*A = B'C'^*C'B' = B'^2 ,$$
$$B' = \sqrt{A^*A} = B \quad \text{(by theorem 4)},$$
$$C' = B^{-1}A = C. \ \square$$

Exercise. Suppose $A = CB$ where $B \in \mathcal{B}$ is a positive selfadjoint operator and $C \in \mathcal{B}$ is an isometric operator.

a) $B=\sqrt{A^*A}$; consequently B is uniquely determined by the stated requirements. (Hint: use § 14 exercise.)

b) C is uniquely determined by the stated requirements iff A is one-to-one. (Hint: if A is one-to-one apply § 17 theorem 6 to B; if A is not one-to-one apply § 24 corollary 2.1 to B.)

§ 33. Step functions of a bounded selfadjoint operator

Let A be a bounded selfadjoint operator on $\mathfrak{H}$. Our next aim is to extend the order preserving homomorphism $\varphi(x)\rightarrow\varphi(A)$ of the algebra $\mathscr{S}$ of continuous functions $\varphi(x)$ on $\mathbf{R}$ into $\mathscr{B}$, restricted now to real-valued functions, to a larger domain, namely an algebra of functions containing the "*step functions*" $\varphi_\lambda(x)$ $(\lambda\in\mathbf{R})$ defined by

$$(1) \qquad \varphi_\lambda(x) = \begin{cases} 1 & \text{for } -\infty < \xi \leqslant \lambda, \\ 0 & \text{for } \lambda < \xi < +\infty \end{cases} \quad \text{(fig. 20)}.$$

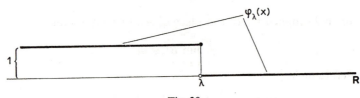

Fig. 20

The pleasant result which we look for is that $\bar{\varphi}_\lambda(x)=\varphi_\lambda(x)$ and $\varphi_\lambda^2(x)= =\varphi_\lambda(x)$ will then imply $[\varphi_\lambda(A)]^*=\varphi_\lambda(A)$ and $[\varphi_\lambda(A)]^2=\varphi_\lambda(A)$, i.e. $\varphi_\lambda(A)$ will then be a projection (cf. § 15 theorem 2). The unpleasant obstacle which we have to overcome lies in the fact that the main tool for defining an operator $\varphi(A)$ as used in § 32 now fails: the function $\varphi_\lambda(x)$ cannot be approximated uniformly by continuous functions, let alone polynomials, on any interval containing λ. Thus in general there is no way to define an operator $\varphi_\lambda(A)$ as a uniform limit of operators $\varphi_{\lambda,n}(A)$ where $\varphi_{\lambda,n}(x)\in\mathscr{S}$.

Fortunately there is still the concept of a strong limit (§ 30 definition 2) which comes to rescue in this situation. Observe that the function $\varphi_\lambda(x)$ may be obtained as a pointwise limit of a decreasing sequence of real-valued continuous functions $\varphi_{\lambda,n}(x)$ defined e.g. by

$$(2) \qquad \varphi_{\lambda,n}(x) = \begin{cases} 1 & \text{for } -\infty < \xi \leqslant \lambda, \\ 1 - n(\xi - \lambda) & \text{for } \lambda \leqslant \xi \leqslant \lambda + 1/n, \\ 0 & \text{for } \lambda + 1/n \leqslant \xi < +\infty \end{cases} \quad \text{(fig. 21)}.$$

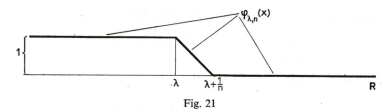

Fig. 21

By § 32 corollary 3.5 the sequence of corresponding selfadjoint operators $\varphi_{\lambda,n}(A)$ is non-increasing and bounded below by zero. It therefore converges strongly to some bounded selfadjoint operator on $\mathfrak{H}$ (cf. § 30 theorem 4 and § 30 remark 2) which, as we hope, will be a good candidate for $\varphi_{\lambda}(A)$.

In order to treat this matter systematically we start with the proper terminology.

DEFINITION 1. A real-valued function $\varphi(x)$ on $\mathbf{R}$ is called *upper semi-continuous* if it is a pointwise limit of a non-increasing sequence of continuous real-valued functions on $\mathbf{R}$.

Remark 1. It can be shown that a real-valued function $\varphi(x)$ on $\mathbf{R}$ is upper semi-continuous iff for every $\xi_0 \in \mathbf{R}$ and for every $\varepsilon > 0$ there exists a $\delta > 0$ such that

$$\varphi(\xi) < \varphi(\xi_0) + \varepsilon \quad \text{for all } \xi \in]\xi_0 - \delta, \xi_0 + \delta[\,.$$

We shall not need and therefore not prove this statement which, however, is sometimes used to define upper semi-continuity of a function $\varphi(x)$.

THEOREM 1. *Let $A \in \mathscr{B}$ be selfadjoint and let $\varphi(x)$ be a non-negative upper semi-continuous function on $\mathbf{R}$. Then there exists a unique positive selfadjoint operator $\varphi(A) \in \mathscr{B}$ such that whenever $\{\varphi_n(x)\}_{n=1}^{\infty}$ is any non-increasing sequence of non-negative functions in $\mathscr{S}$, pointwise converging to $\varphi(x)$ on $\Sigma(A)$, then*

$$\varphi(A) = (\text{s}) \lim \varphi_n(A) \,.$$

PROOF. If $\{\varphi_n(x)\}_{n=1}^{\infty} \subset \mathscr{S}$ is any sequence as described in theorem 1, then $\{\varphi_n(A)\}_{n=1}^{\infty}$ is a non-increasing sequence of positive selfadjoint operators (cf. § 32 corollary 3.5) which converges strongly to some positive selfadjoint operator $\varphi(A) \in \mathscr{B}$ by § 30 theorem 4 and § 30 remark 2.

Suppose $\{\varphi'_n(x)\}_{n=1}^{\infty}$ is any other sequence enjoying the same properties as stated in theorem 1 for the sequence $\{\varphi_n(x)\}_{n=1}^{\infty}$ and therefore supplying a sequence $\{\varphi'_n(A)\}_{n=1}^{\infty} \subset \mathscr{B}$ which converges strongly to some positive selfadjoint operator $\varphi'(A)$. Then for any $m \geqslant 1$ we have

(3)
$$\lim_{n \to \infty} \varphi_n(\xi) \leqslant \varphi'_m(\xi) \quad \text{for all } \xi \in \Sigma(A).$$

We assert that for all $n \geqslant n(m)$ even

(4)
$$\varphi_n(\xi) \leqslant \varphi'_m(\xi) + \frac{1}{m} \quad \text{for all } \xi \in \Sigma(A)$$

(the reader familiar with topology is referred to DINI's theorem which replaces the proof of assertion (4) given below; cf. [6] (13.40), [7] § 6.2.III). Otherwise there would exist a sequence $\{\xi_n\}_{n=1}^{\infty} \subset \Sigma(A)$ such that

(5)
$$\varphi_n(\xi_n) > \varphi'_m(\xi_n) + \frac{1}{m} \quad \text{for all } n \geqslant 1.$$

Without loss of generality (replacing $\{\xi_n\}_{n=1}^{\infty}$ by a subsequence if necessary) we may assume that the sequence $\{\xi_n\}_{n=1}^{\infty}$ converges to some $\xi_0 \in \Sigma(A)$. By (3) we can choose an index k such that

$$\varphi_k(\xi_0) < \varphi'_m(\xi_0) + \frac{1}{4m}.$$

For all sufficiently large indices $n \geqslant k$ the numbers ξ_n are then close enough to ξ_0 in order to guarantee the following chain of inequalities:

$$\varphi_k(\xi_0) < \varphi'_m(\xi_0) + \frac{1}{4m}$$

$$< \varphi'_m(\xi_n) + \frac{1}{2m} \quad \text{(by continuity of } \varphi'_m(x) \text{ in } \xi_0)$$

$$< \varphi_n(\xi_n) - \frac{1}{2m} \quad \text{(by (5))}$$

$$\leqslant \varphi_k(\xi_n) - \frac{1}{2m} \quad \text{(since } \varphi_n(x) \leqslant \varphi_k(x))$$

$$< \varphi_k(\xi_0) - \frac{1}{4m} \quad \text{(by continuity of } \varphi_k(x) \text{ in } \xi_0).$$

The inequality $\varphi_k\{\xi_0\} < \varphi_k(\xi_0) - 1/4m$, however, is impossible.
This proves our assertion (4) which in turn implies

$$\varphi(A) \leqslant \varphi_n(A) \leqslant \varphi'_m(A) + \frac{1}{m} I$$

by § 30 theorem 4, § 30 remark 2, and § 32 corollary 3.5. We conclude

$$\varphi(A) \leqslant (\text{s}) \lim_{m \to \infty} \left[\varphi'_m(A) + \frac{1}{m} I \right] = \varphi'(A).$$

Interchanging the roles of $\varphi(A)$ and $\varphi'(A)$ we get by the same argument

$$\varphi'(A) \leqslant \varphi(A)$$

and therefore

$$\varphi(A) = \varphi'(A) = (\text{s}) \lim_{m \to \infty} \varphi'_m(A).$$

At the same time this proves the statement about uniqueness in theorem 1. □

Remark 2. If $\varphi(x)$ is continuous, then the operator $\varphi(A)$ corresponding to $\varphi(x)$ by theorem 1 coincides with the one corresponding to $\varphi(x)$ by § 32 theorem 1, as can be seen by taking $\varphi_n(x) = \varphi(x)$ for all $n \geqslant 1$.

THEOREM 2. *Let $A \in \mathcal{B}$ be selfadjoint, let $\varphi(x)$ and $\psi(x)$ be non-negative upper semi-continuous functions on $\mathbf{R}$, and let $\alpha > 0$ be given. Then the functions $(\varphi + \psi)(x)$, $(\alpha\varphi)(x)$, and $(\varphi\psi)(x)$ are non-negative upper semi-continuous and*

$$(\varphi + \psi)(A) = \varphi(A) + \psi(A),$$
$$(\alpha\varphi)(A) = \alpha\varphi(A),$$
$$(\varphi\psi)(A) = \varphi(A)\psi(A).$$

If $\varphi(\xi) \leqslant \psi(\xi)$ for all $\xi \in \Sigma(A)$, in particular if $\varphi(x) \leqslant \psi(x)$, then $\varphi(A) \leqslant \psi(A)$.
PROOF. If $\{\varphi_n(x)\}_{n=1}^{\infty}$ and $\{\psi_n(x)\}_{n=1}^{\infty}$ are non-increasing sequences of real-valued functions in $\mathcal{S}$ converging pointwise on $\Sigma(A)$ to $\varphi(x)$ and $\psi(x)$ respectively, then the non-increasing sequences $\{(\varphi_n + \psi_n)(x)\}_{n=1}^{\infty}$, $\{(\alpha\varphi_n)(x)\}_{n=1}^{\infty}$, and $\{(\varphi_n\psi_n)(x)\}_{n=1}^{\infty}$ in $\mathcal{S}$ converge pointwise on $\Sigma(A)$ to $(\varphi + \psi)(x)$, $(\alpha\varphi)(x)$, and $(\varphi\psi)(x)$ respectively and we have

$$(\varphi + \psi)(A) = (\text{s}) \lim_{n \to \infty} (\varphi_n + \psi_n)(A) = (\text{s}) \lim_{n \to \infty} \varphi_n(A) + (\text{s}) \lim_{n \to \infty} \psi_n(A)$$

$$= \varphi(A) + \psi(A),$$

$$(\alpha\varphi)(A) = (\text{s}) \lim_{n \to \infty} (\alpha\varphi_n)(A) = \alpha \left[(\text{s}) \lim_{n \to \infty} \varphi_n(A) \right]$$

$$= \alpha\varphi(A),$$

while the strong convergence of $\{(\varphi_n\psi_n)(A)\}_{n=1}^{\infty}$ to $\varphi(A)\psi(A)$ follows from

$$\| \varphi_n(A)\psi_n(A) f - \varphi(A)\psi(A) f \| \leqslant$$
$$\leqslant \| \varphi_n(A)[\psi_n(A) f - \psi(A) f] \| + \| [\varphi_n(A) - \varphi(A)]\psi(A) f \|$$
$$\leqslant \| \varphi_n(x) \|_A \| \psi_n(A) f - \psi(A) f \| + \| [\varphi_n(A) - \varphi(A)]\psi(A) f \|,$$

$$\|\varphi_n(x)\|_A \leqslant \|\varphi_1(x)\|_A,$$

(s) $\lim\limits_{n \to \infty} \varphi_n(A) = \varphi(A),$

(s) $\lim\limits_{n \to \infty} \psi_n(A) = \psi(A).$

If $\varphi(\xi) \leqslant \psi(\xi)$ for all $\xi \in \Sigma(A)$, then the argument used in the proof of theorem 1 (4) shows that for all $n \geqslant n(m)$ we have

$$\varphi_n(\xi) \leqslant \psi_m(\xi) + \frac{1}{m} \quad \text{for all } \xi \in \Sigma(A),$$

$$\varphi(A) \leqslant \varphi_n(A) \leqslant \psi_m(A) + \frac{1}{m} I$$

and therefore

$$\varphi(A) \leqslant (\text{s}) \lim_{m \to \infty} \left[\psi_m(A) + \frac{1}{m} I \right] = \psi(A). \ \Box$$

COROLLARY 2.1. *Let A, $\varphi(x)$, and $\psi(x)$ be as in theorem 2. If $\varphi(\xi) = \psi(\xi)$ for all $\xi \in \Sigma(A)$, then $\varphi(A) = \psi(A)$.*
PROOF. From $\varphi(A) \leqslant \psi(A)$ and $\varphi(A) \geqslant \psi(A)$ follows $\varphi(A) = \psi(A)$ by § 30 theorem 1 c). $\Box$

For the benefit of a smoother formulation and application of the preceding results we enlarge the class of non-negative upper semi-continuous functions to an algebra by defining

$$\mathscr{R} = \{\varphi(x) = \varphi_1(x) - \varphi_2(x): \varphi_k(x) \text{ is non-negative}$$
$$\text{and upper semi-continuous on } \mathbf{R} \text{ for } k = 1, 2\}.$$

THEOREM 3. *The class $\mathscr{R}$ of real-valued functions on $\mathbf{R}$ is a real algebra if sum, scalar multiple, and product in $\mathbf{R}$ are defined pointwise.*
PROOF. Let $\varphi_k(x)$ and $\psi_k(x)$ $(k = 1, 2)$ be non-negative upper semi-continuous functions on $\mathbf{R}$. The assertion then follows from theorem 2 and from the identities

$$(6) \quad \begin{cases} [\varphi_1(x) - \varphi_2(x)] + [\psi_1(x) - \psi_2(x)] \\ \qquad = (\varphi_1 + \psi_1)(x) - (\varphi_2 + \psi_2)(x), \\ \alpha[\varphi_1(x) - \varphi_2(x)] = \begin{cases} (\alpha\varphi_1)(x) - (\alpha\varphi_2)(x) & \text{for } \alpha \geqslant 0, \\ (-\alpha\varphi_2)(x) - (-\alpha\varphi_1)(x) & \text{for } \alpha < 0, \end{cases} \\ [\varphi_1(x) - \varphi_2(x)][\psi_1(x) - \psi_2(x)] \\ \qquad = (\varphi_1\psi_1 + \varphi_2\psi_2)(x) - (\varphi_1\psi_2 + \varphi_2\psi_1)(x). \ \Box \end{cases}$$

THEOREM 4. *Let the selfadjoint operator $A \in \mathscr{B}$ and the function $\varphi(x) \in \mathscr{R}$ be given. Then there exists a unique selfadjoint operator $\varphi(A) \in \mathscr{B}$ such that if*

$$\varphi(x) = \varphi_1(x) - \varphi_2(x) \tag{7}$$

where $\varphi_k(x)$ is non-negative and upper-continuous on $\mathbf{R}$ for $k = 1, 2$, then

$$\varphi(A) = \varphi_1(A) - \varphi_2(A). \tag{8}$$

The mapping $\varphi(x) \to \varphi(A)$ is a homomorphism of $\mathscr{R}$ into $\mathscr{B}$ which is order preserving in the following sense: if $\varphi(x) \in \mathscr{R}$ and $\psi(x) \in \mathscr{R}$ have the property that

$$\varphi(\xi) \geqslant \psi(\xi) \quad \text{for all } \xi \in \Sigma(A),$$

then $\varphi(A) \geqslant \psi(A)$.

PROOF. From

$$\varphi(x) = \varphi_1(x) - \varphi_2(x) = \varphi_1'(x) - \varphi_2'(x),$$

where $\varphi_k(x)$ and $\varphi_k'(x)$ are non-negative upper semi-continuous functions on $\mathbf{R}$ for $k = 1, 2$, we conclude

$$\varphi_1(x) + \varphi_2'(x) = \varphi_1'(x) + \varphi_2(x),$$
$$\varphi_1(A) + \varphi_2'(A) = \varphi_1'(A) + \varphi_2(A) \quad \text{(by theorem 2)},$$
$$\varphi_1(A) - \varphi_2(A) = \varphi_1'(A) - \varphi_2'(A).$$

The operator $\varphi(A)$ as defined in (8) is therefore independent of the particular representation (7) of $\varphi(x)$ as a difference of two non-negative upper semi-continuous functions on $\mathbf{R}$. At the same time this implies the uniqueness of $\varphi(A)$. The operator $\varphi(A)$ is selfadjoint since

$$[\varphi(A)]^* = [\varphi_1(A)]^* - [\varphi_2(A)]^* = \varphi_1(A) - \varphi_2(A) = \varphi(A).$$

If (by means of an analogous decomposition for $\psi(x) \in \mathscr{R}$)

$$\varphi(\xi) = \varphi_1(\xi) - \varphi_2(\xi) \geqslant \psi_1(\xi) - \psi_2(\xi) = \psi(\xi) \quad \text{for all } \xi \in \Sigma(A),$$

then we conclude

$$\varphi_1(\xi) + \psi_2(\xi) \geqslant \psi_1(\xi) + \varphi_2(\xi) \quad \text{for all } \xi \in \Sigma(A),$$
$$\varphi_1(A) + \psi_2(A) \geqslant \psi_1(A) + \varphi_2(A)$$
$$\varphi(A) = \varphi_1(A) - \varphi_2(A) \geqslant \psi_1(A) - \psi_2(A) = \psi(A).$$

That the mapping $\varphi(x) \to \varphi(A)$ is a homomorphism of $\mathscr{R}$ into $\mathscr{B}$ can be shown by means of the identities (6). Replacing there x by A we get for instance

$$(\varphi\psi)(A) = (\varphi_1\psi_1 + \varphi_2\psi_2)(A) - (\varphi_1\psi_2 + \varphi_2\psi_1)(A)$$
$$= [\varphi_1(A)\psi_1(A) + \varphi_2(A)\psi_2(A)] - [\varphi_1(A)\psi_2(A) + \varphi_2(A)\psi_1(A)]$$
$$\text{(by theorem 2)}$$
$$= [\varphi_1(A) - \varphi_2(A)][\psi_1(A) - \psi_2(A)] = \varphi(A)\psi(A). \ \square$$

THEOREM 5. *Let the selfadjoint operator $A \in \mathscr{B}$ and the function $\varphi(x) \in \mathscr{R}$ be given and suppose the operator $B \in \mathscr{B}$ satisfies $AB = BA$. Then $\varphi(A)B = B\varphi(A)$.*
PROOF. It suffices to prove the assertion for a non-negative upper semi-continuous function $\varphi(x)$. Suppose

$$\varphi(A) = (s)\lim_{n \to \infty} \varphi_n(A), \quad \varphi_n(x) \in \mathscr{S} \text{ for all } n \geqslant 1$$

(cf. theorem 1). By § 32 theorem 2 we have

$$\varphi(A)Bf = \lim_{n \to \infty} \varphi_n(A)Bf = \lim_{n \to \infty} B\varphi_n(A)f$$
$$= B\lim_{n \to \infty} \varphi_n(A)f = B\varphi(A)f \quad \text{for all } f \in \mathfrak{H}. \ \square$$

Example 1. Let $\varphi(x) = c_{]\alpha, \beta]}$ be the characteristic function of the interval $]\alpha, \beta]$ $(-\infty \leqslant \alpha < \beta < +\infty)$, given by

$$\varphi(\xi) = \begin{cases} 0 & \text{for } -\infty < \xi \leqslant \alpha \quad (\text{if } -\infty < \alpha), \\ 1 & \text{for } \alpha < \xi \leqslant \beta, \\ 0 & \text{for } \beta < \xi < +\infty. \end{cases}$$

Then we have $\varphi(x) \in \mathscr{R}$ (in fact $\varphi(x) = \varphi_\beta(x) - \varphi_\alpha(x)$ for $-\infty < \alpha$ where $\varphi_\alpha(x)$ and $\varphi_\beta(x)$ are defined as in (1)), and $\varphi^2(x) = \varphi(x)$. From $[\varphi(A)]^2 = \varphi(A)$ we conclude that $\varphi(A) = \varphi_\beta(A) - \varphi_\alpha(A)$ is a projection (cf. § 15 theorem 2). By theorem 5 this projection commutes with every operator $B \in \mathscr{B}$ which commutes with A. By § 21 theorem 3 the corresponding subspace reduces every such operator B, in particular it reduces A.

The following exercise exhibits a decomposition of a bounded selfadjoint operator on $\mathfrak{H}$ analogous to the familiar decomposition of real-valued functions into positive and negative parts (cf. appendix B6 definition 5).

Exercise. Let $A \in \mathscr{B}$ be selfadjoint and let $P = I - \varphi_0(A)$ where $\varphi_0(x) \in \mathscr{R}$ is defined as in (1).

a) If $B \in \mathscr{B}$ is positive selfadjoint and commutes with A, then $O \leqslant PB \leqslant B$. (Hint: B then also commutes with P and $P^\perp$).

b) The operators $A^+ = PA$ and $A^- = -P^\perp A$ are positive selfadjoint, commute with A, and $A = A^+ - A^-$. (Hint: consider the functions $x[1 - \varphi_0(x)]$ and $-x\varphi_0(x)$.)

c) If the operators $B_1 \in \mathscr{B}$ and $B_2 \in \mathscr{B}$ are positive selfadjoint, commute with A, and satisfy $A = B_1 - B_2$, then $B_1 \geqslant A^+$ and $B_2 \geqslant A^-$. (Hint: use a).)

§ 34. The spectral decomposition of a bounded selfadjoint operator

Let A be a bounded selfadjoint operator on $\mathfrak{H}$. Recall that we wanted to split the operator A into a sum (or something similar) of simpler operators, namely of scalar multiples of projections.

In fact we already have a family of projections at hand which seems to be closely related to A: if the functions $\varphi_\lambda(x)$ are defined for $\lambda \in \mathbf{R}$ as in § 33 (1) (fig. 20), then for every $\lambda \in \mathbf{R}$ the operator

$$(1) \qquad\qquad P(\lambda) = \varphi_\lambda(A)$$

is a projection which reduces A (cf. § 33 example 1). Let us have a closer look at the properties of this one-parameter family of projections $\{P(\lambda): \lambda \in \mathbf{R}\}$.

(a) From $\varphi_\lambda(x) \leqslant \varphi_{\lambda'}(x)$ for $\lambda \leqslant \lambda'$ we conclude

$$P(\lambda) \leqslant P(\lambda') \quad \text{for } \lambda \leqslant \lambda'$$

(cf. § 33 theorem 2).

(b) Let θ be any positive real number and let $\alpha_1 = \min \Sigma(A)$, $\alpha_2 = \max \Sigma(A)$ (cf. § 30 corollary 8.1). From

$$\varphi_{\alpha_1 - \theta}(\xi) = 0 \quad \text{for all } \xi \in \Sigma(A)$$

we conclude

$$(2) \qquad\qquad P(\alpha_1 - \theta) = O \quad \text{for } \theta > 0$$

(cf. § 33 corollary 2.1). On the other hand we have

$$\varphi_{\alpha_2}(\xi) = 1 \quad \text{for all } \xi \in \Sigma(A)$$

and therefore, again by § 33 corollary 2.1,

$$P(\alpha_2) = I.$$

By (a) this also implies

$$P(\lambda) = O \quad \text{for } -\infty < \lambda < \alpha_1,$$
$$P(\lambda) = I \quad \text{for } \alpha_2 \leqslant \lambda < +\infty.$$

(c) Let the continuous function $\varphi_{\lambda,n}(x) \in \mathscr{S}$ be defined for $\lambda \in \mathbf{R}$ and $n \geqslant 1$ as in § 33 (2) (fig. 21) and let $\{\theta_n\}_{n=1}^{\infty}$ be any decreasing sequence of positive real numbers converging to zero. The sequence $\{\varphi_{\lambda+\theta_n,n}(x)\}_{n=1}^{\infty}$ also decreases and converges pointwise to $\varphi_{\lambda}(x)$ (fig. 22).

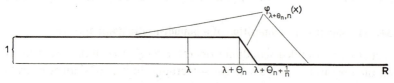

Fig. 22

From $\varphi_{\lambda}(x) \leqslant \varphi_{\lambda+\theta_n}(x) \leqslant \varphi_{\lambda+\theta_n,n}(x)$ we deduce

$$P(\lambda) \leqslant (\text{s}) \lim_{n \to \infty} P(\lambda + \theta_n) \leqslant (\text{s}) \lim_{n \to \infty} \varphi_{\lambda+\theta_n,n}(A) = P(\lambda),$$

$$P(\lambda) = (\text{s}) \lim_{n \to \infty} P(\lambda + \theta_n).$$

Since this limit only depends on λ but not on the particular choice of the sequence $\{\theta_n\}_{n=1}^{\infty}$ we shall write the last equation in the form

$$P(\lambda) = P(\lambda + 0) \quad \text{for all } \lambda \in \mathbf{R}$$

and express this statement in words by saying that the one-parameter family $\{P(\lambda): \lambda \in \mathbf{R}\}$ is continuous from the right.

In fact, for every $\lambda \in \mathbf{R}$ there also exists a unique projection denoted by $P(\lambda - 0)$ with the property that

$$P(\lambda - 0) = (\text{s}) \lim_{n \to \infty} P(\lambda - \theta_n)$$

for every sequence $\{\theta_n\}_{n=1}^{\infty}$ as described above. Indeed, given two such sequences $\{\theta_n\}_{n=1}^{\infty}$ and $\{\theta_n'\}_{n=1}^{\infty}$, the operators

$$P = (\text{s}) \lim_{n \to \infty} P(\lambda - \theta_n),$$

$$P' = (\text{s}) \lim_{n \to \infty} P(\lambda - \theta_n')$$

are projections by § 30 theorem 5. Moreover, from

$$P(\lambda - \theta_m) \leqslant (\text{s}) \lim_{n \to \infty} P(\lambda - \theta_n') = P',$$

$$P(\lambda - \theta_m') \leqslant (\text{s}) \lim_{n \to \infty} P(\lambda - \theta_n) = P$$

we conclude

$$P = (s) \lim_{m \to \infty} P(\lambda - \theta_m) \leqslant P',$$

$$P' = (s) \lim_{m \to \infty} P(\lambda - \theta'_m) \leqslant P,$$

$$P = P'.$$

Note that $P(\lambda - 0) \leqslant P(\lambda)$ but $P(\lambda - 0)$ need not coincide with $P(\lambda)$ (cf. theorem 5).

Using this notation we may replace (2) by the statement

$$P(\alpha_1 - 0) = O.$$

Indeed this statement is not only implied by (2) but also implies (2) since by (a) for $\lambda < \alpha_1$ we have

$$O \leqslant P(\lambda) \leqslant (s) \lim_{n \to \infty} P(\alpha_1 - \theta_n) = P(\alpha_1 - 0) = O.$$

(d) Let the function $\varphi(x)$ be continuous and real-valued on **R**. If we fix any $\theta > 0$, then $\varphi(x)$ is uniformly continuous on the interval $[\alpha_1 - \theta, \alpha_2]$, i.e. for every $\varepsilon > 0$ there exists a $\delta > 0$ such that

$$|\varphi(\xi) - \varphi(\eta)| \leqslant \varepsilon$$

as soon as $\xi \in [\alpha_1 - \theta, \alpha_2]$, $\eta \in [\alpha_1 - \theta, \alpha_2]$ and $|\xi - \eta| \leqslant \delta$. Choose any subdivision

(3) $$\alpha_1 - \theta = \lambda_0 < \lambda_1 < \cdots < \lambda_{n-1} < \lambda_n = \alpha_2$$

of the interval $[\alpha_1 - \theta, \alpha_2]$ such that

$$\max \{\lambda_k - \lambda_{k-1} : 1 \leqslant k \leqslant n\} \leqslant \delta$$

and for $1 \leqslant k \leqslant n$ choose any real number $\lambda'_k \in [\lambda_{k-1}, \lambda_k]$. We obtain

$$-\varepsilon \leqslant \varphi(\xi) - \varphi(\lambda'_k) \leqslant \varepsilon \quad \text{for } \xi \in [\lambda_{k-1}, \lambda_k], \quad 1 \leqslant k \leqslant n,$$

$$\varphi(\lambda'_k) [\varphi_{\lambda_k}(\xi) - \varphi_{\lambda_{k-1}}(\xi)] = \begin{cases} \varphi(\lambda'_k) & \text{for } \xi \in]\lambda_{k-1}, \lambda_k], \\ 0 & \text{for } \xi \notin]\lambda_{k-1}, \lambda_k], \end{cases} \quad 1 \leqslant k \leqslant n,$$

and therefore (fig. 23)

$$-\varepsilon \leqslant \varphi(\xi) - \sum_{k=1}^{n} \varphi(\lambda'_k) [\varphi_{\lambda_k}(\xi) - \varphi_{\lambda_{k-1}}(\xi)] \leqslant \varepsilon \quad \text{for } \xi \in]\alpha_1 - \theta, \alpha_2].$$

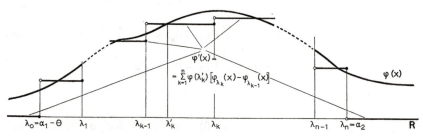

Fig. 23

By § 33 theorem 4 we conclude

$$- \varepsilon I \leqslant \varphi(A) - \sum_{k=1}^{n} \varphi(\lambda_k') \left[P(\lambda_k) - P(\lambda_{k-1}) \right] \leqslant \varepsilon I.$$

Applying § 30 theorem 6 to the selfadjoint operator

$$\varphi(A) - \sum_{k=1}^{n} \varphi(\lambda_k') \left[P(\lambda_k) - P(\lambda_{k-1}) \right]$$

we get

(4) $$\left\| \varphi(A) - \sum_{k=1}^{n} \varphi(\lambda_k') \left[P(\lambda_k) - P(\lambda_{k-1}) \right] \right\| \leqslant \varepsilon.$$

Let us reconsider the meaning of this last statement. A first look at formula (4) reveals that we can at least come close to the operator $\varphi(A)$ by finite linear combinations of projections. Indeed, the operator $P(\lambda_k) - P(\lambda_{k-1})$ is a projection since $P(\lambda_k) \geqslant P(\lambda_{k-1})$ (cf. § 15 theorem 6 e), f) and § 33 example 1). But even more is true: if the "mesh" of the subdivision (3) is taken smaller and smaller (by letting δ tend to zero), then the sum

$$\sum_{k=1}^{n} \varphi(\lambda_k') \left[P(\lambda_k) - P(\lambda_{k-1}) \right]$$

approaches the operator $\varphi(A)$, independently of the particular choice of the subdivisions (3) and the numbers $\lambda_k' \in [\lambda_{k-1}, \lambda_k]$. A similar situation occurs if the non-decreasing projection-valued function $P(x)$ is replaced by some non-decreasing real-valued function $\psi(x)$: it is well known (and shown by a similar argument) that the "*Riemann-Stieltjes sums*"

$$\sum_{k=1}^{n} \varphi(\lambda_k') \left[\psi(\lambda_k) - \psi(\lambda_{k-1}) \right]$$

then approach a real number which is called the *Riemann-Stieltjes integral* of $\varphi(x)$ (with respect to the function $\psi(x)$) over $[\alpha_1 - \theta, \alpha_2]$ and which is denoted by

$$\int\limits_{\alpha_1-\theta}^{\alpha_2} \varphi(\lambda)\, d\psi(\lambda)$$

(cf. appendix B2). Since we are dealing here with a perfectly analogous situation there is no reason why we should not use the same notation and write

(5)
$$\varphi(A) = \int\limits_{\alpha_1-\theta}^{\alpha_2} \varphi(\lambda)\, dP(\lambda)$$

in order to describe the limiting process in question. The meaning of this notation is underlined by saying that the integral in (5) is of *Riemann-Stieltjes type*.

Formula (5) still incorporates a somewhat strange feature, namely the positive but otherwise arbitrary number θ in the lower limit of the integral. The reason why θ is arbitrary lies in the fact that every term

$$\varphi(\lambda_k')\, [P(\lambda_k) - P(\lambda_{k-1})], \quad \lambda_{k-1} \leqslant \lambda_k' \leqslant \lambda_k < \alpha_1,$$

in the Riemann-Stieltjes sums approaching the integral in (5) simply vanishes. Indeed, by (a) we have $P(\lambda) = O$ for $\lambda < \alpha_1$. The reason why the number θ cannot completely be done away with lies in the fact that, choosing $\lambda_1' = \lambda_1 = \alpha_1$, the term

(6) $\varphi(\alpha_1)\, [P(\alpha_1) - P(\lambda_0)] = \varphi(\alpha_1)\, P(\alpha_1), \quad \lambda_0 < \lambda_1' = \lambda_1 = \alpha_1,$

might insist to make a non-zero contribution to the Riemann-Stieltjes sums approaching the integral in (5). Indeed, this will be the case if $\varphi(\alpha_1) \neq 0$ and $P(\alpha_1) > O$. Still, we may indicate (if also somewhat imprecisely) the fact that $\theta > 0$ may be chosen arbitrarily small by writing (5) in the form

$$\varphi(A) = \int\limits_{\alpha_1-0}^{\alpha_2} \varphi(\lambda)\, dP(\lambda).$$

In particular, for $\varphi(x) = x$ we obtain

$$A = \int\limits_{\alpha_1-0}^{\alpha_2} \lambda\, dP(\lambda).$$

Furthermore, in the Riemann-Stieltjes sums approaching the integral we may take the last mentioned fact into account by choosing the subdivision without loss of generality in the form

$$\lambda_0 < \alpha_1 = \lambda_1 < \cdots < \lambda_{n-1} < \lambda_n = \alpha_2.$$

Finally, let the function $\varphi(x) = \varphi^{(1)}(x) + i\varphi^{(2)}(x)$ be continuous and complex-valued on $\mathbf{R}$ with real and imaginary parts $\varphi^{(j)}(x)$ $(j=1, 2)$. Choosing the number $\delta > 0$ as above we get

$$|\varphi^{(j)}(\xi) - \varphi^{(j)}(\eta)| \leqslant \varepsilon \qquad (j = 1, 2)$$

if $\xi \in [\alpha_1 - \delta, \alpha_2]$, $\eta \in [\alpha_1 - \delta, \alpha_2]$, and $|\xi - \eta| \leqslant \delta$, and therefore

$$\left\| \varphi^{(j)}(A) - \sum_{k=1}^{n} \varphi^{(j)}(\lambda'_k) \left[P(\lambda_k) - P(\lambda_{k-1}) \right] \right\| \leqslant \varepsilon \quad \text{for } j = 1, 2,$$

$$\left\| \varphi(A) - \sum_{k=1}^{n} \varphi(\lambda'_k) \left[P(\lambda_k) - P(\lambda_{k-1}) \right] \right\| \leqslant$$

$$\leqslant \sum_{j=1}^{2} \left\| \varphi^{(j)}(A) - \sum_{k=1}^{n} \varphi^{(j)}(\lambda'_k) \left[P(\lambda_k) - P(\lambda_{k-1}) \right] \right\| \leqslant 2\varepsilon.$$

We sum up these results in a theorem.

THEOREM 1 (SPECTRAL THEOREM FOR BOUNDED SELFADJOINT OPERATORS). *Let A be a bounded selfadjoint operator on $\mathfrak{H}$ and let $\alpha_1 = \min \Sigma(A)$, $\alpha_2 = \max \Sigma(A)$. Then there exists a family of projections $\{P(\lambda): \lambda \in \mathbf{R}\}$, called the spectral family of A, with the following properties:*

  a) $P(\lambda) \leqslant P(\lambda')$ *for* $\lambda \leqslant \lambda'$;
  b) $P(\alpha_1 - 0) = O$,
        $P(\alpha_2) = I$;
  c) $P(\lambda + 0) = P(\lambda)$ *for all* $\lambda \in \mathbf{R}$;
  d) $A = \displaystyle\int_{\alpha_1 - 0}^{\alpha_2} \lambda \, dP(\lambda).$

More generally, for every continuous complex-valued function $\varphi(x)$ on $\mathbf{R}$ and for every $\varepsilon > 0$ there exists a $\delta > 0$ such that

(7)
$$\left\| \varphi(A) - \sum_{k=1}^{n} \varphi(\lambda'_k) \left[P(\lambda_k) - P(\lambda_{k-1}) \right] \right\| \leqslant \varepsilon$$

whenever

(8)
$$\begin{cases} \lambda_0 < \alpha_1 = \lambda_1 < \cdots < \lambda_{n-1} < \lambda_n = \alpha_2, \\ \lambda_k - \lambda_{k-1} \leqslant \delta \quad \text{for } 1 \leqslant k \leqslant n, \\ \lambda'_k \in [\lambda_{k-1}, \lambda_k] \quad \text{for } 1 \leqslant k \leqslant n. \end{cases}$$

In other words

$$\varphi(A) = \int\limits_{\alpha_1-0}^{\alpha_2} \varphi(\lambda)\,\mathrm{d}P(\lambda)$$

where the integral is of Riemann-Stieltjes type.

Remark 1. From what has been said in the discussion of (d) it follows that in order to guarantee (7) it suffices to choose δ so small that $|\varphi(\xi)-\varphi(\eta)| \leqslant \frac{1}{2}\varepsilon$ (or $|\varphi(\xi)-\varphi(\eta)| \leqslant \varepsilon$ in case of a real-valued function $\varphi(x)$) whenever $\xi\in[\alpha_1-\delta,\alpha_2]$, $\eta\in[\alpha_1-\delta,\alpha_2]$, and $|\xi-\eta| \leqslant \delta$.

COROLLARY 1.1. *Let A and $\{P(\lambda): \lambda\in\mathbf{R}\}$ be as in theorem 1. If $\varphi(x)$ is a continuous complex-valued function on $\mathbf{R}$, then*

$$\varphi(A)f = \int\limits_{\alpha_1-0}^{\alpha_2} \varphi(\lambda)\,\mathrm{d}P(\lambda)\,f \qquad \text{for all } f\in\mathfrak{H},$$

$$\langle\varphi(A)f,g\rangle = \int\limits_{\alpha_1-0}^{\alpha_2} \varphi(\lambda)\,\mathrm{d}\langle P(\lambda)f,g\rangle \quad \text{for all } f\in\mathfrak{H} \text{ and all } g\in\mathfrak{H}.$$

More explicitly, for every $\varepsilon>0$ there exists a $\delta>0$ such that

$$\left\|\varphi(A)f - \sum_{k=1}^{n} \varphi(\lambda'_k)\left[P(\lambda_k)f - P(\lambda_{k-1})f\right]\right\| \leqslant \varepsilon\|f\|\,\text{for all } f\in\mathfrak{H},$$

$$\left|\langle\varphi(A)f,g\rangle - \sum_{k=1}^{n} \varphi(\lambda'_k)\left[\langle P(\lambda_k)f,g\rangle - \langle P(\lambda_{k-1})f,g\rangle\right]\right| \leqslant \varepsilon\|f\|\,\|g\|$$

$$\text{for all } f\in\mathfrak{H} \text{ and all } g\in\mathfrak{H}$$

whenever (8) is satisfied.

COROLLARY 1.2. *Let A and $\{P(\lambda): \lambda\in\mathbf{R}\}$ be as in theorem 1. If $\varphi(x)$ is a continuous complex-valued function on $\mathbf{R}$, then*

$$\|\varphi(A)f\|^2 = \int\limits_{\alpha_1-0}^{\alpha_2} |\varphi(\lambda)|^2\,\mathrm{d}\,\|P(\lambda)f\|^2 \quad \text{for all } f\in\mathfrak{H}.$$

More explicitly, for every $\varepsilon>0$ there exists a $\delta>0$ such that

$$\left|\|\varphi(A)f\|^2 - \sum_{k=1}^{n} |\varphi(\lambda'_k)|^2 \left[\|P(\lambda_k)f\|^2 - \|P(\lambda_{k-1})f\|^2\right]\right| \leqslant \varepsilon\|f\|^2$$

$$\text{for all } f\in\mathfrak{H}$$

whenever (8) is satisfied.

PROOF. Applying corollary 1.1 to the function $|\varphi(x)|^2 = \bar\varphi(x)\,\varphi(x)$ we get in particular

$$|\langle [\varphi(A)]^*\,\varphi(A)\,f, f\rangle - \sum_{k=1}^{n} |\varphi(\lambda'_k)|^2\,[\langle P(\lambda_k)\,f, f\rangle - \langle P(\lambda_{k-1})\,f, f\rangle]| =$$

$$= |\,\|\varphi(A)\,f\|^2 - \sum_{k=1}^{n} |\varphi(\lambda'_k)|^2\,[\|P(\lambda_k)\,f\|^2 - \|P(\lambda_{k-1})\,f\|^2|\, \leqslant \varepsilon\,\|f\|^2.\ \square$$

Note that corollary 1.2 and the second formula in corollary 1.1 contain ordinary Riemann-Stieltjes integrals (cf. appendix B2) of which the one in corollary 1.1 still displays a somewhat unusual feature: while the function $\psi(x) = \|P(x)f\|^2$ is real-valued (even non-negative) and non-decreasing (cf. § 15 theorem 6 d)), in general the function $\psi'(x) = \langle P(x)f, g\rangle = \langle P(x)f, P(x)g\rangle$ is complex-valued. Still, the definition of the Riemann-Stieltjes integral

$$\int_{\alpha_1 - 0}^{\alpha_2} \varphi(\lambda)\,\mathrm{d}\langle P(\lambda)\,f, g\rangle$$

by means of Riemann-Stieltjes sums is carried through without any difficulty. This will appear less surprising if one realizes that by the polar identity (§ 13 theorem 1) the function $\psi'(x)$ is a linear combination of four real-valued non-decreasing functions.

The next theorem shows that it is legitimate to talk about "*the*" spectral family of A since it is uniquely determined by the requirements a), b), c), d) in theorem 1.

THEOREM 2. *Let A be a bounded selfadjoint operator on $\mathfrak{H}$ and let $\{P'(\lambda):$ $\lambda \in \mathbf{R}\}$ be a family of projections satisfying the requirements* a), b), c), d) *in theorem 1. Then $P'(\lambda) = P(\lambda)$ where $P(\lambda)$ is defined as in* (1).
PROOF. We first show

$$(9)\qquad P(\lambda)\,P'(\mu) = P'(\mu)\,P(\lambda)\quad\text{for all } \lambda \in \mathbf{R} \text{ and all } \mu \in \mathbf{R}$$

which implies that $P(\lambda)\,P'(\mu)$ is again a projection (cf. § 15 theorem 4). Indeed, $P'(\mu)$ commutes with all projections $P'(\lambda)$ $(\lambda \in \mathbf{R})$ by property a) and by § 15 theorem 6 b), c), e). As a consequence, $P'(\mu)$ commutes with all Riemann-Stieltjes sums approximating the integral

$$A = \int_{\alpha_1 - 0}^{\alpha_2} \lambda\,\mathrm{d}P'(\lambda)$$

and therefore also with A. Applying § 33 theorem 5 (with $B = P'(\mu)$ and $\varphi(A) = \varphi_\lambda(A) = P(\lambda)$) we obtain (9).

Assume now there were a $\mu \in \mathbf{R}(\alpha_1 \leqslant \mu < \alpha_2)$ such that $P(\mu) \neq P'(\mu)$. The projection $P'(\mu) P(\mu)$ cannot both equal $P(\mu)$ and $P'(\mu)$. Without loss of generality we assume

$$P'(\mu) P(\mu) \neq P(\mu),$$

hence

$$[I - P'(\mu)] P(\mu) \neq O.$$

Note that also $[I - P'(\mu)]$ commutes with $P(\mu)$ and therefore $[I - P'(\mu)] P(\mu)$ is a non-zero projection. Let f be a non-zero vector in the corresponding subspace. We obtain

$$f = [I - P'(\mu)] P(\mu) f,$$

(10) $\qquad P(\mu) f = P(\mu) [I - P'(\mu)] P(\mu) f = f,$

(11) $\qquad P'(\mu) f = P'(\mu) [I - P'(\mu)] P(\mu) f = o \quad$ (since $f \in [I - P'(\mu)] \mathfrak{H}$).

From property b) and d) we infer

$$(A - \mu I) = \int\limits_{\alpha_1 - 0}^{\alpha_2} (\lambda - \mu) \, dP(\lambda) = \int\limits_{\alpha_1 - 0}^{\alpha_2} (\lambda - \mu) \, dP'(\lambda).$$

This again as in corollary 1.1 implies

$$\langle (A - \mu I) f, f \rangle = \int\limits_{\alpha_1 - 0}^{\alpha_2} (\lambda - \mu) \, d \langle P(\lambda) f, f \rangle = \int\limits_{\alpha_1 - 0}^{\alpha_2} (\lambda - \mu) \, d \langle P'(\lambda) f, f \rangle.$$

From (10) we conclude

$$\langle [P(\lambda_k) - P(\lambda_{k-1})] f, f \rangle = 0 \quad \text{for } \lambda_{k-1} \geqslant \mu.$$

Using μ as a subdivision point for all our Riemann-Stieltjes sums we therefore get

(12) $\qquad \langle (A - \mu I) f, f \rangle = \int\limits_{\alpha_1 - 0}^{\mu} (\lambda - \mu) \, d \langle P(\lambda) f, f \rangle \leqslant 0.$

On the other hand, from (11) we conclude

$$\langle [P'(\lambda_k) - P'(\lambda_{k-1})] f, f \rangle = 0 \quad \text{for } \lambda_k \leqslant \mu.$$

Moreover, by property c) we may choose $\theta > 0$ so small that

$$\langle P'(\mu + \theta) f, f \rangle \leqslant \tfrac{1}{2} \|f\|^2$$

and therefore

$$\langle [I - P'(\mu + \theta)] f, f \rangle \geqslant \tfrac{1}{2} \|f\|^2 .$$

Using μ and $\mu + \theta$ as subdivision points for all our Riemann-Stieltjes sums we therefore obtain

$$\langle (A - \mu I) f, f \rangle = \int_{\mu}^{\mu+\theta} (\lambda - \mu) \, d \langle P'(\lambda) f, f \rangle + \int_{\mu+\theta}^{\alpha_2} (\lambda - \mu) \, d \langle P'(\lambda) f, f \rangle$$

$$\geqslant \int_{\mu+\theta}^{\alpha_2} \theta \, d \langle P'(\lambda) f, f \rangle = \theta \langle [I - P'(\mu + \theta)] f, f \rangle$$

$$\geqslant \tfrac{1}{2} \theta \|f\|^2 > 0 .$$

Together with (12) this supplies the looked-for contradiction. $\square$

Example 1. Let $\mathfrak{H} = \mathfrak{L}_2(\alpha, \beta)$ $(-\infty < \alpha < \beta < +\infty)$ and let $A = E$ be the multiplication operator in $\mathfrak{H}$ (cf. § 19 theorem 1) defined by

$$Ef = xf \quad \text{for all } f \in \mathfrak{H}.$$

Recall that $\Sigma(E) = [\alpha, \beta]$ and that therefore $\alpha_1 = \min \Sigma(E) = \alpha$, $\alpha_2 = \max \Sigma(E) = \beta$ (cf. § 23 example 1). Define the projection $P'(\lambda)$ by

$$P'(\lambda) = O \quad \text{for } \lambda \leqslant \alpha ,$$
$$P'(\lambda) = I \quad \text{for } \lambda \geqslant \beta ,$$

and let $P'(\lambda)$ for $\alpha < \lambda < \beta$ be the projection upon $\mathfrak{L}_2(\alpha, \lambda)$ considered as a subspace of $\mathfrak{H} = \mathfrak{L}_2(\alpha, \beta)$, given by

$$P'(\lambda) f(\xi) = \begin{cases} f(\xi) & \text{for } \alpha < \xi < \lambda \\ 0 & \text{for } \lambda \leqslant \xi < \beta \end{cases}$$

(cf. § 15 example 1, § 21 example 1). The family $\{P'(\lambda) : \lambda \in \mathbf{R}\}$ obviously has the properties a), b), c) mentioned in theorem 1. We even have $P'(\lambda) = P'(\lambda - 0) = P'(\lambda + 0)$ for all $\lambda \in \mathbf{R}$, i.e. the family $\{P'(\lambda) : \lambda \in \mathbf{R}\}$ is everywhere continuous. As a consequence, if $\varphi(x)$ is any complex-valued continuous function on $\mathbf{R}$, then in each of the Riemann-Stieltjes sums entering into the definition of the integral

$$\int_{\alpha_1 - 0}^{\alpha_2} \varphi(\lambda) \, dP'(\lambda)$$

the starting term $\varphi(\lambda_1') [P'(\alpha_1) - P'(\lambda_0)]$ vanishes (cf. (6)). In other words, we may simply forget about the numbers λ_0, λ_1' and replace the lower limit $\alpha_1 - 0$ by $\alpha_1 = \alpha$.

In order to check property d) let $\varepsilon > 0$ be given and suppose

$$\alpha = \lambda_1 < \lambda_2 < \cdots < \lambda_n = \beta ,$$
$$\lambda_k - \lambda_{k-1} \leqslant \varepsilon \quad \text{for } 2 \leqslant k \leqslant n ,$$
$$\lambda_k' \in [\lambda_{k-1}, \lambda_k] \quad \text{for } 2 \leqslant k \leqslant n .$$

Observe that $P'(\lambda_k) - P'(\lambda_{k-1})$ is the projection upon $\mathfrak{L}_2(\lambda_{k-1}, \lambda_k)$ considered as a subspace of $\mathfrak{H} = \mathfrak{L}_2(\alpha, \beta)$ for $2 \leqslant k \leqslant n$. We obtain

$$\left\| Ef - \sum_{k=2}^{n} \lambda_k' [P'(\lambda_k) - P'(\lambda_{k-1})] f \right\|^2 = \sum_{k=2}^{n} \int_{\lambda_{k-1}}^{\lambda_k} |\xi - \lambda_k'|^2 \, |f(\xi)|^2 \, d\xi$$

$$\leqslant \sum_{k=2}^{n} \varepsilon^2 \int_{\lambda_{k-1}}^{\lambda_k} |f(\xi)|^2 \, d\xi = \varepsilon^2 \, \|f\|^2 ,$$

$$\left\| E - \sum_{k=2}^{n} \lambda_k' [P'(\lambda_k) - P'(\lambda_{k-1})] \right\| \leqslant \varepsilon ,$$

$$(13) \qquad E = \int_{\alpha}^{\beta} \lambda \, dP'(\lambda) .$$

By theorem 2 we conclude that $\{P'(\lambda) : \lambda \in \mathbf{R}\}$ is indeed the spectral family of the operator E.

Formula (13) admits a simple intuitive interpretation in this case: multiplication by x within $\mathfrak{L}_2(\alpha, \beta)$ may be "approximated" (in the operator norm) by multiplication by step functions (fig. 24). Moreover, a similar reasoning

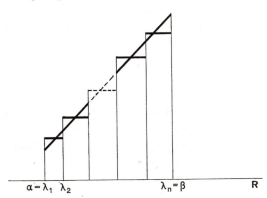

Fig. 24

with Riemann-Stieltjes sums as carried through above reveals that the oper-
ator $\varphi(E)$ coincides with the multiplication (within $\mathfrak{L}_2(\alpha, \beta)$) by the function
$\varphi(x)$ (cf. § 19 exercise).

Example 2. Let A be a compact selfadjoint operator on $\mathfrak{H}$. In order to fix
our ideas and to avoid notational complications we only consider a special
case, the general case being treated similarly. Suppose therefore that A is not
of finite rank, that all non-zero eigenvalues $\lambda_k(1 \leqslant k \leqslant \infty)$ are negative ($-A$
is then a positive selfadjoint operator) and that

(14)
$$A = \sum_{k=1}^{\infty} \lambda_k P_k,$$
$$\lambda_1 < \lambda_2 < \cdots \leqslant 0, \quad \lim_{k \to \infty} \lambda_k = 0$$

is the spectral decomposition of A (cf. § 28 theorem 2). The operator P_k is
the projection upon the eigenspace $\mathfrak{M}_k$ corresponding to the eigenvalue λ_k
of A. Let $\mathfrak{M}_0$ be the eigenspace corresponding to the "eigenvalue" $\lambda_0 = 0$
(we formally admit the case $\mathfrak{M}_0 = \mathfrak{O}$). We then have

$$\Sigma(A) = \{\lambda_k\}_{k=0}^{\infty},$$
$$\alpha_1 = \min \Sigma(A) = \lambda_1,$$
$$\alpha_2 = \max \Sigma(A) = 0$$

(cf. § 27 theorem 7 and § 27 theorem 8). For $-\infty < \lambda < +\infty$ let $P'(\lambda)$ be the
projection upon the subspace

$$\mathfrak{M}(\lambda) = \bigvee_{\lambda_k \leqslant \lambda} \mathfrak{M}_k.$$

It is left to the reader to check that the family $\{P'(\lambda): \lambda \in \mathbf{R}\}$ indeed satisfies
the requirements a), b), c), d) mentioned in theorem 1 and that the integral

$$A = \int_{\alpha_1 - 0}^{\alpha_2} \lambda \, dP'(\lambda)$$

reduces to the series (14) (observe that $P'(\lambda_k) - P'(\lambda_{k-1}) = P_k$ for $1 \leqslant k < \infty$
and that $\|A - \sum_{k=1}^{n} \lambda_k P_k\| = |\lambda_{n+1}|$).

Moreover, if $\varphi(x)$ is a complex-valued continuous function on $\mathbf{R}$ vanish-
ing at zero, then formula (5) reduces to

(15)
$$\varphi(A) = \sum_{k=1}^{\infty} \varphi(\lambda_k) P_k,$$

which at the same time implies that $\varphi(A)$ is again a compact operator (cf. § 28 theorem 5). On the other hand, if $\varphi(0)\neq 0$, then the series $\sum_{k=0}^{\infty}\varphi(\lambda_k)P_k$ diverges in $\mathscr{B}$. While formula (5) still holds it does not reduce to (15). This agrees with the fact that the operator $\varphi(A)$ (in the case under consideration) will not any more be compact: its eigenvalues $\varphi(\lambda_k)$ may have $\varphi(0)$ as a non-zero accumulation point or $\varphi(0)$ may be an eigenvalue with an infinite-dimensional corresponding eigenspace (cf. § 27 corollary 2.2 and § 27 corollary 2.1).

By theorem 1 and theorem 2, the spectral family $\{P(\lambda): \lambda\in\mathbf{R}\}$ uniquely determines and in turn is uniquely determined by the operator A. Moreover, the spectral family also reflects in a direct way the properties of the operator A (cf. also § 28 theorem 3).

THEOREM 3. *Let* $\{P(\lambda): \lambda\in\mathbf{R}\}$ *be the spectral family of the bounded selfadjoint operator* A *on* $\mathfrak{H}$. *If* B *is a bounded linear operator on* $\mathfrak{H}$, *then* $AB=BA$ *iff*

(16) $P(\lambda)\,B = BP(\lambda)$ *for all* $\lambda\in\mathbf{R}$.

In particular,
$$P(\lambda)\,A = AP(\lambda)   \text{ for all } \lambda\in\mathbf{R}.$$

PROOF. From $AB=BA$ follows (16) by § 33 theorem 5 and (1). Conversely, if B satisfies (16), then B commutes with all Riemann-Stieltjes sums approximating the integral $\int_{\alpha_1-0}^{\alpha_2}\lambda dP(\lambda)$ and therefore also with A (cf. the proof of § 32 theorem 2). $\square$

THEOREM 4. *Let* $\{P(\lambda): \lambda\in\mathbf{R}\}$ *be the spectral family of the bounded self-adjoint operator* A *on* $\mathfrak{H}$. *A number* $\mu\in\mathbf{R}$ *is a regular value for* A *iff there exists a* $\theta>0$ *such that* $P(\mu-\theta)=P(\mu+\theta)$.

PROOF. a) Suppose μ is a regular value for A. Since $\Sigma(A)$ is a closed subset of $\mathbf{R}$, there exists an open interval of positive radius 2θ about μ which does not contain any point of $\Sigma(A)$. Let the continuous functions $\varphi_{\mu-\theta,n}(x)$ and $\varphi_{\mu+\theta,n}(x)$ be defined as in § 33 (2) (fig. 25).

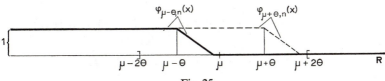

Fig. 25

For sufficiently large n (i.e. for $1/n < \theta$) we then have

$$\varphi_{\mu-\theta,n}(\xi) = \varphi_{\mu+\theta,n}(\xi) \quad \text{for all } \xi \in \Sigma(A),$$
$$\varphi_{\mu-\theta,n}(A) = \varphi_{\mu+\theta,n}(A) \quad \text{(by § 32 corollary 3.3)},$$
$$P(\mu - \theta) = \text{(s)} \lim_{n\to\infty} \varphi_{\mu-\theta,n}(A) = \text{(s)} \lim_{n\to\infty} \varphi_{\mu+\theta,n}(A) = P(\mu + \theta).$$

b) Suppose $P(\mu-\theta) = P(\mu+\theta)$ for some $\theta > 0$. Choosing $\mu - \theta$ and $\mu + \theta$ as subdivision points for all Riemann-Stieltjes sums approximating the integral in (5) we get

$$\varphi(A) = \int_{\alpha_1 - 0}^{\mu - \theta} \varphi(\lambda)\, dP(\lambda) + \int_{\mu + \theta}^{\alpha_2} \varphi(\lambda)\, dP(\lambda)$$

for any complex-valued continuous function $\varphi(x)$ on **R**. Let in particular the function $\psi(x)$ be defined by

$$\psi(\xi) = \begin{cases} \dfrac{1}{\xi - \mu} & \text{for } |\xi - \mu| \geqslant \theta, \\[2mm] \dfrac{1}{\theta^2}(\xi - \mu) & \text{for } |\xi - \mu| \leqslant \theta, \end{cases} \qquad \text{(fig. 26)}.$$

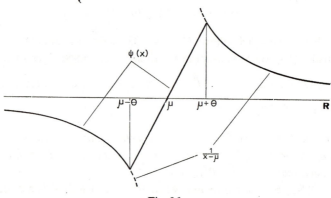

Fig. 26

Putting $\varphi(x) = (x - \mu)\,\psi(x)$ we get

$$\varphi(A) = (A - \mu I)\,\psi(A) = \psi(A)\,(A - \mu I)$$
$$= \int_{\alpha_1 - 0}^{\mu - \theta} (\lambda - \mu)\,\psi(\lambda)\, dP(\lambda) + \int_{\mu + \theta}^{\alpha_2} (\lambda - \mu)\,\psi(\lambda)\, dP(\lambda)$$
$$= [P(\mu - \theta) - O] + [I - P(\mu + \theta)] = I.$$

Therefore $(A - \mu I)$ is invertible and μ is a regular value for A. $\square$

COROLLARY 4.1. *Let A and $\{P(\lambda): \lambda \in \mathbf{R}\}$ be as in theorem 4. Then $\mu \in \Sigma(A)$ iff $P(\mu-\theta) < P(\mu+\theta)$ for all $\theta > 0$.*

THEOREM 5. *Let $\{P(\lambda): \lambda \in \mathbf{R}\}$ be the spectral family of the bounded self-adjoint operator A on $\mathfrak{H}$ and let $\mu \in \mathbf{R}$ be given. Then $P(\mu)-P(\mu-0)$ is the projection upon the subspace*

$$\mathfrak{M}_\mu = \{f : Af = \mu f\}.$$

As a consequence, μ is an eigenvalue of A iff $P(\mu-0) < P(\mu)$.

PROOF. Because of $P(\mu-0) \leqslant P(\mu)$ the operator $P(\mu)-P(\mu-0)$ is a projection upon some subspace $\mathfrak{M}$ (cf. the discussion (c) of property c) and § 15 theorem 6 e), f)). For every $f \in \mathfrak{M}$ we have

$$P(\lambda) f = o \quad \text{for } \lambda < \mu$$
$$P(\lambda) f = f \quad \text{for } \lambda \geqslant \mu.$$

Choosing μ as a subdivision point $\lambda_k = \lambda'_k$ for the Riemann-Stieltjes sums approximating the integral

$$\int_{\alpha_1-0}^{\alpha_2} \lambda \, dP(\lambda) f$$

every such sum therefore reduces to the single term $\mu P(\mu) f = \mu f$ (cf. corollary 1.1). As a consequence we get

$$Af = \int_{\alpha_1-0}^{\alpha_2} \lambda \, dP(\lambda) f = \mu f.$$

Conversely, from $(A-\mu I) f = o$ we conclude

$$(17) \qquad \|(A - \mu I) f\|^2 = \int_{\alpha_1-0}^{\alpha_2} (\lambda - \mu)^2 \, d\|P(\lambda) f\|^2 = 0$$

(cf. corollary 1.2), where $\|P(x)f\|^2$ is a non-decreasing function on $\mathbf{R}$, bounded below by 0 and bounded above by $\|f\|^2$. If $\|P(\lambda')f\|^2 < \|f\|^2$ for some $\lambda' > \mu$, then because of $(\lambda - \mu)^2 \geqslant 0$ we would have

$$\int_{\alpha_1-0}^{\alpha_2} (\lambda - \mu)^2 \, d\|P(\lambda) f\|^2 \geqslant \int_{\lambda'}^{\alpha_2} (\lambda - \mu)^2 \, d\|P(\lambda) f\|^2$$
$$\geqslant (\lambda' - \mu)^2 \left[\|f\|^2 - \|P(\lambda') f\|^2\right] > 0$$

which would contradict (17). Similarly, the inequality $0 < \|P(\lambda')f\|^2$ for some

$\lambda' < \mu$ would imply

$$\int\limits_{\alpha_1 - 0}^{\alpha_2} (\lambda - \mu)^2 \, \mathrm{d} \, \|P(\lambda) f\|^2 \geqslant \int\limits_{\alpha_1 - 0}^{\lambda'} (\lambda - \mu)^2 \, \mathrm{d} \, \|P(\lambda) f\|^2$$

$$\geqslant (\lambda' - \mu)^2 \, \|P(\lambda') f\|^2 > 0,$$

which again would contradict (17). Taking into account the right continuity of the spectral family we conclude

$$\|P(\lambda) f\|^2 = \begin{cases} \|f\|^2 & \text{for } \lambda \geqslant \mu \\ 0 & \text{for } \lambda < \mu \end{cases}$$

and therefore

$$[P(\mu) - P(\mu - 0)] f = f$$

(cf. § 15 theorem 1). $\square$

Exercise. Let $\{P(\lambda): \lambda \in \mathbf{R}\}$ be the spectral family of the bounded self-adjoint operator A on $\mathfrak{H}$. For any vector $f_0 \neq o$ define

$$\mathfrak{M}_0 = \vee \{A^n f_0\}_{n=0}^{\infty}.$$

a) The subspace $\mathfrak{M}_0$ is invariant under A and therefore reduces A.

b) If P_0 is the projection upon $\mathfrak{M}_0$, then

$$P(\lambda)P_0 = P_0 P(\lambda) \quad \text{for all } \lambda \in \mathbf{R}.$$

(Hint: cf. § 21 theorem 3.)

c) If A_0 is the restriction of A to $\mathfrak{M}_0$, then $\{P_0 P(\lambda): \lambda \in \mathbf{R}\}$ is the spectral family of A_0. (Hint: cf. theorem 2.)

d) $\Sigma(A_0) \subset \Sigma(A)$.

Remark 2. A bounded selfadjoint operator A on $\mathfrak{H}$ is said to have *simple spectrum* if there exists a vector f_0 as above, called *generating vector*, such that $\mathfrak{M}_0 = \mathfrak{H}$. The assertions of the exercise above are connected with the concept of *multiplicity* of the spectrum (or parts thereof) which generalizes the concept of multiplicity of an eigenvalue. The interested reader is referred to e.g. [1] sections 69–73 or [2] chapter III.

§ 35. Functions of a unitary operator

Having obtained a satisfactory decomposition for bounded selfadjoint operators on $\mathfrak{H}$ by means of the spectrum we shall now try to obtain a similar spectral decomposition for the other important class of bounded operators on $\mathfrak{H}$, namely the unitary operators.

In order to achieve this goal we parallel the construction of functions of an operator, in particular of a bounded selfadjoint operator, as described in § 31, § 32 and § 33, now taking advantage of the property $U^* = U^{-1}$ characterizing a unitary operator U. We start with the algebra $\mathscr{P}'$ of all polynomials in x and x^{-1} with complex coefficients, considered as functions on the complex plane from which the point 0 has been deleted:

$$\mathscr{P}' = \{\varphi(x) = \sum_{k=-n}^{n} \alpha_k x^k \colon \alpha_k \in \mathbf{C} \text{ for } -n \leqslant k \leqslant n, n \geqslant 0\}.$$

THEOREM 1. *Let U be a unitary operator on $\mathfrak{H}$ and for*

$$\varphi(x) = \sum_{k=-n}^{n} \alpha_k x^k \in \mathscr{P}' \quad (x^0 = 1)$$

define

$$\varphi(U) = \sum_{k=-n}^{n} \alpha_k U^k \in \mathscr{B} \quad (U^0 = I),$$

$$\bar{\varphi}(U) = \sum_{k=-n}^{n} \bar{\alpha}_k U^{-k} \in \mathscr{B}.$$

The mapping $\varphi(x) \to \varphi(U)$ is a homomorphism of the algebra $\mathscr{P}'$ into $\mathscr{B}$ with the additional property that

$$[\varphi(U)]^* = \bar{\varphi}(U) = \sum_{k=-n}^{n} \bar{\alpha}_k U^{-k}.$$

PROOF. By § 14 theorem 2 we get

$$[\varphi(U)]^* = \left[\sum_{k=-n}^{n} \alpha_k U^k\right]^* = \sum_{k=-n}^{n} \bar{\alpha}_k U^{-k} = \bar{\varphi}(U).$$

The other assertions are trivial. $\square$

Recall that $|\xi| = 1$ for $\xi \in \Sigma(U)$ (cf. § 23 theorem 7).

THEOREM 2. *Let U be a unitary operator on $\mathfrak{H}$ and let $\varphi(x) \in \mathscr{P}'$ be given. Then*

$$\Sigma(\varphi(U)) = \{\varphi(\xi) \colon \xi \in \Sigma(U)\}$$
$$= \varphi(\Sigma(U)).$$

PROOF. If $\varphi(x) = \sum_{k=-n}^{n} \alpha_k x^k$, then we have $x^n \varphi(x) = \psi(x) \in \mathscr{P}$ (as defined in § 31). The complex number η belongs to $\Sigma(\varphi(U))$ iff the operator $[\varphi(U) - \eta I] = = U^{-n}[\psi(U) - \eta U^n]$ is not invertible. This is the case iff $\psi(U) - \eta U^n$ is not invertible or, equivalently, iff 0 belongs to the spectrum of $\psi(U) - \eta U^n$. Again, by § 31 theorem 2 this is the case iff $\psi(\xi) - \eta \xi^n = 0$ for some $\xi \in \Sigma(U)$ or, equivalently, iff $\eta = \psi(\xi)/\xi^n = \varphi(\xi)$ for some $\xi \in \Sigma(U)$. $\square$

COROLLARY 2.1. *Let U and $\varphi(x)$ be as in theorem 2. If $\varphi(\xi) \in \mathbf{R}$ for $|\xi| = 1$, then $\varphi(U)$ is selfadjoint and*

$$\|\varphi(U)\| = \|\varphi(x)\|_U \quad (= \sup\{|\varphi(\xi)| : \xi \in \Sigma(U)\}).$$

PROOF. By assumption we have

$$\overline{\varphi(\xi)} = \sum_{k=-n}^{n} \bar{\alpha}_k \xi^{-k} = \sum_{k=-n}^{n} \alpha_k \xi^k = \varphi(\xi) \quad \text{for } |\xi| = 1,$$

$$\sum_{k=-n}^{n} (\alpha_k - \bar{\alpha}_{-k}) \xi^k = 0 \qquad \text{for } |\xi| = 1,$$

$$\sum_{k=-n}^{n} (\alpha_k - \bar{\alpha}_{-k}) e^{2\pi i k \eta} = 0 \qquad \text{for all } \eta \in \mathbf{R}.$$

Since we already know the functions $e^{2\pi i k y}$ $(-\infty < k < +\infty)$ on $\mathbf{R}$ to be linearly independent (cf. § 8 example 3) we obtain

$$\alpha_k = \bar{\alpha}_{-k} \quad \text{for } -n \leqslant k \leqslant n,$$

$$[\varphi(U)]^* = \bar{\varphi}(U) = \sum_{k=-n}^{n} \bar{\alpha}_k U^{-k} = \sum_{k=-n}^{n} \alpha_k U^k = \varphi(U). \quad \square$$

We shall now restrict the variable x to the unit circle in the complex plane. The reason for this departure from what has been done in § 32 is clear: while the spectrum of a bounded selfadjoint operator on $\mathfrak{H}$ lies on $\mathbf{R}$, the spectrum of a unitary operator lies on the complex unit circle. The further reasoning as followed in § 32, however, remains the same also in the present case and allows to extend the homomorphism of $\mathscr{P}'$ into $\mathscr{B}$ to the larger algebra $\mathscr{S}'$ of all continuous complex-valued functions on the unit circle:

$$\mathscr{S}' = \{\varphi(x) : \varphi(x) \text{ is a continuous complex-valued function}$$
$$\text{on the unit circle in } \mathbf{C}\}.$$

Note that by means of the substitution

$$x = e^{iy},$$
$$\varphi'(y) = \varphi(e^{iy})$$

to every function $\varphi(x) \in \mathscr{S}'$ there corresponds a periodic function $\varphi'(y)$ on $\mathbf{R}$ with period 2π. Via this substitution $\mathscr{S}'$ may also be identified with the algebra of all periodic continuous complex-valued functions on $\mathbf{R}$ with period 2π.

The proofs of the following theorems exactly parallel those of the corresponding theorems in § 32 and § 33. The interval $[\alpha_1 = \min \Sigma(A), \alpha_2 =$

$=\max \Sigma(A)]$ now has to be replaced by the complex unit circle, subintervals have to be replaced by arcs, and WEIERSTRASS' approximation theorem has to be applied in the third formulation as given in appendix B1. We omit a mere repetition of arguments and only state the theorems. By way of illustration a proof is given for corollary 3.1.

THEOREM 3. *Let the unitary operator* $U \in \mathcal{B}$ *and the function* $\varphi(x) \in \mathcal{S}'$ *be given. Then there exists a unique operator* $\varphi(U) \in \mathcal{B}$ *with the following property: whenever* $\{\varphi_n(x)\}_{n=1}^{\infty}$ *is a sequence in* $\mathcal{P}'$ *such that*

(1)
$$\lim_{n \to \infty} \|\varphi(x) - \varphi_n(x)\|_U = 0,$$

then
$$\varphi(U) = \lim_{n \to \infty} \varphi_n(U).$$

The mapping $\varphi(x) \to \varphi(U)$ *is a homomorphism of the algebra* $\mathcal{S}'$ *into* $\mathcal{B}$ *with the additional properties*

$$[\varphi(U)]^* = \bar{\varphi}(U) \quad (= \psi(U) \text{ where } \psi(x) = \bar{\varphi}(x) \in \mathcal{S}'),$$
$$\|\varphi(U)\| \leqslant 2 \|\varphi(x)\|_U.$$

PROOF. Analogous to the proof of § 32 theorem 1. □

COROLLARY 3.1. *Let* U *and* $\varphi(x)$ *be as in theorem 3. Then* $\varphi(U)$ *is normal. If* $\varphi(x)$ *is real-valued, then* $\varphi(U)$ *is selfadjoint.*
PROOF. Let $\psi(x) = \bar{\varphi}(x)$. Then $\psi(x) \in \mathcal{S}'$ and

$$\varphi(U) [\varphi(U)]^* = \varphi(U) \bar{\varphi}(U) = \varphi(U) \psi(U) = (\varphi\psi)(U)$$
$$= (\psi\varphi)(U) = \psi(U) \varphi(U) = [\varphi(U)]^* \varphi(U).$$

For a direct proof of the second assertion let $\varphi(x)$ be real-valued and let $\{\varphi_n(x)\}_{n=1}^{\infty}$ be a sequence in $\mathcal{P}'$ satisfying (1). Then also the sequence $\{\frac{1}{2}[\varphi_n(x) + \bar{\varphi}_n(x)]\}_{n=1}^{\infty} \subset \mathcal{P}'$ satisfies

$$\lim_{n \to \infty} \|\varphi(x) - \frac{1}{2}[\varphi_n(x) + \bar{\varphi}_n(x)]\|_U = 0.$$

Since every term of the sequence $\{\frac{1}{2}[\varphi_n(U) + \bar{\varphi}_n(U)]\}_{n=1}^{\infty}$ is selfadjoint, so is its limit $\varphi(U)$. □

THEOREM 4. *Let the unitary operator* $U \in \mathcal{B}$ *and the function* $\varphi(x) \in \mathcal{S}'$ *be given and let the operator* $B \in \mathcal{B}$ *satisfy* $UB = BU$. *Then* $\varphi(U) B = B\varphi(U)$.
PROOF. Analogous to the proof of § 32 theorem 2. □

THEOREM 5. *Let the unitary operator $U \in \mathcal{B}$ and the function $\varphi(x) \in \mathcal{S}'$ be given. Then*

$$\Sigma(\varphi(U)) = \{\varphi(\xi) : \xi \in \Sigma(U)\}$$
$$= \varphi(\Sigma(U)).$$

PROOF. Analogous to the proof of § 32 theorem 3. $\square$

COROLLARY 5.1. *Let U and $\varphi(x)$ be as in theorem 5.*

a) *The operator $\varphi(U)$ is selfadjoint iff*

$$\varphi(\xi) \in \mathbf{R} \quad \text{for all } \xi \in \Sigma(U).$$

b) *The operator $\varphi(U)$ is unitary iff*

$$|\varphi(\xi)| = 1 \quad \text{for all } \xi \in \Sigma(U).$$

c) *The operator $\varphi(U)$ is invertible iff*

$$\varphi(\xi) \neq 0 \quad \text{for all } \xi \in \Sigma(U).$$

PROOF. Analogous to the proof of § 32 corollary 3.1. $\square$

COROLLARY 5.2. *Let U and $\varphi(x)$ be as in theorem 5. If $\varphi(U)$ is selfadjoint, then*

$$\|\varphi(U)\| = \|\varphi(x)\|_U.$$

PROOF. Analogous to the proof of § 32 corollary 3.2. $\square$

COROLLARY 5.3. *Let the unitary operator $U \in \mathcal{B}$ and the functions $\varphi(x) \in \mathcal{S}'$, $\psi(x) \in \mathcal{S}'$ be given. Then $\varphi(U) = \psi(U)$ iff $\|\varphi(x) - \psi(x)\|_U = 0$.*
PROOF. Analogous to the proof of § 32 corollary 3.3. $\square$

COROLLARY 5.4. *Let U and $\varphi(x)$ be as in theorem 5. If $\varphi(x)$ is real-valued and non-negative on $\Sigma(U)$ (in particular if $\varphi(x) \geqslant 0$ on the complex unit circle), then $\varphi(U) \geqslant O$.*
PROOF. Analogous to the proof of § 32 corollary 3.4. $\square$

COROLLARY 5.5. *Let $U \in \mathcal{B}$ be unitary. The homomorphism $\varphi(x) \to \varphi(U)$ of $\mathcal{S}'$ into $\mathcal{B}$ is order preserving in the following sense: if $\varphi(x) \in \mathcal{S}'$ and $\psi(x) \in \mathcal{S}'$ are real-valued on $\Sigma(U)$ and*

$$\varphi(\xi) \geqslant \psi(\xi) \quad \text{for all } \xi \in \Sigma(U)$$

(in particular if $\varphi(x) \geqslant \psi(x)$ on the complex unit circle), then $\varphi(U) \geqslant \psi(U)$.
PROOF. Analogous to the proof of § 32 corollary 3.5. $\square$

As in the case of a bounded selfadjoint operator on $\mathfrak{H}$ the last step before obtaining the spectral decomposition of a unitary operator consists in

a) restricting the attention to *real-valued* functions on the unit circle and
b) extending the homomorphism $\varphi(x) \to \varphi(U)$ to an algebra of real-valued functions on the complex unit circle which also contains step functions.

DEFINITION 1. A real-valued function $\varphi(x)$ on the complex unit circle is called *upper semi-continuous* if it is a pointwise limit of a non-increasing sequence of continuous real-valued functions on the complex unit circle.

THEOREM 6. *Let $U \in \mathscr{B}$ be unitary and let $\varphi(x)$ be a non-negative upper semi-continuous function on the complex unit circle. Then there exists a unique positive selfadjoint operator $\varphi(U) \in \mathscr{B}$ such that whenever $\{\varphi_n(x)\}_{n=1}^{\infty}$ is any non-increasing sequence of non-negative functions in $\mathscr{S}'$, pointwise converging to $\varphi(x)$ on $\Sigma(U)$, then*

$$\varphi(U) = (s) \lim \varphi_n(U).$$

PROOF. Analogous to the proof of § 33 theorem 1. □

THEOREM 7. *Let $U \in \mathscr{B}$ be unitary, let $\varphi(x)$ and $\psi(x)$ be non-negative upper semi-continuous functions on the complex unit circle and let $\alpha > 0$ be given. Then the functions $(\varphi + \psi)(x)$, $(\alpha\varphi)(x)$, and $(\varphi\psi)(x)$ are non-negative and upper semi-continuous and*

$$(\varphi + \psi)(U) = \varphi(U) + \psi(U),$$
$$(\alpha\varphi)(U) = \alpha\varphi(U),$$
$$(\varphi\psi)(U) = \varphi(U)\,\psi(U).$$

If $\varphi(\xi) \leqslant \psi(\xi)$ for all $\xi \in \Sigma(U)$, in particular if $\varphi(x) \leqslant \psi(x)$ on the complex unit circle, then $\varphi(U) \leqslant \psi(U)$.
PROOF. Analogous to the proof of § 33 theorem 2. □

COROLLARY 7.1. *Let U, $\varphi(x)$, and $\psi(x)$ be as in theorem 7. If $\varphi(\xi) = \psi(\xi)$ for all $\xi \in \Sigma(U)$, then $\varphi(U) = \psi(U)$.*
PROOF. Analogous to the proof of § 33 corollary 2.1. □

Again we enlarge the class of non-negative upper semi-continuous functions on the complex unit circle to an algebra by defining

$$\mathscr{R}' = \{\varphi(x) = \varphi_1(x) - \varphi_2(x) : \varphi_k(x) \text{ is non-negative and upper semi-continuous}$$
$$\text{on the complex unit circle for } k = 1, 2\}.$$

THEOREM 8. *The class $\mathscr{R}'$ of real-valued functions on the complex unit circle is a real algebra if sum, scalar multiple, and product in $\mathscr{R}'$ are defined pointwise.*
PROOF. Analogous to the proof of § 33 theorem 3. □

THEOREM 9. *Let the unitary operator $U \in \mathscr{B}$ and the function $\varphi(x) \in \mathscr{R}'$ be given.*

Then there exists a unique selfadjoint operator $\varphi(U) \in \mathscr{B}$ such that if

$$\varphi(x) = \varphi_1(x) - \varphi_2(x)$$

where $\varphi_k(x)$ is non-negative and upper semi-continuous on the complex unit circle for $k = 1, 2$, then

$$\varphi(U) = \varphi_1(U) - \varphi_2(U).$$

The mapping $\varphi(x) \to \varphi(U)$ is a homomorphism of $\mathscr{R}'$ into $\mathscr{B}$ which is order preserving in the following sense: if $\varphi(x) \in \mathscr{R}'$ and $\psi(x) \in \mathscr{R}'$ have the property that

$$\varphi(\xi) \geqslant \psi(\xi) \quad \text{for all } \xi \in \Sigma(U),$$

then $\varphi(U) \geqslant \psi(U)$.

PROOF. Analogous to the proof of § 33 theorem 4. □

THEOREM 10. *Let the unitary operator $U \in \mathscr{B}$ and the function $\varphi(x) \in \mathscr{R}'$ be given and suppose the operator $B \in \mathscr{B}$ satisfies $UB = BU$. Then $\varphi(U)B = B\varphi(U)$.*
PROOF. Analogous to the proof of § 33 theorem 5. □

Exercise. Let the operator U be defined on $\mathfrak{L}_2(\alpha, \beta)$ $(-\infty \leqslant \alpha < \beta \leqslant +\infty)$ for given $\gamma \in \mathbf{R}$ by

$$Uf = e^{i\gamma x} f(x).$$

   a) U is unitary.
   b) $\Sigma(U) = \{e^{i\gamma\xi} : \alpha_{(\leqq)}\xi_{(\leqq)}\beta\}$ (the equality signs in parentheses have to be omitted if α or β is infinite). (Hint: apply directly § 23 definition 1.)
   c) If $\varphi(x) \in \mathscr{R}'$ is given, then $\varphi(U)$ is given on $\mathfrak{L}_2(\alpha, \beta)$ by

$$\varphi(U)f = \varphi(e^{i\gamma x}) f(x).$$

(Hint: prove this first for $\varphi(x) \in \mathscr{P}'$ and $\varphi(x) \in \mathscr{S}'$.)

§ 36. The spectral decomposition of a unitary operator

Similarly as in § 34 we shall now use a special class of step functions on the complex unit circle in order to define a family of projections which in turn will serve to recapture the initial unitary operator U and its "functions" in the form of Riemann-Stieltjes integrals.

To this end, for $0 \leqslant \lambda \leqslant 2\pi$ define the functions $\varphi_\lambda(x)$ on the complex unit circle in the following way (fig. 27):

$$\varphi_0(x) \equiv 0,$$

$$\varphi_\lambda(e^{i\eta}) = \begin{cases} 1 & \text{for } 0 < \eta \leqslant \lambda, \\ 0 & \text{for } \lambda < \eta \leqslant 2\pi, \end{cases} \quad (0 < \lambda < 2\pi),$$

$$\varphi_{2\pi}(x) \equiv 1.$$

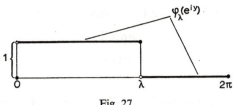

Fig. 27

The functions $\varphi_\lambda(x)$ $(0 \leqslant \lambda \leqslant 2\pi)$ belong to the algebra $\mathscr{R}'$ defined in § 35. Indeed, if we furthermore for $0 \leqslant \lambda < 2\pi$ define the functions $\psi_\lambda(x)$ on the complex unit circle (fig. 28) by

$$\psi_0(e^{i\eta}) = \begin{cases} 0 & \text{for } 0 < \eta < 2\pi, \\ 1 & \text{for } \eta = 2\pi, \end{cases}$$

$$\psi_\lambda(x) = \varphi_\lambda(x) + \psi_0(x) \quad (0 \leqslant \lambda < 2\pi),$$

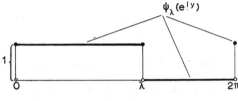

Fig. 28

then the functions $\psi_\lambda(x)$ are real-valued and upper semi-continuous and we have

(1) $\varphi_\lambda(x) = \psi_\lambda(x) - \psi_0(x) \in \mathscr{R}'$ for $0 \leqslant \lambda < 2\pi$.

Define the selfadjoint operator $P(\lambda) \in \mathscr{B}$ by

(2) $P(\lambda) = \varphi_\lambda(U)$ for $0 \leqslant \lambda \leqslant 2\pi$

(cf. § 35 theorem 9). Since $\varphi_\lambda^2(x) = \varphi_\lambda(x)$ the operator $P(\lambda)$ satisfies $P^2(\lambda) = = P(\lambda)$ and therefore is a projection (cf. § 15 theorem 2). Moreover, the family $\{P(\lambda): 0 \leqslant \lambda \leqslant 2\pi\}$ displays features similar to those described in (a), (b), (c), (d) of § 34:

(a) From $\varphi_\lambda(x) \leqslant \varphi_{\lambda'}(x)$ for $0 \leqslant \lambda \leqslant \lambda' \leqslant 2\pi$ we conclude

$$P(\lambda) \leqslant P(\lambda') \quad \text{for } 0 \leqslant \lambda \leqslant \lambda' \leqslant 2\pi$$

(cf. § 35 theorem 9).

(b) From $\varphi_0(x) \equiv 0$, $\varphi_{2\pi}(x) \equiv 1$ we conclude

$$P(0) = O,$$
$$P(2\pi) = I$$

(cf. § 35 theorem 1).

(c) Let $\{\theta_n\}_{n=1}^{\infty}$ be any non-increasing sequence of positive real numbers converging to zero and consider the family of functions $\psi_\lambda(x)$ $(0 \leqslant \lambda < 2\pi)$. Exactly as in § 34 (c) we obtain

$$\psi_\lambda(U) = \text{(s)} \lim_{n \to \infty} \psi_{\lambda+\theta_n}(U) \quad (0 \leqslant \lambda < 2\pi)$$

(we define $\psi_{\lambda+\theta_n}(U) = P(\lambda+\theta_n) \doteq I$ for $\lambda+\theta_n \geqslant 2\pi$). Using (1) we get

$$\begin{aligned}
\text{(s)} \lim_{n \to \infty} P(\lambda + \theta_n) &= \text{(s)} \lim_{n \to \infty} \varphi_{\lambda+\theta_n}(U) \\
&= \text{(s)} \lim_{n \to \infty} [\psi_{\lambda+\theta_n}(U) - \psi_0(U)] \\
&= \psi_\lambda(U) - \psi_0(U) = \varphi_\lambda(U) \\
&= P(\lambda) \quad (0 \leqslant \lambda < 2\pi).
\end{aligned}$$

Using the notation introduced in § 34 (c) we conclude

$$P(\lambda + 0) = P(\lambda) \quad \text{for } 0 \leqslant \lambda < 2\pi.$$

In other words, the family $\{P(\lambda) : 0 \leqslant \lambda \leqslant 2\pi\}$ is continuous from the right. A similar reasoning as in § 34 (c) also shows that for every $\lambda \in]0, 2\pi]$ there exists a unique projection denoted by $P(\lambda - 0)$ and less than or equal to $P(\lambda)$ with the property that

$$P(\lambda - 0) = \text{(s)} \lim_{n \to \infty} P(\lambda - \theta_n)$$

for every sequence $\{\theta_n\}_{n=1}^{\infty}$ as above.

(d) Let $\varphi(x)$ be a continuous complex-valued function on the complex unit circle. Then $\varphi(e^{iy})$ is a continuous complex-valued function on $[0, 2\pi]$. A similar reasoning as in § 34 (d) shows that for every $\varepsilon > 0$ there exists a $\delta > 0$ such that

$$\left\| \varphi(U) - \sum_{k=2}^{n} \varphi(e^{i\lambda_k'}) [P(\lambda_k) - P(\lambda_{k-1})] \right\| \leqslant \varepsilon$$

whenever

$$0 = \lambda_1 < \lambda_2 < \cdots < \lambda_{n-1} < \lambda_n = 2\pi,$$
$$\lambda_k - \lambda_{k-1} \leqslant \delta \quad \text{for } 2 \leqslant k \leqslant n,$$
$$\lambda_k' \in [\lambda_{k-1}, \lambda_k] \quad \text{for } 2 \leqslant k \leqslant n.$$

Again this statement may briefly be written in the form

$$\varphi(U) = \int\limits_0^{2\pi} \varphi(e^{i\lambda})\, dP(\lambda).$$

In particular for $\varphi(x) = x$ we get

$$U = \int\limits_0^{2\pi} e^{i\lambda}\, dP(\lambda).$$

We sum up the result in a theorem corresponding to § 34 theorem 1.

THEOREM 1 (SPECTRAL THEOREM FOR UNITARY OPERATORS). *Let U be a unitary operator on $\mathfrak{H}$. Then there exists a family of projections $\{P(\lambda): 0 \leqslant \lambda \leqslant 2\pi\}$, called the spectral family of U, with the following properties:*

a) $P(\lambda) \leqslant P(\lambda')$ *for* $0 \leqslant \lambda \leqslant \lambda' \leqslant 2\pi$;

b) $P(0) = O$,

        $P(2\pi) = I$;

c) $P(\lambda + 0) = P(\lambda)$ *for* $0 \leqslant \lambda < 2\pi$;

d) $U = \int\limits_0^{2\pi} e^{i\lambda}\, dP(\lambda).$

More generally, for every continuous complex-valued function $\varphi(x)$ on the complex unit circle and for every $\varepsilon > 0$ there exists a $\delta > 0$ such that

$$\left\| \varphi(U) - \sum_{k=2}^{n} \varphi(e^{i\lambda_k'}) \left[P(\lambda_k) - P(\lambda_{k-1}) \right] \right\| \leqslant \varepsilon$$

whenever

(3) $\begin{cases} 0 = \lambda_1 < \cdots < \lambda_{n-1} < \lambda_n = 2\pi, \\ \lambda_k - \lambda_{k-1} \leqslant \delta \quad \textit{for } 2 \leqslant k \leqslant n, \\ \lambda_k' \in [\lambda_{k-1}, \lambda_k] \quad \textit{for } 2 \leqslant k \leqslant n. \end{cases}$

In other words,

$$\varphi(U) = \int\limits_0^{2\pi} \varphi(e^{i\lambda})\, dP(\lambda)$$

where the integral is of Riemann-Stieltjes type.

COROLLARY 1.1. *Let U and $\{P(\lambda): 0 \leqslant \lambda \leqslant 2\pi\}$ be as in theorem 1. If $\varphi(x)$ is* *a continuous complex-valued function on the complex unit circle, then*

$$\varphi(U)f = \int_0^{2\pi} \varphi(e^{i\lambda})\,dP(\lambda)f \qquad \text{for all } f \in \mathfrak{H},$$

$$\langle \varphi(U)f, g \rangle = \int_0^{2\pi} \varphi(e^{i\lambda})\,d\langle P(\lambda)f, g \rangle \quad \text{for all } f \in \mathfrak{H} \text{ and all } g \in \mathfrak{H}.$$

More explicitly, for every $\varepsilon>0$ there exists a $\delta>0$ such that

$$\|\varphi(U)f - \sum_{k=2}^{n} \varphi(e^{i\lambda_k'})[P(\lambda_k)f - P(\lambda_{k-1})f]\| \leqslant \varepsilon \|f\| \text{ for all } f \in \mathfrak{H},$$

$$|\langle \varphi(U)f, g \rangle - \sum_{k=2}^{n} \varphi(e^{i\lambda_k'})[\langle P(\lambda_k)f, g \rangle - \langle P(\lambda_{k-1})f, g \rangle]| \leqslant \varepsilon \|f\| \|g\|$$

$$\text{for all } f \in \mathfrak{H} \text{ and all } g \in \mathfrak{H}$$

whenever (3) *is satisfied.*

COROLLARY 1.2. *Let U and $\{P(\lambda):0\leqslant\lambda\leqslant2\pi\}$ be as in theorem 1. If $\varphi(x)$ is a continuous complex-valued function on the complex unit circle, then*

$$\|\varphi(U)f\|^2 = \int_0^{2\pi} |\varphi(e^{i\lambda})|^2\,d\|P(\lambda)f\|^2 \quad \text{for all } f \in \mathfrak{H}.$$

More explicitly, for every $\varepsilon>0$ there exists a $\delta>0$ such that

$$|\|\varphi(U)f\|^2 - \sum_{k=2}^{n} |\varphi(e^{i\lambda_k'})|^2 [\|P(\lambda_k)f\|^2 - \|P(\lambda_{k-1})f\|^2]| \leqslant \varepsilon \|f\|^2$$

$$\text{for all } f \in \mathfrak{H}$$

whenever (3) *is satisfied.*

PROOF. Analogous to the proof of § 34 corollary 1.2. $\square$

The spectral family of U is again uniquely determined by the properties a), b), c), d) mentioned in theorem 1.

THEOREM 2. *Let U be a unitary operator on $\mathfrak{H}$ and let $\{P'(\lambda):0\leqslant\lambda\leqslant2\pi\}$ be a family of projections satisfying the requirements a), b), c), d) in theorem 1. Then $P'(\lambda)=P(\lambda)$ where $P(\lambda)$ is defined as in (2).*

PROOF. Analogous to the proof of § 34 theorem 2 (α_1-0 and α_2 have to be replaced now by 0 and 2π respectively). $\square$

If we extend the definition of the spectral family $\{P(\lambda):0\leqslant\lambda\leqslant2\pi\}$ by defining $P(\lambda)=O$ for $\lambda<0$, $P(\lambda)=I$ for $\lambda>2\pi$, then the family $\{P(\lambda):\lambda\in\mathbf{R}\}$

very much resembles the spectral family of a bounded selfadjoint operator as described in § 34 theorem 1, although we do not yet know precisely where in the interval $[0, 2\pi]$ to look for α_1 and α_2. In fact, for a sequence of successively refined subdivisions of type (3) the corresponding Riemann-Stieltjes sums

$$\sum_{k=2}^{n} \lambda'_k [P(\lambda_k) - P(\lambda_{k-1})]$$

converge in the operator norm to a unique bounded selfadjoint operator

$$A = \int_0^{2\pi} \lambda \, dP(\lambda)$$

on $\mathfrak{H}$, although here (as opposed to § 34) there has not been any such operator to start with. This fact will be obvious to the reader familiar with Riemann-Stieltjes integrals and it will hardly surprise the reader not so familiar with these objects but having followed the discussion in § 34. We therefore omit further details. If we put

$$\alpha_1 = \sup \{\lambda : P(\lambda) = O\},$$
$$\alpha_2 = \inf \{\lambda : P(\lambda) = I\},$$

then § 34 theorem 2 asserts that the family $\{P(\lambda) : \lambda \in \mathbf{R}\}$ is indeed the spectral family of the bounded selfadjoint operator A. Moreover, by § 34 theorem 1 we get

$$U = \int_0^{2\pi} e^{i\lambda} \, dP(\lambda) = \int_{\alpha_1 - 0}^{\alpha_2} e^{i\lambda} \, dP(\lambda) = e^{iA}.$$

It turns out that starting with the unitary operator U we have constructed a bounded selfadjoint operator A on $\mathfrak{H}$ such that $U = e^{iA}$ (moreover satisfying $\Sigma(A) \subset [0, 2\pi]$ and not having 0 as an eigenvalue, cf. § 34 theorem 5). The spectral family of U in fact coincides with the spectral family of A. According to § 34 theorem 1 this spectral family allows to represent the operator U as the Riemann-Stieltjes integral

(4) $$U = e^{iA} = \int_0^{2\pi} e^{i\lambda} \, dP(\lambda).$$

Recall that also conversely, if A is any bounded selfadjoint operator on $\mathfrak{H}$, then e^{iA} is unitary (cf. § 32 example 1). A similar relation between unitary operators and selfadjoint operators in $\mathfrak{H}$ in general will lead to a spectral decomposition of unbounded selfadjoint operators in $\mathfrak{H}$ (cf. § 38 theorem 1 and § 38 theorem 2).

Example 1. Let F be the Fourier-Plancherel operator in $\mathfrak{L}_2(-\infty, +\infty)$ (cf. § 16 theorem 3). For reasons of notational convenience we shall consider its inverse F^{-1} in place of F. We already know that the spectrum of F^{-1} (as also the spectrum of F) only consists of the four eigenvalues $i^k (1 \leqslant k \leqslant 4)$ and that

$$(5) \qquad\qquad F^{-1} = \sum_{k=1}^{4} i^k P_k$$

where P_k is the projection upon the eigenspace corresponding to the eigenvalue i^k (cf. § 22 example 2 (5); we have changed the notation by putting $P_0 = P_4$). It follows that the spectral family of the unitary operator F^{-1} is defined by

$$P(\lambda) = \begin{cases} O & \text{for } 0 \leqslant \lambda < \tfrac{1}{2}\pi, \\ P_1 & \text{for } \tfrac{1}{2}\pi \leqslant \lambda < \pi, \\ P_1 + P_2 & \text{for } \pi \leqslant \lambda < \tfrac{3}{2}\pi, \\ P_1 + P_2 + P_3 & \text{for } \tfrac{3}{2}\pi \leqslant \lambda < 2\pi, \\ I & \text{for } \lambda = 2\pi. \end{cases}$$

Indeed, since the projections P_k are mutually orthogonal (cf. § 22 remark 4), the operators $P(\lambda)$ as defined above are projections for $0 \leqslant \lambda \leqslant 2\pi$ (cf. § 15 theorem 5 a), f)), and the family $\{P(\lambda) : 0 \leqslant \lambda \leqslant 2\pi\}$ satisfies the requirements a), b), c), d) in theorem 2, while the Riemann-Stieltjes integral (4) reduces to the sum (5). The spectral family of F^{-1} coincides with the spectral family of the bounded selfadjoint operator

$$A = \sum_{k=1}^{4} \tfrac{1}{2}\pi k P_k$$

and we have

$$F^{-1} = e^{iA}.$$

Note, however, that the selfadjoint operator A is not compact since the projections P_k are not of finite rank (cf. § 28 theorem 2).

After all what has been said so far in this section, it is to be expected that a unitary operator and its spectral family show the same close relationship as already observed in the case of a bounded selfadjoint operator on $\mathfrak{H}$.

THEOREM 3. *Let* $\{P(\lambda) : 0 \leqslant \lambda \leqslant 2\pi\}$ *be the spectral family of the unitary oper-*

ator U on $\mathfrak{H}$. If B is a bounded linear operator on $\mathfrak{H}$, then $UB = BU$ iff

$$P(\lambda) B = BP(\lambda) \quad for \ 0 \leqslant \lambda \leqslant 2\pi.$$

In particular,

$$P(\lambda) U = UP(\lambda) \quad for \ 0 \leqslant \lambda \leqslant 2\pi.$$

PROOF. Analogous to the proof of § 34 theorem 3. $\square$

THEOREM 4. Let $\{P(\lambda): 0 \leqslant \lambda \leqslant 2\pi\}$ be the spectral family of the unitary oper-ator U on $\mathfrak{H}$. The number $e^{i\mu}$ $(0 < \mu \leqslant 2\pi)$ is a regular value for U iff there exists a $\theta > 0$ such that $P(\mu - \theta) = P(\mu + \theta)$ (defining $P(\lambda) = I$ for $\lambda > 2\pi$).
PROOF. Analogous to the proof of § 34 theorem 4. $\square$

COROLLARY 4.1. Let U and $\{P(\lambda): 0 \leqslant \lambda \leqslant 2\pi\}$ be as in theorem 4. The number $e^{i\mu}$ $(0 < \mu \leqslant 2\pi)$ belongs to $\Sigma(U)$ iff $P(\mu - \theta) < P(\mu + \theta)$ for all $\theta > 0$ (defining $P(\lambda) = O$ for $\lambda < 0$, $P(\lambda) = I$ for $\lambda > 2\pi$).

THEOREM 5. Let $\{P(\lambda): 0 \leqslant \lambda \leqslant 2\pi\}$ be the spectral family of the unitary oper-ator U on $\mathfrak{H}$ and let the number $\mu \in]0, 2\pi]$ be given. Then $P(\mu) - P(\mu - 0)$ is the projection upon the subspace

$$\mathfrak{M}_\mu = \{f : Uf = e^{i\mu} f\}.$$

As a consequence, $e^{i\mu}$ is an eigenvalue of U iff $P(\mu - 0) < P(\mu)$.
PROOF. Analogous to the proof of § 34 theorem 5. $\square$

Assertion c) of the following exercise is the so-called *statistical ergodic theorem* of VON NEUMANN (cf. [3], [9] section 144).

Exercise. Let $\{P(\lambda): 0 \leqslant \lambda \leqslant 2\pi\}$ be the spectral family of the unitary oper-ator U on $\mathfrak{H}$ and define $P_1 = I - P(2\pi - 0)$.

a) $P_1 f = f$ iff $Uf = f$. As a consequence, we have $P_1 f = UP_1 f$ for all $f \in \mathfrak{H}$, and P_1 is the projection upon the subspace of all $f \in \mathfrak{H}$ which are invariant under U.

b) $P(\lambda) P_1 f = \begin{cases} o & for \ 0 \leqslant \lambda < 2\pi, \\ P_1 f & for \quad \lambda = 2\pi. \end{cases}$

c) $\lim\limits_{n \to \infty} \left\| \dfrac{1}{n} \sum\limits_{k=0}^{n-1} U^k f - P_1 f \right\| = 0 \quad for \ all \ f \in \mathfrak{H}.$

(Hint: use

$$P_1 f = \frac{1}{n} \sum_{k=0}^{n-1} U^k P_1 f$$

and

$$\left\| \frac{1}{n} \sum_{k=0}^{n-1} U^k (f - P_1 f) \right\|^2 = \int_0^\theta \frac{1}{n^2} \left| \sum_{k=0}^{n-1} e^{ik\lambda} \right|^2 d \, \|P(\lambda)(f - P_1 f)\|^2 +$$

$$+ \int_\theta^{2\pi-\theta} \frac{1}{n^2} \left| \frac{e^{in\lambda} - 1}{e^{i\lambda} - 1} \right|^2 d \, \|P(\lambda)(f - P_1 f)\|^2 +$$

$$+ \int_{2\pi-\theta}^{2\pi} \frac{1}{n^2} \left| \sum_{k=0}^{n-1} e^{ik\lambda} \right|^2 d \, \|P(\lambda)(f - P_1 f)\|^2$$

where the first summand is majorized by $\|P(\theta)f\|^2$ and where the third summand is majorized by $\|P(2\pi-0)f\|^2 - \|P(2\pi-\theta)f\|^2$.)

§ 37. The spectral decomposition of a bounded normal operator

From the spectral decomposition of bounded selfadjoint operators on $\mathfrak{H}$ we can deduce a similar decomposition of bounded normal operators on $\mathfrak{H}$. The following reasoning parallels the one having been applied in the case of compact normal operators in § 28 (cf. § 28 theorem 4).

Let A be a bounded normal operator on $\mathfrak{H}$ and let

$$B = \tfrac{1}{2}(A + A^*) \quad (\|B\| \leqslant \|A\|),$$

$$C = \frac{1}{2i}(A - A^*) \quad (\|C\| \leqslant \|A\|).$$

The operators B and C are selfadjoint and satisfy

(1) $$A = B + iC,$$

(2) $$BC = CB$$

(cf. § 14 theorem 7 (6), (7)). Furthermore, let $\{P(\xi) : \xi \in \mathbf{R}\}$ and $\{Q(\eta) : \eta \in \mathbf{R}\}$ be the spectral families of B and C respectively (cf. § 34 theorem 1). From (1) we conclude

$$A = \int_{-\|A\|-0}^{\|A\|} \xi \, dP(\xi) + i \int_{-\|A\|-0}^{\|A\|} \eta \, dQ(\eta)$$

$$= \int_{-\|A\|-0}^{\|A\|} \xi \, d[P(\xi) + iQ(\xi)]$$

(we take into account the possibility that $-\|A\|$ may be an eigenvalue of B or C). This representation of the operator A as a Riemann-Stieltjes integral, however, does not quite correspond to the spectral decompositions dealt with so far: the family $\{P(\xi)+iQ(\xi):\xi\in\mathbf{R}\}$ is not any more a family of projections, increasing monotonically from O to I. Also, the integration is not performed over a set a priori containing the spectrum of A, since in general this spectrum will not any more lie on the real axis. Thus there is no reason to expect that the family of operators $\{P(\xi)+iQ(\xi):\xi\in\mathbf{R}\}$ will display the nice features and the close relationship with A which would entitle it to be called a spectral family of A.

We therefore proceed in a different way. From (2) we conclude

$$P(\xi)\,Q(\eta) = Q(\eta)\,P(\xi) \quad \text{for all } \xi\in\mathbf{R} \text{ and all } \eta\in\mathbf{R}$$

by § 34 theorem 3. As a consequence, the operator $R(\xi,\eta)=P(\xi)\,Q(\eta)$ is a projection for all pairs of real numbers $(\xi,\eta)\in\mathbf{R}^2$ by § 15 theorem 4. From (2) we further conclude that

$$P(\xi)\,C = CP(\xi) \quad \text{for all } \xi\in\mathbf{R},$$
$$Q(\eta)\,B = BQ(\eta) \quad \text{for all } \eta\in\mathbf{R}$$

(cf. § 34 theorem 3) and therefore

$$R(\xi,\eta)\,A = AR(\xi,\eta) \quad \text{for all } (\xi,\eta)\in\mathbf{R}^2.$$

As a consequence, the subspace corresponding to the projection $R(\xi,\eta)$ reduces A (cf. § 21 theorem 3). We shall now investigate to which extent the family $\{R(\xi,\eta):(\xi,\eta)\in\mathbf{R}^2\}$ still displays features similar to those characteristic for spectral families in the cases already dealt with.

(a) For any two points (ξ,η) and (ξ',η') in $\mathbf{R}^2$ we have

$$R(\xi,\eta)\,R(\xi',\eta') = R(\min\{\xi,\xi'\},\,\min\{\eta,\eta'\})$$

and therefore in particular

$$R(\xi,\eta)\,R(\xi',\eta') = R(\xi,\eta) \quad \text{for } \xi\leqslant\xi' \text{ and } \eta\leqslant\eta'.$$

We conclude

$$R(\xi,\eta) \leqslant R(\xi',\eta') \quad \text{for } \xi\leqslant\xi' \text{ and } \eta\leqslant\eta'$$

(cf. § 15 theorem 6 c), e)). Intuitively speaking, the projection-valued function $R(x,y)$ is non-decreasing with respect to transition to points "above to the right" in $\mathbf{R}^2$.

(b) The corresponding properties of the spectral families $\{P(\xi): \xi \in \mathbf{R}\}$ and $\{Q(\eta): \eta \in \mathbf{R}\}$ imply

$$R(\xi, \eta) = P(\xi)Q(\eta) = O \quad \text{for } \xi < -\|A\| \text{ or } \eta < -\|A\|,$$
$$R(\xi, \eta) = P(\xi)Q(\eta) = I \quad \text{for } \xi \geqslant \|A\| \text{ and } \eta \geqslant \|A\|.$$

(c) If $\{\theta_n\}_{n=1}^{\infty}$ and $\{\varepsilon_n\}_{n=1}^{\infty}$ are non-increasing sequences of positive numbers, converging to zero, then for every $f \in \mathfrak{H}$ we get

$$\|R(\xi, \eta) f - R(\xi + \theta_n, \eta + \varepsilon_n) f\| \leqslant \|[P(\xi) - P(\xi + \theta_n)] Q(\eta) f\| +$$
$$+ \|P(\xi + \theta_n)\| \|Q(\eta) f - Q(\eta + \varepsilon_n) f\|.$$

Taking into account that $P(\xi+0)=P(\xi)$ and $Q(\eta+0)=Q(\eta)$ we see that

$$(s) \lim_{n \to \infty} R(\xi + \theta_n, \eta + \varepsilon_n) = R(\xi, \eta).$$

Again this fact may briefly be written in the form

$$R(\xi + 0, \eta + 0) = R(\xi, \eta) \quad \text{for all } (\xi, \eta) \in \mathbf{R}^2.$$

In the same way as for the individual spectral families $\{P(\xi): \xi \in \mathbf{R}\}$ and $\{Q(\eta): \eta \in \mathbf{R}\}$ we find that there also exists a unique projection $R(\xi-0, \eta-0) \leqslant R(\xi, \eta)$ such that

$$R(\xi - 0, \eta - 0) = (s) \lim_{n \to \infty} R(\xi - \theta_n, \eta - \varepsilon_n)$$

for every pair of sequences $\{\theta_n\}_{n=1}^{\infty}$, $\{\varepsilon_n\}_{n=1}^{\infty}$ as above.

(d) We now approximate the operator A by means of Riemann-Stieltjes sums ultimately defining a double integral of Riemann-Stieltjes type with respect to the projection-valued function $R(x, y)$ on $\mathbf{R}^2$. To this end, let two sequences of real numbers $\{\xi_k\}_{k=0}^{n}$, $\{\eta_l\}_{l=0}^{m}$ be given such that

(3)
$$\begin{cases} \xi_0 < -\|A\| = \xi_1 < \cdots < \xi_{n-1} < \xi_n = \|A\|, \\ \quad \xi_k - \xi_{k-1} \leqslant \delta \quad \text{for } 1 \leqslant k \leqslant n, \\ \eta_0 < -\|A\| = \eta_1 < \cdots < \eta_{m-1} < \eta_m = \|A\|, \\ \quad \eta_l - \eta_{l-1} \leqslant \delta \quad \text{for } 1 \leqslant l \leqslant m. \end{cases}$$

Identifying $\mathbf{R}^2$ with the complex plane we further choose in every rectangle $\{\lambda = \xi + i\eta: \xi_{k-1} \leqslant \xi \leqslant \xi_k, \eta_{l-1} \leqslant \eta \leqslant \eta_l\}$ a complex number $\lambda_{k,l}$ (fig. 29).

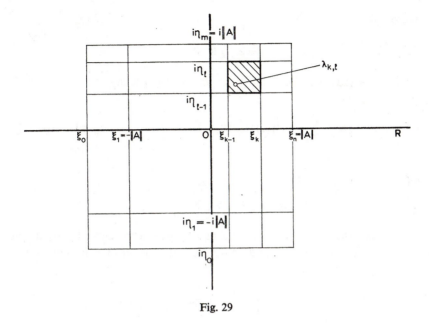

Fig. 29

Observe that the operator

$$R_{k,l} = [P(\xi_k) - P(\xi_{k-1})] [Q(\eta_l) - Q(\eta_{l-1})]$$
$$= R(\xi_k, \eta_l) - R(\xi_k, \eta_{l-1}) - R(\xi_{k-1}, \eta_l) + R(\xi_{k-1}, \eta_{l-1})$$

associated with this rectangle is again a projection (cf. § 15 theorem 4) and that

$$R_{k,l} \perp R_{k',l'} \quad \text{if } k \neq k' \text{ or } l \neq l',$$

$$\sum_{k=1}^{n} \sum_{l=1}^{m} R_{k,l} = \sum_{k=1}^{n} [P(\xi_k) - P(\xi_{k-1})] \sum_{l=1}^{m} [Q(\eta_l) - Q(\eta_{l-1})] = I.$$

Consider now the double Riemann-Stieltjes sums

(4) $$S = \sum_{k=1}^{n} \sum_{l=1}^{m} \lambda_{k,l} R_{k,l},$$

$$S' = \sum_{k=1}^{n} \sum_{l=1}^{m} (\xi_k + i\eta_l) R_{k,l}$$

$$= \sum_{k=1}^{n} \xi_k [P(\xi_k) - P(\xi_{k-1})] + i \sum_{l=1}^{m} \eta_l [Q(\eta_l) - Q(\eta_{l-1})].$$

We obtain

$$\|S' - S\| = \|\sum_{k=1}^{n} \sum_{l=1}^{m} (\xi_k + i\eta_l - \lambda_{k,l}) R_{k,l}\|$$

$$\leqslant \max\{|\xi_k + i\eta_l - \lambda_{k,l}| : 1 \leqslant k \leqslant n, 1 \leqslant l \leqslant m\}$$

$$< 2\delta,$$

$$\|A - S'\| \leqslant \|B - \sum_{k=1}^{n} \xi_k [P(\xi_k) - P(\xi_{k-1})]\| + \|C - \sum_{l=1}^{m} \eta_l [Q(\eta_l) - Q(\eta_{l-1})]\|$$

$$\leqslant 2\delta,$$

(cf. § 34 remark 1)

$$\|A - S\| \leqslant \|A - S'\| + \|S' - S\| \leqslant 4\delta.$$

The fact that A may be approximated arbitrarily close by Riemann-Stieltjes sums of type (4) by choosing the subdivision (3) sufficiently fine is again expressed in the form

$$A = \int_{-\infty}^{+\infty} \int_{-\infty}^{+\infty} (\xi + i\eta) \, dR(\xi, \eta)$$

where the double integral is of Riemann-Stieltjes type. In fact the reasoning applied above shows that the integration may be restricted to a square of the form

$$[-\|A\| - \theta, \|A\|] \times [-\|A\| - \theta, \|A\|] \subset \mathbf{R}^2$$

where $\theta > 0$ is arbitrary. Indeed, outside of any such a square the contributions $\lambda_{k,l} R_{k,l}$ to the sum S vanish because of $P(\xi_k) - P(\xi_{k-1}) = O$ or $Q(\eta_l) - Q(\eta_{l-1}) = O$.

Applying a similar reasoning to the operator $A^* = B - iC$ we get at the same time

$$A^* = \int_{-\infty}^{+\infty} \int_{-\infty}^{+\infty} (\xi - i\eta) \, dR(\xi, \eta).$$

We sum up these results in the following theorem where we also switch back from the notation $\{R(\xi, \eta) : (\xi, \eta) \in \mathbf{R}^2\}$ to the more familiar notation $\{P(\xi, \eta) : (\xi, \eta) \in \mathbf{R}^2\}$ for a spectral family.

THEOREM 1 (SPECTRAL THEOREM FOR BOUNDED NORMAL OPERATORS). *Let A be a bounded normal operator on $\mathfrak{H}$. Then there exists a family of projections $\{P(\xi, \eta) : (\xi, \eta) \in \mathbf{R}^2\}$ which commute with A, called a spectral family of A,* *with the following properties:*

a) $P(\xi, \eta) P(\xi', \eta') = P(\min \{\xi, \xi'\}, \min \{\eta, \eta'\})$
$$\text{for all } (\xi, \eta) \in \mathbf{R}^2 \text{ and all } (\xi', \eta') \in \mathbf{R}^2;$$

b) $\qquad\qquad P(\xi, \eta) = O \qquad$ for $\xi < - \|A\|$ or $\eta < - \|A\|$,
$\qquad\qquad P(\xi, \eta) = I \qquad$ for $\xi \geqslant \|A\|$ and $\eta \geqslant \|A\|$;

c) $\quad P(\xi+0, \eta+0) = P(\xi, \eta) \quad$ for all $(\xi, \eta) \in \mathbf{R}^2$;

d) $\qquad\qquad\qquad A = \displaystyle\int_{-\infty}^{+\infty} \int_{-\infty}^{+\infty} (\xi + i\eta) \, dP(\xi, \eta).$

More explicitly, we have

$$\left\| A - \sum_{k=1}^{n} \sum_{l=1}^{m} \lambda_{k,l} [P(\xi_k, \eta_l) - P(\xi_k, \eta_{l-1}) - P(\xi_{k-1}, \eta_l) + P(\xi_{k-1}, \eta_{l-1})] \right\| \leqslant 4\delta$$

whenever

$$\xi_0 < - \|A\| = \xi_1 < \cdots < \xi_{n-1} < \xi_n = \|A\|,$$
$$\xi_k - \xi_{k-1} \leqslant \delta \quad \text{for } 1 \leqslant k \leqslant n,$$
$$\eta_0 < - \|A\| = \eta_1 < \cdots < \eta_{m-1} < \eta_m = \|A\|,$$
$$\eta_l - \eta_{l-1} \leqslant \delta \quad \text{for } 1 \leqslant l \leqslant m,$$
$$\lambda_{k,l} \in \{\lambda = \xi + i\eta : \xi_{k-1} \leqslant \xi \leqslant \xi_k, \eta_{l-1} \leqslant \eta \leqslant \eta_l\}.$$

In the same sense the operator A^* is represented in the form

$$A^* = \int_{-\infty}^{+\infty} \int_{-\infty}^{+\infty} (\xi - i\eta) \, dP(\xi, \eta).$$

We shall not enter upon a discussion of the fact that for a selfadjoint or a unitary operator this theorem reduces to the spectral theorems established in § 34 and § 36 respectively. We also omit a discussion of corollaries and theorems which, after all we know about spectral decompositions so far, might be expected to accompany this spectral theorem.

Example 1. Let A be a compact normal operator on $\mathfrak{H}$, not of finite rank, and let

(5) $$A = \sum_{k=1}^{\infty} \lambda_k P_k$$

be its spectral decomposition (cf. § 28 theorem 4). Suppose

$$\lambda_k = \xi_k + i\eta_k, \quad \xi_k \in \mathbf{R}, \quad \eta_k \in \mathbf{R} \quad \text{for } 1 \leqslant k < \infty,$$

and let $\mathfrak{M}_k$ be the corresponding eigenspace of A. Define $P(\xi, \eta)$ to be the projection upon the subspace

$$\mathfrak{M}(\xi, \eta) = \vee \{\mathfrak{M}_k : \xi_k \leqslant \xi, \eta_k \leqslant \eta\}.$$

The family $\{P(\xi, \eta) : (\xi, \eta) \in \mathbf{R}^2\}$ then satisfies the requirements a), b), c), d) in theorem 1, and d) reduces to (5).

If A is any (not necessary normal) bounded linear operator on $\mathfrak{H}$, then the operators

$$B = \tfrac{1}{2}(A + A^*),$$

$$C = \frac{1}{2i}(A - A^*)$$

are still selfadjoint (cf. § 14 theorem 7). The spectral families of B and C will, however, not any more commute in general and the preceding reasoning breaks down. Still, if R and S are corresponding Riemann-Stieltjes sums approximating the operators B and C respectively within ε, then because of $A = B + iC$ we have

$$\|A - (R + iS)\| \leqslant \|B - R\| + \|C - S\| \leqslant 2\varepsilon.$$

Since $R + iS$ is a linear combination of projections we see that every bounded linear operator on $\mathfrak{H}$ can be approximated arbitrarily close by linear operators of this special type.

THEOREM 2. *The set of all linear combinations of projections is everywhere dense in $\mathscr{B}$ (in the sense of the operator norm).*

The following exercise generalizes § 34 theorem 5 and § 36 theorem 5 to the case of bounded normal operators in general.

Exercise. Let A be a bounded normal operator on $\mathfrak{H}$, let $\{P(\xi) : \xi \in \mathbf{R}\}$ and $\{Q(\eta) : \eta \in \mathbf{R}\}$ be the spectral families of the commuting selfadjoint operators $B = \tfrac{1}{2}(A + A^*)$ and $C = -\tfrac{1}{2}i(A - A^*)$ respectively and define

$$R(\xi, \eta) = P(\xi) Q(\eta) \quad \text{for all } (\xi, \eta) \in \mathbf{R}^2.$$

a) Show that for every $(\xi, \eta) \in \mathbf{R}^2$

$$R(\xi, \eta - 0) = P(\xi) Q(\eta - 0),$$
$$R(\xi - 0, \eta) = P(\xi - 0) Q(\eta),$$
$$R(\xi - 0, \eta - 0) = P(\xi - 0) Q(\eta - 0).$$

(Recall that e.g. $R(\xi, \eta - 0) = (s) \lim_{n \to \infty} R(\xi, \eta - \theta_n)$ by definition, where

$\{\theta_n\}_{n=1}^{\infty}$ is a non-increasing sequence of positive numbers which converges to zero.)

b) If $Af=(\xi+i\eta)f$ $(\xi\in\mathbf{R}, \eta\in\mathbf{R})$, then $Bf=\xi f$ and $Cf=\eta f$. (Hint: prove and use the fact that $A^*f=(\xi-i\eta)f$; cf. § 14 theorem 6 and § 23 exercise b).)

c) If $Af=(\xi+i\eta)f$ $(\xi\in\mathbf{R}, \eta\in\mathbf{R})$, then $R(\xi, \eta)f=f$ and $R(\xi, \eta-0)f=$ $=R(\xi-0, \eta)f=R(\xi-0, \eta-0)f=o$. (Hint: cf. § 34 theorem 5.)

d) If $(\xi+i\eta)$ is an eigenvalue of A for some $\xi\in\mathbf{R}$, $\eta\in\mathbf{R}$, then $[R(\xi, \eta)-R(\xi, \eta-0)-R(\xi-0, \eta)+R(\xi-0, \eta-0)]$ is the projection upon the corresponding eigenspace.

e) If $[R(\xi, \eta)-R(\xi, \eta-0)-R(\xi-0, \eta)+R(\xi-0, \eta-0)]\neq O$, then $(\xi+i\eta)$ is an eigenvalue of A. (Hint: the operator in brackets is a projection and every non-zero vector in the corresponding subspace is an eigenvector of B and C.)

CHAPTER VII

SPECTRAL ANALYSIS OF UNBOUNDED
SELFADJOINT OPERATORS

§ 38. The Cayley transform

The result we still are looking for is a spectral decomposition of a (not necessarily bounded) selfadjoint operator A in $\mathfrak{H}$. It seems reasonable to try to use the decomposition theorems already available to us in order to achieve this goal.

In order to explain the idea which we shall follow suppose for a moment that the selfadjoint operator A is bounded. Then for every continuous complex-valued function $\varphi(x)$ on $\mathbf{R}$ the operator $\varphi(A)$ is well defined (cf. § 32 theorem 1). In particular, consider the continuous function on $\mathbf{R}$

$$\varphi(x) = \frac{x - \mathrm{i}}{x + \mathrm{i}} = (x - \mathrm{i})(x + \mathrm{i})^{-1}.$$

We have $|\varphi(\xi)| = 1$ for all $\xi \in \mathbf{R}$ and the function $\varphi(x)$ maps the real axis into the complex unit circle. In fact the mapping is onto if we admit x to assume the "value" $\xi = \infty$ which then is mapped into 1. We conclude that the operator

$$U = \varphi(A) = (A - \mathrm{i}I)(A + \mathrm{i}I)^{-1}$$

is unitary (cf. § 32 example 2 and § 32 corollary 3.1 b)), and its spectrum $\Sigma(U) = \Sigma(\varphi(A)) = \varphi(\Sigma(A))$ is a closed subset of the complex unit circle not containing the point 1 (cf. § 32 theorem 3).

Let $\{Q(\mu) : 0 \leqslant \mu \leqslant 2\pi\}$ be the spectral family of U. We now reverse the mapping considered above, defining the function $\psi(x)$ on the complex unit circle by

$$\psi(x) = \mathrm{i}\,\frac{1 + x}{1 - x} = \mathrm{i}(1 + x)(1 - x)^{-1}.$$

Although the function $\psi(x)$ is not continuous on the whole complex unit circle (it maps 1 into the "point" ∞) it suffices to define the operator

$$\psi(U) = \mathrm{i}(I + U)(I - U)^{-1}$$

since $1 \notin \Sigma(U)$ (this and the following somewhat ruthless reasoning could – and would have to – be made watertight if its ambitions were greater than just to give an intuitive picture of what is going on in this and the following section). Since $x = \psi[\varphi(x)]$ we get

$$A = \psi(U) = \int_0^{2\pi} \psi(e^{i\mu}) \, dQ(\mu)$$

where

$$\psi(e^{i\mu}) = i \frac{1 + e^{i\mu}}{1 - e^{i\mu}} = i \frac{e^{-\frac{1}{2}i\mu} + e^{\frac{1}{2}i\mu}}{e^{-\frac{1}{2}i\mu} - e^{\frac{1}{2}i\mu}} = -\frac{\cos\frac{1}{2}\mu}{\sin\frac{1}{2}\mu} = -\cot\frac{1}{2}\mu$$

increases from $-\infty$ to $+\infty$ as μ increases from 0 to 2π. Substituting

$$\lambda = -\cot\tfrac{1}{2}\mu$$

we get

$$A = \int_{-\infty}^{+\infty} \lambda \, dQ(2 \operatorname{arccot}(-\lambda)).$$

The conclusion seems to be that defining $P(\lambda) = Q(2 \operatorname{arccot}(-\lambda))$ we have again obtained the spectral family of A, but now on a detour via the spectral family of the unitary operator $U = \varphi(A)$.

We shall now substantiate the foregoing more or less intuitive reasoning, even without the assumption that A be bounded, by first constructing explicitly the operator U as a function of A and then conversely A as a function of U. The construction of a spectral family of A and the corresponding spectral decomposition of A will follow in the next section.

THEOREM 1. *Let A be a selfadjoint operator in $\mathfrak{H}$. Then the operators $A \pm iI$ have bounded inverses defined on all of $\mathfrak{H}$ and the operator*

$$U = (A - iI)(A + iI)^{-1}$$

is a unitary operator on $\mathfrak{H}$, called the Cayley transform of A.

PROOF. Since A is selfadjoint we have $\Sigma(A) \subset \mathbf{R}$ and $\pm i \notin \Sigma(A)$ (cf. § 24 theorem 3). As a consequence, the operators $A \pm iI$ have bounded inverses defined on all of $\mathfrak{H}$ and ($\mathfrak{D}_A$ denoting the domain of A)

$$(A + iI)^{-1} \mathfrak{H} = \mathfrak{D}_{A+iI} = \mathfrak{D}_A,$$
$$(A - iI) \mathfrak{D}_A = \mathfrak{H}.$$

Combining these two statements we see that the linear operator

$$U = (A - iI)(A + iI)^{-1}$$

maps $\mathfrak{H}$ in one-to-one fashion onto $\mathfrak{H}$.

It remains to show that U preserves the norm (cf. § 13 corollary 1.2). To this end, observe that for every $g \in \mathfrak{D}_A$ we have

$$\|(A \pm iI)g\|^2 = \langle Ag, Ag \rangle \pm i\langle g, Ag \rangle \mp i\langle Ag, g \rangle + \langle g, g \rangle$$
$$= \|Ag\|^2 + \|g\|^2,$$
$$\|(A + iI)g\| = \|(A - iI)g\|.$$

Substituting $g = (A + iI)^{-1}f$ we conclude

$$\|Uf\| = \|(A - iI)(A + iI)^{-1}f\| = \|(A + iI)(A + iI)^{-1}f\|$$
$$= \|f\| \quad \text{for all } f \in \mathfrak{H}. \ \square$$

THEOREM 2. *If U is the Cayley transform of the selfadjoint operator A in $\mathfrak{H}$, then $I - U$ maps $\mathfrak{H}$ in one-to-one fashion onto $\mathfrak{D}_A$ and*

$$A = i(I + U)(I - U)^{-1}.$$

PROOF. We already know that the operator $A + iI$ maps $\mathfrak{D}_A$ in one-to-one fashion onto $\mathfrak{H}$. For every corresponding pair of vectors $f \in \mathfrak{D}_A$ and $g \in \mathfrak{H}$ we get

$$g = (A + iI)f,$$
$$Ug = (A - iI)f,$$

and therefore

(1) $$\qquad\qquad (I + U)g = 2Af,$$

(2) $$\qquad\qquad (I - U)g = 2if.$$

The last equation shows that $I - U$ maps $\mathfrak{H}$ onto $\mathfrak{D}_A$. This mapping is one-to-one since $(I - U)g = o$ implies $f = o$ which in turn implies $g = (A + iI)f = o$. Finally, for every given $f \in \mathfrak{D}_A$ we conclude by (1) and (2)

$$Af = \tfrac{1}{2}(I + U)g = \tfrac{1}{2}(I + U)(I - U)^{-1}(2if)$$
$$= i(I + U)(I - U)^{-1}f. \ \square$$

COROLLARY 2.1. *Let U be the Cayley transform of the selfadjoint operator A in $\mathfrak{H}$. Then 1 is not an eigenvalue of U. Moreover, 1 is a regular value for U iff A is bounded.*

PROOF. If A is bounded, then $U=\varphi(A)$ where $\varphi(x)=(x-\mathrm{i})(x+\mathrm{i})^{-1}$, and $\Sigma(U)=\varphi(\Sigma(A))$. However, as remarked earlier, we have $\varphi(\xi)\neq1$ for all $\xi\in\mathbf{R}$ and therefore $1\notin\Sigma(U)$. The other assertions immediately follow from theorem 2. $\square$

THEOREM 3. *Let U be the Cayley transform of a selfadjoint operator A in $\mathfrak{H}$ and let B be a bounded linear operator on $\mathfrak{H}$. Then $BU=UB$ iff B commutes with A, in other words iff $B\mathfrak{D}_A\subset\mathfrak{D}_A$ and $BA\subset AB$ (cf. § 21 definition 3).*
PROOF. Suppose B commutes with A. For every $f\in\mathfrak{H}$ we have

$$(A+\mathrm{i}I)^{-1}f\in\mathfrak{D}_A,$$
$$B(A+\mathrm{i}I)^{-1}f\in\mathfrak{D}_A,$$
$$(A+\mathrm{i}I)B(A+\mathrm{i}I)^{-1}f=B(A+\mathrm{i}I)(A+\mathrm{i}I)^{-1}f=Bf,$$
$$B(A+\mathrm{i}I)^{-1}f=(A+\mathrm{i}I)^{-1}Bf,$$
$$BUf=B(A-\mathrm{i}I)(A+\mathrm{i}I)^{-1}f=(A-\mathrm{i}I)B(A+\mathrm{i}I)^{-1}f$$
$$=(A-\mathrm{i}I)(A+\mathrm{i}I)^{-1}Bf=UBf.$$

Conversely, if $BU=UB$ and if the vector $f\in\mathfrak{D}_A=\mathfrak{D}_{(I-U)^{-1}}$ is given, then putting $g=(I-U)^{-1}f$ we get

$$f=(I-U)g,$$
$$Bf=B(I-U)g=(I-U)Bg\in\mathfrak{D}_{(I-U)^{-1}}=\mathfrak{D}_A,$$
$$(I-U)^{-1}Bf=Bg=B(I-U)^{-1}f,$$
$$ABf=\mathrm{i}(I+U)(I-U)^{-1}Bf=\mathrm{i}(I+U)B(I-U)^{-1}f$$
$$=B\mathrm{i}(I+U)(I-U)^{-1}f=BAf.$$

This proofs the assertion. $\square$

COROLLARY 3.1. *Let A and U be as in theorem 3 and let $\{Q(\mu):0\leqslant\mu\leqslant2\pi\}$ be the spectral family of U. If $0\leqslant\mu<\mu'\leqslant2\pi$, then the subspace corresponding to the projection $Q(\mu')-Q(\mu)$ reduces A.*
PROOF. Cf. § 21 theorem 3 and § 36 theorem 3. $\square$

The chain of assertions in the following exercise leads up to the statement that every unitary operator on $\mathfrak{H}$ which does not have 1 as an eigenvalue is the Cayley transform of a selfadjoint operator in $\mathfrak{H}$.

Exercise. Let U be a unitary operator on $\mathfrak{H}$ which does not have 1 as an eigenvalue and define an operator A in $\mathfrak{H}$ by $A=\mathrm{i}(I+U)(I-U)^{-1}$.

a) $\mathfrak{D}_A=\mathfrak{D}_{(I-U)^{-1}}=\mathfrak{H}$. (Hint: cf. § 22 theorem 5.)

b) 1 is not an eigenvalue of U^{-1}, moreover $[(I-U)^{-1}]^*=(I-U^{-1})^{-1}$. (Hint: cf. § 17 theorem 5.)

c) $A \subset A^*$. (Hint: let $f=(I-U)f_1 \in \mathfrak{D}_A$ and $g=(I-U)g_1 \in \mathfrak{D}_A$ be given and show $\langle Af, g \rangle = \langle f, Ag \rangle$.)

d) $\mathfrak{D}_{A^*} \subset \mathfrak{D}_A$; consequently A is selfadjoint. (Hint: if $\langle Af, g \rangle = \langle f, g^* \rangle$ for all $f \in \mathfrak{D}_A$ let $f=(I-U)f_1$ where f_1 runs through $\mathfrak{H}$ and show

$$g = (I - U)[\tfrac{1}{2}(g - ig^*)] .)$$

e) $U=(A-iI)(A+iI)^{-1}$; consequently U is the Cayley transform of A. (Hint: for every vector $f_1 \in \mathfrak{H}$ we have $f=(I-U)f_1 \in \mathfrak{D}_A$ and $Af=i(I+U)f_1$.)

§ 39. The spectral decomposition of an unbounded selfadjoint operator

In what follows let A be a fixed, possibly unbounded, selfadjoint operator in $\mathfrak{H}$, let U be its Cayley transform (cf. § 38 theorem 1 and § 38 theorem 2), and let $\{Q(\mu):0 \leqslant \mu \leqslant 2\pi\}$ be the spectral family of U. Then we have

$$U = (A - iI)(A + iI)^{-1},$$
$$A = i(I + U)(I - U)^{-1},$$

and for every continuous complex-valued function $\varphi(x)$ on the complex unit circle there exists a well-defined bounded operator

$$\varphi(U) = \int_0^{2\pi} \varphi(e^{i\mu}) \, dQ(\mu)$$

on $\mathfrak{H}$ (cf. § 36 theorem 1).

The introduction of the preceding section already suggested that A, as a function of U, may perhaps also be represented as a Riemann-Stieltjes integral with respect to the projection-valued function $Q(y)$ on $[0, 2\pi]$, and that the substitution $\lambda = -\cot \tfrac{1}{2}\mu$ might even lead to an integral representation of the familiar form

$$A = \int_{-\infty}^{+\infty} \lambda \, dP(\lambda).$$

The obstacle, however, which we still have to overcome lies in the fact that, as far as we know, this procedure is only applicable to continuous functions $\varphi(x)$ on the complex unit circle and then produces bounded linear operators $\varphi(U)$. Accordingly, if A is unbounded there is no hope of obtaining A from U via such a continuous function. The way to master this difficulty is to cut

$\mathfrak{H}$ into pieces, more precisely into a sum of mutually orthogonal reducing subspaces, in each of which A is restricted to a bounded selfadjoint operator.

THEOREM 1. *For every integer m let $\mu_m = 2 \operatorname{arccot}(-m)$ (fig. 30) and let $\mathfrak{H}_m$ be the subspace corresponding to the projection $P_m = Q(\mu_m) - Q(\mu_{m-1})$. The subspaces $\mathfrak{H}_m$ are pairwise orthogonal and span $\mathfrak{H}$:*

$$\text{(1)} \qquad \mathfrak{H} = \sum_{m=-\infty}^{+\infty} \mathfrak{H}_m .$$

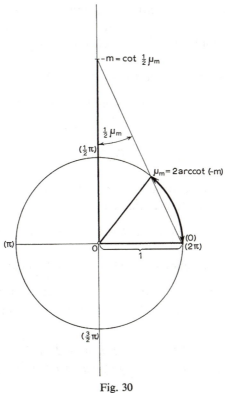

Fig. 30

PROOF. For $m < n$ we have $Q(\mu_m) \leqslant Q(\mu_n)$ and therefore

$$P_m P_n = [Q(\mu_m) - Q(\mu_{m-1})] [Q(\mu_n) - Q(\mu_{n-1})]$$
$$= [Q(\mu_m) - Q(\mu_{m-1})] - [Q(\mu_m) - Q(\mu_{m-1})] = O$$

(cf. § 15 theorem 6 c), e)). We conclude $\mathfrak{H}_m \perp \mathfrak{H}_n$ (cf. § 15 theorem 5 a), d)).

Recall that $Q(2\pi-0)=Q(2\pi)=I$ since $1=e^{2\pi i}$ is not an eigenvalue of U (cf. § 36 theorem 5 and § 38 corollary 2.1). For every $f \in \mathfrak{H}$ we get

$$\sum_{m=-\infty}^{+\infty} P_m f = \lim_{m \to +\infty} Q(\mu_m) f - \lim_{n \to -\infty} Q(\mu_n) f$$
$$= Q(2\pi - 0) f - Q(0 + 0) f = f - o = f.$$

This proves the assertion (cf. § 7 theorem 4). □

The reader is warned that there is no reason to be afraid of a "two-sided" orthogonal series (=vector sum) of subspaces of the form (1) in place of the "one-sided" series considered in § 7. In fact (1) could also be written (if we would prefer the trouble to do so) in the form

$$\mathfrak{H} = \mathfrak{H}_0 + \mathfrak{H}_1 + \mathfrak{H}_{-1} + \mathfrak{H}_2 + \mathfrak{H}_{-2} + \cdots,$$

since changing the order of summation does not afflict the convergence or divergence of the series

$$\sum_{m=-\infty}^{+\infty} \|f_m\|^2 \quad (f_m \in \mathfrak{H}_m \text{ for } -\infty < m < +\infty).$$

Since formally we have arrived here in a natural way at a series of two-sided type we shall also stick to it in the sequel.

THEOREM 2. *For every integer m, the subspace $\mathfrak{H}_m$ as defined in theorem* 1 *reduces A.*
PROOF. The assertion follows immediately from § 38 corollary 3.1. □

By § 21 theorem 1 the operator A induces in every subspace $\mathfrak{H}_m$ $(-\infty < m < +\infty)$ a selfadjoint operator A_m which we now proceed to study in some more detail.

THEOREM 3. *Let $\mathfrak{H}_m$ be defined as in theorem* 1. *Then $\mathfrak{H}_m \subset \mathfrak{D}_A$ and the restriction A_m of A to $\mathfrak{H}_m$ is a bounded selfadjoint operator on $\mathfrak{H}_m$ given by*

$$A_m f = Af = \int_{\mu_{m-1}}^{\mu_m} (-\cot\tfrac{1}{2}\mu) \, dQ(\mu) f \qquad \text{for all } f \in \mathfrak{H}_m.$$

PROOF. Let $f \in \mathfrak{H}_m$ be given. Then $Q(\mu)f=f$ for $\mu \geqslant \mu_m$ and $Q(\mu)f=o$ for $\mu \leqslant \mu_{m-1}$ by the definition of $\mathfrak{H}_m$. Using μ_{m-1} and μ_m as fixed subdivision points for our Riemann-Stieltjes sums we get for any continuous function $\varphi(x)$ on the complex unit circle

$$\varphi(U) f = \int_{\mu_{m-1}}^{\mu_m} \varphi(e^{i\mu}) \, dQ(\mu) \, f \in \mathfrak{H}_m$$

(cf. § 36 corollary 1.1). In particular, let us define for every integer m

$$\xi_m = e^{i\mu_m},$$

$$\varphi_m(\xi) = \begin{cases} (1-\xi)^{-1} & \text{for } \xi = e^{i\mu}, \mu_{m-1} \leqslant \mu \leqslant \mu_m, \\ (\xi_m - \xi_{m-1})^{-1} [(1-\xi_{m-1})^{-1} (\xi_m - \xi) + (1-\xi_m)^{-1} (\xi - \xi_{m-1})] \end{cases}$$
$$\text{for all other } \xi \text{ on the complex unit circle}.$$

The function $\varphi_m(x)$ is continuous on the complex unit circle and we get

$$(I-U) \, \varphi_m(U) \, f = \int_{m-1}^{\mu_m} (1 - e^{i\mu}) \, \varphi_m(e^{i\mu}) \, dQ(\mu) \, f$$

$$= \int_{\mu_{m-1}}^{\mu_m} dQ(\mu) \, f = Q(\mu_m) \, f - Q(\mu_{m-1}) \, f = f.$$

We conclude $f \in (I-U) \, \mathfrak{H} = \mathfrak{D}_{(I-U)^{-1}} = \mathfrak{D}_A$. This proves $\mathfrak{H}_m \subset \mathfrak{D}_A$. Since the operator A_m is selfadjoint and defined on all of $\mathfrak{H}_m$ it is also bounded (cf. § 21 theorem 1 and § 17 corollary 8.2). Moreover we get

$$(I-U)^{-1} f = \varphi_m(U) \, f$$
$$A_m f = Af = i(I+U)(I-U)^{-1} f = i(I+U) \, \varphi_m(U) \, f$$

$$= \int_{\mu_{m-1}}^{\mu_m} i \frac{1+e^{i\mu}}{1-e^{i\mu}} \, dQ(\mu) \, f = \int_{\mu_{m-1}}^{\mu_m} (-\cot \tfrac{1}{2}\mu) \, dQ(\mu) \, f . \ \square$$

COROLLARY 3.1. *Let $\mathfrak{H}_m$ and A_m be defined as in theorem 3. Then*

$$\|A_m f\|^2 = \int_{\mu_{m-1}}^{\mu_m} |\cot \tfrac{1}{2}\mu|^2 \, d \, \|Q(\mu) \, f\|^2 \quad \text{for all } f \in \mathfrak{H}_m.$$

PROOF. Applying § 36 corollary 1.2 to the function $\varphi(x) = i(1+x) \, \varphi_m(x)$ and to the vector $f \in \mathfrak{H}_m$ we get

$$\|A_m f\|^2 = \|Af\|^2 = \|i(I+U) \, \varphi_m(U) \, f\|^2$$

$$= \int_{\mu_{m-1}}^{\mu_m} |\cot \tfrac{1}{2}\mu|^2 \, d \, \|Q(\mu) \, f\|^2 . \ \square$$

COROLLARY 3.2. *Let μ_m and P_m be defined as in theorem 1. Then AP_m is a bounded selfadjoint operator on $\mathfrak{H}$ and*

$$AP_m = \int_{\mu_{m-1}}^{\mu_m} (-\cot \tfrac{1}{2}\mu) \, dQ(\mu).$$

PROOF. If the continuous function $\varphi_m(x)$ on the complex unit circle is defined as in the proof of theorem 3, then we have

$$i(I + U) \, \varphi_m(U) = \int_0^{2\pi} i(1 + e^{i\mu}) \, \varphi_m(e^{i\mu}) \, dQ(\mu),$$

$$(2) \qquad i(I + U) \, \varphi_m(U) \, P_m = \int_{\mu_{m-1}}^{\mu_m} i \frac{1 + e^{i\mu}}{1 - e^{i\mu}} \, dQ(\mu)$$

$$= \int_{\mu_{m-1}}^{\mu_m} (-\cot \tfrac{1}{2}\mu) \, dQ(\mu),$$

$$i(I + U) \, \varphi_m(U) \, P_m f = \int_{\mu_{m-1}}^{\mu_m} (-\cot \tfrac{1}{2}\mu) \, dQ(\mu) \, f \quad \text{for all } f \in \mathfrak{H}.$$

The bounded linear operator $i(I+U) \, \varphi_m(U) \, P_m$ on $\mathfrak{H}$ therefore coincides with A_m on $\mathfrak{H}_m$ (cf. theorem 3), while it annihilates the subspace $\mathfrak{H}_m^{\perp}$. As a consequence it is selfadjoint (this could also already have been deduced from (2)) and it coincides with the operator AP_m. $\square$

COROLLARY 3.3. *Let P_m and A_m be defined as in theorem 1 and theorem 3 and define*

$$P(\lambda) = Q(2 \arccot(-\lambda)) \quad \text{for all } \lambda \in \mathbf{R}.$$

The family of projections $\{P(\lambda): \lambda \in \mathbf{R}\}$ has the following properties:

a) $\qquad P(\lambda) \leqslant P(\lambda') \quad for \ \lambda \leq \lambda';$

b) (s) $\lim_{\lambda \to -\infty} P(\lambda) = O,$

 (s) $\lim_{\lambda \to +\infty} P(\lambda) = I;$

c) $\qquad P(\lambda + 0) = P(\lambda) \quad for \ all \ \lambda \in \mathbf{R};$

d) *for every integer m the following equalities hold:*

$$AP_m = \int_{m-1}^{m} \lambda dP(\lambda),$$

$$A_m f = \int_{m-1}^{m} \lambda dP(\lambda) f \quad \text{for all } f \in \mathfrak{H}_m,$$

$$\|A_m f\|^2 = \int_{m-1}^{m} |\lambda|^2 \, d\|P(\lambda) f\|^2 \quad \text{for all } f \in \mathfrak{H}_m.$$

Moreover, a bounded linear operator B on $\mathfrak{H}$ commutes with A iff $P(\lambda) B = = BP(\lambda)$ for all $\lambda \in \mathbf{R}$.

PROOF. The assertions a), b), c) immediately follow from the corresponding properties of the spectral family $\{Q(\mu): 0 \leqslant \mu \leqslant 2\pi\}$ of the Cayley transform U (for b) cf. § 38 corollary 2.1). Assertion d) follows from theorem 3, corollary 3.1, and corollary 3.2 after the substitution $-\cot \frac{1}{2}\mu = \lambda$ has been applied in the Riemann-Stieltjes sums approximating the corresponding integrals. The last assertion follows from § 38 theorem 3 and § 36 theorem 3. $\square$

It now remains to glue the pieces of A together again. That this can be done is the content of the following theorem.

THEOREM 4 (RIESZ-LORCH). *Suppose $\{\mathfrak{H}_m\}_{m=-\infty}^{+\infty}$ is a sequence of pairwise orthogonal subspaces such that*

(3) $$\mathfrak{H} = \sum_{m=-\infty}^{+\infty} \mathfrak{H}_m.$$

For every integer m let P_m be the projection upon $\mathfrak{H}_m$ and let A_m be a bounded selfadjoint operator on $\mathfrak{H}_m$. Then there exists a unique selfadjoint operator A in $\mathfrak{H}$ which is reduced by every subspace $\mathfrak{H}_m$ and the restriction of which to $\mathfrak{H}_m$ coincides with A_m for $-\infty < m < +\infty$. Moreover,

(4) $$\mathfrak{D}_A = \{f \in \mathfrak{H}: \sum_{m=-\infty}^{+\infty} \|A_m P_m f\|^2 < \infty\}$$

and

(5) $$Af = \sum_{m=-\infty}^{+\infty} A_m P_m f \quad \text{for all } f \in \mathfrak{D}_A.$$

PROOF. In order to simplify the notation, for every $f \in \mathfrak{H}$ write $f_m = P_m f$ $(-\infty < m < +\infty)$. The set $\mathfrak{D}_A$ as defined in (4) is a linear manifold (the proof of this statement parallels the proof of § 4 theorem 1). Since

$$\sum_{m=-n}^{n} \mathfrak{H}_m \subset \mathfrak{D}_A \quad \text{for all } n \geqslant 1$$

the hypothesis (3) implies that $\mathfrak{D}_A$ is everywhere dense in $\mathfrak{H}$. For $f \in \mathfrak{D}_A$ let Af be defined as in (5) (cf. § 7 theorem 3). It follows that A is a linear operator with domain $\mathfrak{D}_A$ and

$$P_m A f = A_m f_m = A f_m = A P_m f \quad \text{for all } f \in \mathfrak{D}_A.$$

Thus $\mathfrak{H}_m$ reduces A (cf. § 21 theorem 3). Moreover, for $f \in \mathfrak{D}_A$ and $g \in \mathfrak{D}_A$ we get

$$\langle Af, g \rangle = \sum_{m=-\infty}^{+\infty} \langle A_m f_m, g_m \rangle = \sum_{m=-\infty}^{+\infty} \langle f_m, A_m g_m \rangle = \langle f, Ag \rangle.$$

We conclude that A is symmetric (cf. § 17 definition 4).

In order to show that A is selfadjoint let $g = \sum_{m=-\infty}^{+\infty} g_m \in \mathfrak{D}_{A^*}$ be given. For every $f \in \mathfrak{H}_m \subset \mathfrak{D}_A$ we get

$$\langle f, A_m g_m \rangle = \langle A_m f, g \rangle = \langle A f, g \rangle = \langle f, A^* g \rangle$$
$$= \langle f, (A^* g)_m \rangle.$$

We conclude

$$A_m g_m = (A^* g)_m,$$

$$\sum_{m=-\infty}^{+\infty} \|A_m g_m\|^2 = \sum_{m=-\infty}^{+\infty} \|(A^* g)_m\|^2 = \|A^* g\|^2 < \infty,$$

$$g \in \mathfrak{D}_A.$$

Let now A' be any other selfadjoint operator in $\mathfrak{H}$ which is reduced by every subspace $\mathfrak{H}_m$ and the restriction of which to $\mathfrak{H}_m$ coincides with A_m for $-\infty < m < +\infty$. For any $f \in \mathfrak{D}_A$ we have

$$\lim_{n \to \infty} \sum_{m=-n}^{n} f_m = f,$$

$$\lim_{n \to \infty} A' \left(\sum_{m=-n}^{n} f_m \right) = \lim_{n \to \infty} \sum_{m=-n}^{n} A_m f_m = A f.$$

Since A' is closed (cf. § 20 corollary 3.1) we conclude $f \in \mathfrak{D}_{A'}$ and $A'f = Af$ (cf. § 20 definition 1). This shows $A' \supset A$ (cf. § 17 definition 2). Taking adjoints we get $A = A^* \supset A'^* = A'$ (cf. § 17 theorem 3). Both inclusions together imply $A = A'$. $\square$

We now apply theorem 4 to the operators A_m $(-\infty < m < +\infty)$ induced by the selfadjoint operator A in the subspaces $\mathfrak{H}_m$ as defined in theorem 1 and theorem 3. Our first conclusion is

(6)
$$\mathfrak{D}_A = \{f \in \mathfrak{H}: \sum_{m=-\infty}^{+\infty} \|A_m P_m f\|^2 < \infty\}$$

$$= \{f \in \mathfrak{H}: \sum_{m=-\infty}^{+\infty} \int_{m-1}^{m} |\lambda|^2 \, d \, \|P(\lambda) f\|^2 < \infty\}$$

(by corollary 3.3). In fact it is legitimate to replace the last series by a single integral with the limits $-\infty$ and $+\infty$ (cf. appendix B2 definition 2). In order to verify this statement observe that for $k-2 \leqslant \mu \leqslant k-1$ and $l \leqslant \nu \leqslant l+1$ we have

$$\sum_{m=k}^{l} \int_{m-1}^{m} |\lambda|^2 \, d \, \|P(\lambda) f\|^2 \leqslant \int_{\mu}^{\nu} |\lambda|^2 \, d \, \|P(\lambda) f\|^2 \leqslant \sum_{m=k-1}^{l+1} \int_{m-1}^{m} |\lambda|^2 \, d \, \|P(\lambda) f\|^2$$

and therefore

$$\sum_{m=-\infty}^{+\infty} \int_{m-1}^{m} |\lambda|^2 \, d \, \|P(\lambda) f\|^2 = \lim_{\substack{\mu \to -\infty \\ \nu \to +\infty}} \int_{\mu}^{\nu} |\lambda|^2 \, d \, \|P(\lambda) f\|^2 = \int_{-\infty}^{+\infty} |\lambda|^2 \, d \, \|P(\lambda) f\|^2.$$

This allows to rewrite (6) in the form

$$\mathfrak{D}_A = \{f \in \mathfrak{H}: \int_{-\infty}^{+\infty} |\lambda|^2 \, d \, \|P(\lambda) f\|^2 < \infty\}.$$

Next we conclude from theorem 4

$$Af = \sum_{m=-\infty}^{+\infty} \int_{m-1}^{m} \lambda \, dP(\lambda) f = \lim_{\substack{k \to -\infty \\ l \to +\infty}} \int_{k}^{l} \lambda \, dP(\lambda) f.$$

Again we would like to replace the second and third member by a single integral from $-\infty$ to $+\infty$. Denoting by $[\mu]$ the greatest integer not exceeding μ (i.e. $[\mu] \leqslant \mu < [\mu] + 1$) we get

$$\int\limits_{\mu}^{\nu} \lambda dP(\lambda) f = \int\limits_{[\mu]}^{[\nu]} \lambda \, dP(\lambda) f + \int\limits_{[\nu]}^{\nu} \lambda \, dP(\lambda) f - \int\limits_{[\mu]}^{\mu} \lambda \, dP(\lambda) f,$$

$$\lim_{\mu \to -\infty} \left\| \int\limits_{[\mu]}^{\mu} \lambda dP(\lambda) f \right\|^2 = \lim_{\mu \to -\infty} \int\limits_{[\mu]}^{\mu} |\lambda|^2 \, d \, \| P(\lambda) f \|^2 \leqslant \lim_{\mu \to -\infty} \| A_{[\mu]+1} f \|^2 = 0,$$

$$\lim_{\nu \to +\infty} \left\| \int\limits_{[\nu]}^{\nu} \lambda \, dP(\lambda) f \right\|^2 = \lim_{\nu \to +\infty} \int\limits_{[\nu]}^{\nu} |\lambda|^2 \, d \, \| P(\lambda) f \|^2 \leqslant \lim_{\nu \to +\infty} \| A_{[\nu]+1} f \|^2 = 0,$$

$$\int\limits_{-\infty}^{+\infty} \lambda dP(\lambda) f = \lim_{\substack{\mu \to -\infty \\ \nu \to +\infty}} \int\limits_{\mu}^{\nu} \lambda dP(\lambda) f = \lim_{\substack{\mu \to -\infty \\ \nu \to +\infty}} \sum_{m=[\mu]+1}^{[\nu]} A_m f = Af.$$

We combine these results with the assertions of corollary 3.3 in the following theorem.

THEOREM 5 (SPECTRAL THEOREM FOR (NOT NECESSARILY BOUNDED) SELFADJOINT OPERATORS). *Let A be a selfadjoint operator in* $\mathfrak{H}$. *Then there exists a family of projections* $\{P(\lambda): \lambda \in \mathbf{R}\}$, *called the spectral family of A, with the following properties*:

a) $P(\lambda) \leqslant P(\lambda')$ *for* $\lambda \leqslant \lambda'$;

b)   (s) $\lim\limits_{\lambda \to -\infty} P(\lambda) = O$,

      (s) $\lim\limits_{\lambda \to +\infty} P(\lambda) = I$;

c) $P(\lambda + 0) = P(\lambda)$ *for all* $\lambda \in \mathbf{R}$;

d) $f \in \mathfrak{D}_A$ *iff* $\int\limits_{-\infty}^{+\infty} \lambda^2 \, d \, \| P(\lambda) f \|^2 < \infty$,

and

$$Af = \int\limits_{-\infty}^{+\infty} \lambda dP(\lambda) f \quad \text{*for all* } f \in \mathfrak{D}_A.$$

Remark 1. The second part of assertion d) in theorem 5 is also written in the form

(7)
$$A = \int_{-\infty}^{+\infty} \lambda dP(\lambda).$$

However, in contrast to the corresponding statement in § 34 theorem 1 d), this Riemann-Stieltjes integral may not be interpreted as a uniform limit of Riemann-Stieltjes sums in $\mathscr{B}$. The proper interpretation of (7) is

$$Af = \lim_{\substack{\mu \to -\infty \\ \nu \to +\infty}} \int_{\mu}^{\nu} \lambda dP(\lambda) f \quad \text{for all } f \in \mathfrak{D}_A$$

where the integral between the finite limits μ and ν is a strong limit of Riemann-Stieltjes sums in $\mathfrak{H}$.

Again the properties a), b), c), d) mentioned in theorem 5 determine the spectral family uniquely.

THEOREM 6. *Let $\{P'(\lambda) : \lambda \in \mathbf{R}\}$ be a family of projections satisfying the requirements a), b), c), d) in theorem 5. Then $P'(\lambda) = P(\lambda)$ where $P(\lambda)$ is defined as in corollary 3.3.*
PROOF. We first show

(8) $P(\lambda) P'(\mu) = P'(\mu) P(\lambda)$ for all $\lambda \in \mathbf{R}$ and all $\mu \in \mathbf{R}$.

Indeed, for every $f \in \mathfrak{D}_A$ and for fixed $\mu \in \mathbf{R}$ we have

$$\int_{-\infty}^{+\infty} |\lambda|^2 \, d \, \| P'(\lambda) P'(\mu) f \|^2 = \int_{-\infty}^{\mu} |\lambda|^2 \, d \, \| P'(\lambda) f \|^2$$
$$\leqslant \int_{-\infty}^{+\infty} |\lambda|^2 \, d \, \| P'(\lambda) f \|^2 < \infty$$

and therefore $P'(\mu) f \in \mathfrak{D}_A$. Furthermore, a look at the Riemann-Stieltjes sums approximating the corresponding integrals with finite limits reveals that

$$AP'(\mu) f = \int_{-\infty}^{+\infty} \lambda \, dP'(\lambda) P'(\mu) f = \int_{-\infty}^{\mu} \lambda \, dP'(\lambda) f = P'(\mu) Af.$$

The last assertion of corollary 3.3 now implies (8).

Suppose now $P'(\mu) \neq P(\mu)$ for some $\mu \in \mathbf{R}$. Without loss of generality we assume $P'(\mu) P(\mu) \neq P(\mu)$ (cf. the proof of § 34 theorem 2) and therefore

$$[I - P'(\mu)] P(\mu) \neq O.$$

From property b) we even conclude that there must be a $v < \mu$ such that

$$[I - P'(\mu)] [P(\mu) - P(v)] \neq O$$

(otherwise we would have

$$[I - P'(\mu)] P(\mu) f = \lim_{v \to -\infty} [I - P'(\mu)] [P(\mu) - P(v)] f = o$$

$$\text{for all } f \in \mathfrak{H}).$$

By (8) this product is again a projection (cf. § 15 theorem 4). Let the non-zero vector f be chosen in the corresponding subspace, i.e. such that

$$[I - P'(\mu)] [P(\mu) - P(v)] f = f \neq o.$$

Then we have

$$P(\mu) f = f,$$
$$P(v) f = P'(\mu) f = o,$$
$$\int_{-\infty}^{+\infty} |\lambda|^2 \, \mathrm{d} \|P(\lambda) f\|^2 = \int_{v}^{\mu} |\lambda|^2 \, \mathrm{d} \|P(\lambda) f\|^2 < \infty$$

and therefore $f \in \mathfrak{D}_A$.

Exactly as in the proof of § 34 theorem 2 we obtain for this vector f

$$\langle (A - \mu I) f, f \rangle = \int_{v}^{\mu} (\lambda - \mu) \, \mathrm{d} \langle P(\lambda) f, f \rangle \leqslant 0,$$

$$\langle (A - \mu I) f, f \rangle = \int_{\mu}^{+\infty} (\lambda - \mu) \, \mathrm{d} \langle P'(\lambda) f, f \rangle > 0,$$

which again supplies the looked-for contradiction. □

Observe that, as a consequence of theorem 6, for a bounded selfadjoint operator A on $\mathfrak{H}$ the spectral families mentioned in theorem 5 and in § 34 theorem 1 coincide and the assertion of theorem 5 is contained in the assertion of § 34 theorem 1.

Example 1. Let $\{e_k\}_{k=1}^{\infty}$ be the standard basis of ℓ_2, let P_k be the projection upon the subspace spanned by e_k, and let $A = \bar{B}$ be the selfadjoint operator in

$\mathfrak{H} = \ell_2$ defined in § 20 example 1 and discussed further in § 22 example 1. Define the projection $P(\lambda)$ by

$$P(\lambda) = \begin{cases} O & \text{for } -\infty < \lambda < 1, \\ \sum\limits_{k=1}^{[\lambda]} P_k & \text{for } 1 \leqslant \lambda < +\infty \end{cases}$$

($[\lambda]$ = greatest integer not exceeding λ). The family $\{P(\lambda) : \lambda \in \mathbf{R}\}$ then satisfies the requirements a), b), c), d) in theorem 5 and therefore (by theorem 6) coincides with the spectral family of A. Assertion d) here reduces to

$$f \in \mathfrak{D}_A \quad \text{iff} \quad \sum_{k=1}^{\infty} k^2 |\langle f, e_k \rangle|^2 < \infty,$$

and

$$Af = \sum_{k=1}^{\infty} k \langle f, e_k \rangle e_k \quad \text{for all } f \in \mathfrak{D}_A.$$

The formal equation (7) which reduces to

$$A = \sum_{k=1}^{\infty} k P_k$$

here expresses the fact that A is the strong limit of the partial sums of this series (cf. § 22 remark 3).

Example 2. Let $A = E$ be the multiplication operator in $\mathfrak{L}_2(-\infty, +\infty)$ (cf. § 19 theorem 1). For every $\lambda \in \mathbf{R}$ let $P(\lambda)$ be the projection upon $\mathfrak{L}_2(-\infty, \lambda)$, considered as a subspace of $\mathfrak{L}_2(-\infty, +\infty)$. The family $\{P(\lambda) : \lambda \in \mathbf{R}\}$ then satisfies the requirements a), b), c), d) in theorem 5 and therefore coincides with the spectral family of A. In fact assertion d) reduces to

$$f \in \mathfrak{D}_E \quad \text{iff} \quad \int_{-\infty}^{+\infty} |\lambda|^2 |f(\lambda)|^2 \, d\lambda < \infty,$$

and

$$Ef = \int_{-\infty}^{+\infty} \lambda \, dP(\lambda) f = xf \quad \text{for all } f \in \mathfrak{D}_A$$

(cf. § 34 example 1).

Example 3. Let F be the Fourier-Plancherel operator (cf. § 16), let E be the multiplication operator in $\mathfrak{L}_2(-\infty, +\infty)$ (cf. example 2), and let $A = D = FEF^{-1}$ be the differentiation operator in $\mathfrak{L}_2(-\infty, +\infty)$ (cf. § 18 theorem 7 and § 19 theorem 3). For every $\lambda \in \mathbf{R}$ define the projection $P(\lambda)$ as in example 2 and define $P'(\lambda) = FP(\lambda) F^{-1}$. The operator $P'(\lambda)$ is again a projection since

$$[P'(\lambda)]^* = FP(\lambda) F^{-1} = P'(\lambda),$$
$$[P'(\lambda)]^2 = F[P(\lambda)]^2 F^{-1} = P'(\lambda).$$

Furthermore, from the characteristic properties of the spectral family $\{P(\lambda) : \lambda \in \mathbf{R}\}$ of E we can deduce the corresponding properties of the family $\{P'(\lambda) : \lambda \in \mathbf{R}\}$.

a) For $\lambda \leqslant \lambda'$ we have

$$\langle [P'(\lambda') - P'(\lambda)] f, f \rangle = \langle [P(\lambda') - P(\lambda)] F^{-1}f, F^{-1}f \rangle \geqslant 0$$
$$\text{for all } f \in \mathfrak{L}_2(-\infty, +\infty).$$

We conclude

$$P'(\lambda) \leqslant P'(\lambda') \quad \text{for } \lambda \leqslant \lambda'.$$

b) For every $f \in \mathfrak{L}_2(-\infty, +\infty)$ we have

$$\lim_{\lambda \to -\infty} P'(\lambda) f = F \lim_{\lambda \to -\infty} P(\lambda) F^{-1}f = o,$$
$$\lim_{\lambda \to +\infty} P'(\lambda) f = F \lim_{\lambda \to +\infty} P(\lambda) F^{-1}f = FF^{-1}f = f.$$

We conclude

$$\text{(s)} \lim_{\lambda \to -\infty} P'(\lambda) = O,$$
$$\text{(s)} \lim_{\lambda \to +\infty} P'(\lambda) = I.$$

c) For any non-increasing sequence of positive numbers $\{\theta_n\}_{n=1}^{\infty}$ converging to zero and for every $\lambda \in \mathbf{R}$ we have

$$\lim_{n \to \infty} P'(\lambda + \theta_n) f = F \lim_{n \to \infty} P(\lambda + \theta_n) F^{-1}f$$
$$= FP(\lambda) F^{-1}f = P'(\lambda) f.$$

We conclude

$$P'(\lambda + 0) = P'(\lambda) \quad \text{for all } \lambda \in \mathbf{R}.$$

d) By § 19 (3) we have $f \in \mathfrak{D}_D$ iff $F^{-1}f \in \mathfrak{D}_E$. This again by what has been said in example 2 is equivalent with

$$(9) \qquad \int_{-\infty}^{+\infty} |\lambda|^2 \, d \, \| P(\lambda) F^{-1}f \|^2 < \infty.$$

Since the operator F is unitary we may insert it under the norm without changing anything and rewrite (9) in the form

$$\int\limits_{-\infty}^{+\infty} |\lambda|^2 \, d\,\|FP(\lambda)\,F^{-1}f\|^2 = \int\limits_{-\infty}^{+\infty} |\lambda|^2 \, d\,\|P'(\lambda)\,f\|^2 < \infty .$$

Finally, from

$$EF^{-1}f = \int\limits_{-\infty}^{+\infty} \lambda \, dP(\lambda)\,F^{-1}f \quad \text{for } F^{-1}f \in \mathfrak{D}_E$$

it follows that

$$Df = FEF^{-1} = \int\limits_{-\infty}^{+\infty} \lambda \, dFP(\lambda)\,\dot{F}^{-1}f$$

$$= \int\limits_{-\infty}^{+\infty} \lambda \, dP'(\lambda)\,f \quad \text{for all } f \in \mathfrak{D}_D .$$

Our conclusion is that $\{P'(\lambda) = FP(\lambda)\,F^{-1} : \lambda \in \mathbf{R}\}$ coincides with the spectral family of the differentiation operator D in $\mathfrak{L}_2(-\infty, +\infty)$.

In order to identify the projection $P'(\nu) - P'(\mu)$ $(-\infty < \mu < \nu < +\infty)$ also in an analytic form suppose the function $f \in \mathfrak{L}_2(-\infty, +\infty)$ is also integrable on $\mathbf{R}$, in other words $f \in \mathfrak{L}_2(-\infty, +\infty) \cap \mathfrak{L}_1(-\infty, +\infty)$ (cf. appendix B6). If $c_{[\mu, \nu]}$ is the characteristic function of the interval $[\mu, \nu]$ (cf. § 15 example 1), then for every $\zeta \in \mathbf{R}$ the function $e^{ix(y-\zeta)}c_{[\mu, \nu]}(x)\,f(y)$ is integrable on $\mathbf{R}^2$. This will entitle us below to apply FUBINI's theorem (cf. appendix B8). We obtain

$$[P'(\nu) - P'(\mu)]\,f = F\,[P(\nu) - P(\mu)]\,F^{-1}f$$

(apply § 16 theorem 4)

$$= F\,[P(\nu) - P(\mu)]\,\frac{1}{\sqrt{2\pi}}\int\limits_{-\infty}^{+\infty} e^{ix\eta}f(\eta)\,d\eta$$

(apply example 2 and § 15 example 1)

$$= \frac{1}{\sqrt{2\pi}}\,Fc_{[\mu, \nu]}(x)\int\limits_{-\infty}^{+\infty} e^{ix\eta}f(\eta)\,d\eta$$

$$= \frac{1}{\sqrt{2\pi}}\,F\int\limits_{-\infty}^{+\infty} e^{ix\eta}c_{[\mu, \nu]}(x)\,f(\eta)\,d\eta$$

(observe that the integral defines a function in $\mathfrak{L}_2(-\infty, +\infty)$ which vanishes outside $[\mu, \nu]$ and therefore is integrable on $\mathbf{R}$, cf. § 5 corollary 2.1; then apply § 16 theorem 4 again)

$$= \frac{1}{2\pi} \int\limits_{-\infty}^{+\infty} e^{-i\xi z} \left[\int\limits_{-\infty}^{+\infty} e^{i\xi \eta} c_{[\mu, \nu]}(\xi) f(\eta) \, d\eta \right] d\xi$$

$$= \frac{1}{2\pi} \int\limits_{-\infty}^{+\infty} \int\limits_{-\infty}^{+\infty} e^{i\xi (\eta - z)} c_{[\mu, \nu]}(\xi) f(\eta) \, d\eta d\xi$$

(apply FUBINI's theorem)

$$= \frac{1}{2\pi} \int\limits_{-\infty}^{+\infty} \left[\int\limits_{\mu}^{\nu} e^{i\xi (\eta - z)} \, d\xi \right] f(\eta) \, d\eta$$

$$= \frac{1}{2\pi} \int\limits_{-\infty}^{+\infty} \frac{e^{i\nu (\eta - z)} - e^{i\mu (\eta - z)}}{i(\eta - z)} f(\eta) \, d\eta .$$

In other words, for every integrable $f \in \mathfrak{L}_2(-\infty, +\infty)$ we have

$$(10) \qquad [P'(\nu) - P'(\mu)] f(\zeta) = \frac{1}{2\pi} \int\limits_{-\infty}^{+\infty} \frac{e^{i\nu (\eta - \zeta)} - e^{i\mu (\eta - \zeta)}}{i(\eta - \zeta)} f(\eta) \, d\eta$$

$$\text{for a.a. } \zeta \in \mathbf{R} .$$

Since for every $\zeta \in \mathbf{R}$ the function

$$g_{[\mu, \nu], \zeta}(y) = \frac{e^{-i\nu (y - \zeta)} - e^{-i\mu (y - \zeta)}}{-i(y - \zeta)}$$

belongs to $\mathfrak{L}_2(-\infty, +\infty)$ (cf. the function g_τ in the proof of § 16 theorem 4) we may write the last equation in the form

$$(11) \qquad [P'(\nu) - P'(\mu)] f(\zeta) = \frac{1}{2\pi} \langle f, g_{[\mu, \nu], \zeta} \rangle \quad \text{for a.a. } \zeta \in \mathbf{R} .$$

Recall that every function $f \in \mathfrak{L}_2(-\infty, +\infty)$ may be approximated (in norm) by *integrable* functions in $\mathfrak{L}_2(-\infty, +\infty)$ (simply cut off its "tails" at $-n$ and $+n$). For a suitably chosen sequence $\{f_n\}_{n=1}^{\infty}$ of integrable functions in $\mathfrak{L}_2(-\infty, +\infty)$ we therefore get

$$[P'(\nu) - P'(\mu)] f = \lim_{n \to \infty} [P'(\nu) - P'(\mu)] f_n$$

and by an extension of § 5 remark 3 to $\mathfrak{L}_2(-\infty, +\infty)$ even

$$[P'(v) - P'(\mu)] f(\zeta) = \lim_{n \to \infty} [P'(v) - P'(\mu)] f_n(\zeta) \quad \text{for a.a. } \zeta \in \mathbf{R}.$$

By the continuity of the inner product we therefore conclude

$$[P'(v) - P'(\mu)] f(\zeta) = \lim_{n \to \infty} \frac{1}{2\pi} \langle f_n, g_{[\mu, v], \zeta} \rangle \qquad \text{(by (11))}$$

$$= \frac{1}{2\pi} \langle f, g_{[\mu, v], \zeta} \rangle \qquad \text{for a.a. } \zeta \in \mathbf{R}.$$

In other words, formula (11) even holds for all functions $f \in \mathfrak{L}_2(-\infty, +\infty)$, and the projection $P'(v) - P'(\mu)$ is given on $\mathfrak{L}_2(-\infty, +\infty)$ by formula (10) even for all $f \in \mathfrak{L}_2(-\infty, +\infty)$.

Exercise. The following chain of assertions leads up to a solution, within $\mathfrak{L}_2(-\infty, +\infty)$, of the "*Schrödinger equation*"

$$\frac{\partial \varphi(x, t)}{\partial t} = \frac{\partial \varphi(x, t)}{\partial x}$$

with the "initial condition" $\varphi(x, 0) = f(x) \in \mathfrak{D}_D \subset \mathfrak{L}_2(-\infty, +\infty)$.

Let $\{P(\lambda): \lambda \in \mathbf{R}\}$ be the spectral family of the (possibly unbounded) self-adjoint operator A in $\mathfrak{H}$.

a) For every $f \in \mathfrak{H}$ and every $\tau \in \mathbf{R}$ there exists the improper Riemann-Stieltjes integral

$$f_\tau = \int_{-\infty}^{+\infty} e^{-i\tau\lambda} \, dP(\lambda) f = \lim_{\substack{\mu \to -\infty \\ v \to +\infty}} \int_\mu^v e^{-i\tau\lambda} \, dP(\lambda) f \in \mathfrak{H}.$$

b) $\|P(\mu) f_\tau\|^2 = \|P(\mu) f\|^2$ for all $f \in \mathfrak{H}$ and all $\mu \in \mathbf{R}$.

c) The operator $e^{-i\tau A}$ defined by

$$e^{-i\tau A} f = f_\tau = \int_{-\infty}^{+\infty} e^{-i\tau\lambda} \, dP(\lambda) f \quad \text{for all } f \in \mathfrak{H}$$

is unitary. (Hint: the operator $e^{-i\tau A}$ is reduced by every subspace $\mathfrak{H}_m$ as defined in theorem 1 and there induces a unitary operator.)

d) $f_\tau \in \mathfrak{D}_A$ iff $f \in \mathfrak{D}_A$. (Hint: use b).)

e) Let $\tau \in \mathbf{R}$ and $\varepsilon < 0$ be given. If $\delta > 0$ is chosen sufficiently small, then

$$\left\| \int_{\lambda_{k-1}}^{\lambda_k} (e^{-i\tau\lambda} - e^{-i\tau\lambda'}) \, dP(\lambda) f \right\|^2 \leqslant \varepsilon^2 \, \| [P(\lambda_k) - P(\lambda_{k-1})] f \|^2 \quad \text{for all } f \in \mathfrak{H}$$

whenever $0 \leqslant \lambda_k - \lambda_{k-1} \leqslant \delta$ and $\lambda' \in [\lambda_{k-1}, \lambda_k]$. (Hint: choose $\delta > 0$ such that $|e^{-i\tau\lambda} - e^{-i\tau\lambda'}| \leqslant \varepsilon$ whenever $|\lambda - \lambda'| \leqslant \delta$.)

f) $\qquad -iAe^{-i\tau A}f = -i \int_{-\infty}^{+\infty} \lambda e^{-i\tau\lambda} \, dP(\lambda) f \quad \text{for all } f \in \mathfrak{D}_A.$

(Hint: use d) and e).)

g) $\qquad -iAe^{-i\tau A}f = \lim_{\theta \to 0} \frac{1}{\theta} [e^{-i(\tau+\theta)A}f - e^{-i\tau A}f] \quad \text{for all } f \in \mathfrak{D}_A.$

(Hint: show

$$\| -iAe^{-i\tau A}f - \frac{1}{\theta}[e^{-i(\tau+\theta)A}f - e^{-i\tau A}f] \|^2 =$$

$$= \int_{-\infty}^{+\infty} \lambda^2 \left| 1 - \frac{e^{-i\theta\lambda} - 1}{-i\theta\lambda} \right|^2 d \|P(\lambda) f\|^2 ;$$

then use the boundedness of the function $(e^{-ix} - 1)/(-ix)$ on $\mathbf{R}$ to determine a $\kappa > 0$ such that

$$\int_{\kappa}^{+\infty} \lambda^2 \left| 1 - \frac{e^{-i\theta\lambda} - 1}{-i\theta\lambda} \right|^2 d \|P(\lambda) f\|^2 < \varepsilon,$$

$$\int_{-\infty}^{-\kappa} \lambda^2 \left| 1 - \frac{e^{-i\theta\lambda} - 1}{-i\theta\lambda} \right|^2 d \|P(\lambda) f\|^2 < \varepsilon,$$

and choose $\delta > 0$ such that

$$\kappa^2 \left| 1 - \frac{e^{-i\theta\kappa} - 1}{-i\theta\kappa} \right|^2 < \frac{\varepsilon}{\|f\|^2} \quad \text{for } 0 < |\theta| < \delta.)$$

Remark 2. Assertion g) may also be written in the form

$$-iAe^{-i\tau A}f = \frac{d(e^{-i\tau A}f)}{d\tau} \quad \text{for all } f \in \mathfrak{D}_A.$$

If in particular we choose for A the differentiation operator D in $\mathfrak{L}_2(-\infty, +\infty)$ and if for any given $f \in \mathfrak{D}_D$ we define

$$\varphi(x, t) = e^{-itD} f(x),$$

then we have

$$\varphi(x, 0) = f(x),$$
$$\varphi(x, t) = e^{-itD} \varphi(x, 0)$$

and

$$\frac{\partial \varphi(x, t)}{\partial x} = -iDe^{-itD} f(x) = \frac{\partial [e^{-itD} f(x)]}{\partial t} = \frac{\partial \varphi(x, t)}{\partial t}.$$

The proper interpretation of this equation, however, is not

$$\frac{\partial \varphi(\xi, \tau)}{\partial \xi} = \frac{\partial \varphi(\xi, \tau)}{\partial \tau} \quad \text{for every } \tau \in \mathbf{R} \text{ and almost all } \xi \in \mathbf{R},$$

but

$$\lim_{\theta \to 0} \left\| \frac{\partial \varphi(x, \tau)}{\partial x} - \frac{\varphi(x, \tau + \theta) - \varphi(x, \tau)}{\theta} \right\| = 0 \quad \text{for every } \tau \in \mathbf{R}.$$

According to § 5 remark 3, extended to $\mathfrak{L}_2(-\infty, +\infty)$, this implies that for every $\tau \in \mathbf{R}$ there exists a sequence $\{\theta_n\}_{n=1}^{\infty}$ such that

$$\lim_{n \to \infty} \theta_n = 0$$

and

$$\frac{\partial \varphi(\xi, \tau)}{\partial \xi} = \lim_{n \to \infty} \frac{\varphi(\xi, \tau + \theta_n) - \varphi(\xi, \tau)}{\theta_n} \quad \text{for a.a. } \xi \in \mathbf{R}.$$

Moreover, using CAUCHY's inequality we see that for every $g \in \mathfrak{L}_2(-\infty, +\infty)$ the inner product $\langle \varphi(x, t), g \rangle$ is everywhere on $\mathbf{R}$ differentiable as a function in t and that

$$\left\langle \frac{\partial \varphi(x, \tau)}{\partial x}, g \right\rangle = \frac{d}{d\tau} \langle \varphi(x, \tau), g \rangle$$

for all $g \in \mathfrak{H}$ and every $\tau \in \mathbf{R}$.

The interested reader is referred to [9] chapter X and to the literature mentioned there.

THE GRAPH OF A LINEAR OPERATOR

Let f be a real-valued function on some domain $\mathfrak{D}_f \subset \mathbf{R}$. An efficient method to obtain an illustrative picture of the behaviour of f is to plot the points $(\xi, f(\xi))$ in the Euclidean plane $\mathbf{R}^2$ considered as the "Cartesian product" $\mathbf{R} \times \mathbf{R}$ of $\mathbf{R}$ with itself. The set of all points $(\xi, f(\xi))$ $(\xi \in \mathfrak{D}_f)$ is then called the *graph* of f (fig. 31).

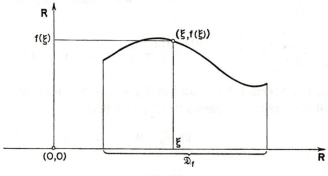

Fig. 31

We shall now imitate this procedure in a more abstract way in order to obtain a more illustrative picture of the behaviour of a linear operator A in a Hilbert space $\mathfrak{H}$. We start with constructing the analogue to the Euclidean plane $\mathbf{R}^2$ in which we shall "plot" the graph of A.

THEOREM 1. *Let* $\mathfrak{H} \times \mathfrak{H}$ *be the set of all ordered pairs* $\mathfrak{f} = (f_1, f_2)$ *of vectors* $f_1 \in \mathfrak{H}, f_2 \in \mathfrak{H}$. *If sum, scalar multiple, and inner product of the vectors* $\mathfrak{f} = (f_1, f_2)$ *and* $\mathfrak{g} = (g_1, g_2)$ *in* $\mathfrak{H} \times \mathfrak{H}$ *are defined by*

$$\mathfrak{f} + \mathfrak{g} = (f_1, f_2) + (g_1, g_2) = (f_1 + g_1, f_2 + g_2),$$
$$\alpha\mathfrak{f} = \alpha(f_1, f_2) = (\alpha f_1, \alpha f_2),$$
$$\langle \mathfrak{f}, \mathfrak{g} \rangle = \langle (f_1, f_2), (g_1, g_2) \rangle = \langle f_1, g_1 \rangle + \langle f_2, g_2 \rangle,$$

then $\mathfrak{H} \times \mathfrak{H}$ *is a Hilbert space.*

PROOF. It is a matter of routine to check that all requirements for $\mathfrak{H} \times \mathfrak{H}$ to be an inner product space are indeed satisfied (cf. § 2 definition 2, § 2 definition 4). For instance we have

$$
\begin{aligned}
\langle \mathfrak{f} + \mathfrak{g}, \mathfrak{h} \rangle &= \langle (f_1, f_2) + (g_1, g_2), (h_1, h_2) \rangle \\
&= \langle (f_1 + g_1, f_2 + g_2), (h_1, h_2) \rangle \\
&= \langle f_1 + g_1, h_1 \rangle + \langle f_2 + g_2, h_2 \rangle \\
&= \langle f_1, h_1 \rangle + \langle f_2, h_2 \rangle + \langle g_1, h_1 \rangle + \langle g_2, h_2 \rangle \\
&= \langle (f_1, f_2), (h_1, h_2) \rangle + \langle (g_1, g_2), (h_1, h_2) \rangle \\
&= \langle \mathfrak{f}, \mathfrak{h} \rangle + \langle \mathfrak{g}, \mathfrak{h} \rangle .
\end{aligned}
$$

It remains to show that $\mathfrak{H} \times \mathfrak{H}$ is complete with respect to the norm defined by

$$
\| \mathfrak{f} \| = \langle (f_1, f_2), (f_1, f_2) \rangle^{\frac{1}{2}} = (\| f_1 \|^2 + \| f_2 \|^2)^{\frac{1}{2}} .
$$

Indeed, if $\{\mathfrak{f}_n\}_{n=1}^{\infty} = \{(f_{n,1}, f_{n,2})\}_{n=1}^{\infty}$ is a fundamental sequence in $\mathfrak{H} \times \mathfrak{H}$, then from

$$
\| \mathfrak{f}_m - \mathfrak{f}_n \|^2 = \| f_{m,1} - f_{n,1} \|^2 + \| f_{m,2} - f_{n,2} \|^2
$$

it follows that also $\{f_{n,1}\}_{n=1}^{\infty}$ and $\{f_{n,2}\}_{n=1}^{\infty}$ are fundamental sequences which therefore converge in $\mathfrak{H}$ to some vectors f_1 and f_2 respectively. Then $\{\mathfrak{f}_n\}_{n=1}^{\infty}$ converges in $\mathfrak{H} \times \mathfrak{H}$ to the vector $\mathfrak{f} = (f_1, f_2)$. $\square$

Note that the subsets $\mathfrak{H}_1 = \{(f, o) : f \in \mathfrak{H}\}$ and $\mathfrak{H}_2 = \{(o, f) : f \in \mathfrak{H}\}$ are mutually orthogonal subspaces of $\mathfrak{H} \times \mathfrak{H}$ and isomorphic images of $\mathfrak{H}$. Furthermore we have

$$
\mathfrak{H} \times \mathfrak{H} = \mathfrak{H}_1 + \mathfrak{H}_2 .
$$

DEFINITION 1. Let A be a (not necessarily linear) operator in $\mathfrak{H}$ defined on a domain $\mathfrak{D}_A \subset \mathfrak{H}$. The *graph* $\mathfrak{G}(A)$ of A is the subset of $\mathfrak{H} \times \mathfrak{H}$ defined by

$$
\mathfrak{G}(A) = \{(f, Af) \in \mathfrak{H} \times \mathfrak{H} : f \in \mathfrak{D}_A\} .
$$

THEOREM 2. *An operator A in $\mathfrak{H}$ is linear iff $\mathfrak{G}(A)$ is a linear manifold in $\mathfrak{H} \times \mathfrak{H}$.*

PROOF. If A is a linear operator in $\mathfrak{H}$ and if $\mathfrak{f} = (f, Af) \in \mathfrak{G}(A)$, $\mathfrak{g} = (g, Ag) \in \mathfrak{G}(A)$, then

$$
\begin{aligned}
\mathfrak{f} + \mathfrak{g} &= (f + g, Af + Ag) = (f + g, A(f + g)) \in \mathfrak{G}(A), \\
\alpha \mathfrak{f} &= (\alpha f, \alpha Af) = (\alpha f, A \alpha f) \in \mathfrak{G}(A).
\end{aligned}
$$

Conversely, if $\mathfrak{G}(A)$ is a linear manifold and if $f \in \mathfrak{D}_A$, $g \in \mathfrak{D}_A$, then we have

$$\mathfrak{f} = (f, Af) \in \mathfrak{G}(A),$$
$$\mathfrak{g} = (g, Ag) \in \mathfrak{G}(A),$$
$$\mathfrak{f} + \mathfrak{g} = (f + g, Af + Ag) \in \mathfrak{G}(A),$$
$$\alpha\mathfrak{f} = (\alpha f, \alpha Af) \in \mathfrak{G}(A).$$

We conclude that $f + g \in \mathfrak{D}_A$ and $\alpha f \in \mathfrak{D}_A$ and moreover $A(f+g) = Af + Ag$ and $A(\alpha f) = \alpha Af$. $\square$

THEOREM 3. *A linear operator A in $\mathfrak{H}$ is closed iff $\mathfrak{G}(A)$ is a subspace of $\mathfrak{H} \times \mathfrak{H}$.*

PROOF. Suppose $\mathfrak{G}(A)$ is a subspace and $\{f_n\}_{n=1}^{\infty}$ is a sequence in $\mathfrak{H}$ with the property that both limits

$$f = \lim_{n \to \infty} f_n,$$
$$g = \lim_{n \to \infty} Af_n$$

exist. Denoting the vectors (f_n, Af_n) and (f, g) in $\mathfrak{H} \times \mathfrak{H}$ by $\mathfrak{f}_n$ and $\mathfrak{f}$ respectively we conclude

$$\mathfrak{f} = \lim_{n \to \infty} \mathfrak{f}_n \in \mathfrak{G}(A)$$

since $\mathfrak{G}(A)$ is closed. As a consequence we get $f \in \mathfrak{D}_A$ and $g = Af$. This shows that A is closed (cf. § 20 definition 1).

Conversely, suppose A is closed and let $\{\mathfrak{f}_n\}_{n=1}^{\infty} = \{(f_n, Af_n)\}_{n=1}^{\infty} \subset \mathfrak{G}(A)$ be a fundamental sequence in $\mathfrak{H} \times \mathfrak{H}$ converging to some vector $\mathfrak{f} = (f, g) \in$ $\in \mathfrak{H} \times \mathfrak{H}$. Then because of

$$\|\mathfrak{f} - \mathfrak{f}_n\|^2 = \|f - f_n\|^2 + \|g - Af_n\|^2$$

we have

$$f = \lim_{n \to \infty} f_n,$$
$$g = \lim_{n \to \infty} Af_n.$$

Since A is closed we conclude that $f \in \mathfrak{D}_A$ and $g = Af$, therefore $\mathfrak{f} = (f, g) \in$ $\in \mathfrak{G}(A)$. This shows that $\mathfrak{G}(A)$ is closed. $\square$

Remark 1. Note that not every subspace, let alone every linear manifold, in $\mathfrak{H} \times \mathfrak{H}$ is the graph of some linear operator in $\mathfrak{H}$ (consider for instance the entire space $\mathfrak{H} \times \mathfrak{H}$). In fact a necessary and sufficient condition for a linear manifold $\mathfrak{G}$ in $\mathfrak{H} \times \mathfrak{H}$ to be a graph of a linear operator in $\mathfrak{H}$ is that $(f_1, f_2) \in$ $\in \mathfrak{G}$ and $(f_1, f_3) \in \mathfrak{G}$ implies $f_2 = f_3$.

We already know that the adjoint of a linear operator in $\mathfrak{H}$ (with every-where dense domain) is closed (cf. § 20 theorem 3). We shall now clarify the connection between the two properties of a linear operator to be closed and to be an adjoint. To this end we first introduce a special unitary operator in $\mathfrak{H} \times \mathfrak{H}$.

THEOREM 4. *Let $I^{\times}$ be the identity operator in $\mathfrak{H} \times \mathfrak{H}$ and let the operator $V^{\times}$ in $\mathfrak{H} \times \mathfrak{H}$ be defined by*

$$V^{\times}(f_1, f_2) = (f_2, -f_1).$$

Then $V^{\times}$ is unitary and $(V^{\times})^2 = -I^{\times}$.

PROOF. It is easy to check that $V^{\times}$ is linear and onto. Furthermore we have

$$\|V^{\times}(f_1, f_2)\|^2 = \|f_2\|^2 + \|f_1\|^2 = \|(f_1, f_2)\|^2,$$
$$(V^{\times})^2(f_1, f_2) = V^{\times}(f_2, -f_1) = (-f_1, -f_2) = -(f_1, f_2)$$

(cf. § 13 corollary 1.2). □

THEOREM 5. *Let A be a linear operator in $\mathfrak{H}$ such that $\mathfrak{D}_A = \mathfrak{H}$. Then*

$$\mathfrak{G}(A^*) = [V^{\times} \mathfrak{G}(A)]^{\perp}.$$

PROOF. Consider the following chain of consecutively equivalent assertions:

a) $g_1 \in \mathfrak{D}_{A^*}$ and $g_2 = A^* g_1$.
b) $\langle Af, g_1 \rangle = \langle f, g_2 \rangle$ for all $f \in \mathfrak{D}_A$.
c) $0 = \langle f, g_2 \rangle - \langle Af, g_1 \rangle$ for all $f \in \mathfrak{D}_A$.
d) $0 = \langle (f, Af), (g_2, -g_1) \rangle$ for all $(f, Af) \in \mathfrak{G}(A)$.
e) $V^{\times}(g_1, g_2) \perp \mathfrak{G}(A)$.
f) $(g_1, g_2) \perp V^{\times} \mathfrak{G}(A)$ (since $(V^{\times})^2 = -I^{\times}$ we   have $(V^{\times})^{-1} \mathfrak{G}(A) = -V^{\times} \mathfrak{G}(A) = V^{\times} \mathfrak{G}(A)$).

The equivalence a)⇔f) coincides with the assertion. □

THEOREM 6. *Let A be a linear operator in $\mathfrak{H}$ such that $\mathfrak{D}_A = \mathfrak{H}$. Then A is closed iff $\mathfrak{D}_{A^*} = \mathfrak{H}$ and $A = A^{**}$.*

PROOF. If $\mathfrak{D}_{A^*} = \mathfrak{H}$ and $A = A^{**}$, then A is closed by § 20 theorem 3.

Conversely, suppose that A is closed. Then $\mathfrak{G}(A)$ is a subspace of $\mathfrak{H} \times \mathfrak{H}$ by theorem 3, and from theorem 5 we conclude

$$\mathfrak{G}(A) = V^{\times}[\mathfrak{G}(A^*)]^{\perp} = [V^{\times} \mathfrak{G}(A^*)]^{\perp}$$

(we have used § 7 theorem 9 and the facts that $(V^{\times})^{-1} = -V^{\times}$ and that $V^{\times}$ is unitary). If $\mathfrak{D}_{A^*}$ were not everywhere dense in $\mathfrak{H}$, then there would exist a non-zero vector $g \perp \mathfrak{D}_{A^*}$ (cf. § 7 corollary 7.1). Under this assumption we have

$$(g, o) \perp \mathfrak{G}(A^*),$$
$$(o, -g) = V^\times (g, o) \perp V^\times \mathfrak{G}(A^*),$$
$$(o, -g) \in \mathfrak{G}(A) = [V^\times \mathfrak{G}(A^*)]^\perp.$$

This, however, contradicts the fact that $Ao = o \neq -g$. Applying now theorem 5 to the operator A^* in place of A we get

$$\mathfrak{G}(A^{**}) = [V^\times \mathfrak{G}(A^*)]^\perp = \mathfrak{G}(A)$$

and therefore $A^{**} = A$. $\square$

Recall that a linear operator B in $\mathfrak{H}$ such that $B \supset A$ is called an extension of A (cf. § 17 definition 2).

COROLLARY 6.1. *Let A be a linear operator in $\mathfrak{H}$ such that $\overline{\mathfrak{D}}_A = \mathfrak{H}$.*

a) *The operator A admits a closed linear extension iff $\overline{\mathfrak{D}}_{A^*} = \mathfrak{H}$.*

b) *If $\overline{\mathfrak{D}}_{A^*} = \mathfrak{H}$, then the operator $\bar{A} = A^{**}$, called the closure of A, is contained in every closed linear extension of A.*

PROOF. a) If there exists a closed linear operator $B \supset A$, then from $B^* \subset A^*$ (cf. § 17 theorem 3) and $\overline{\mathfrak{D}}_{B^*} = \mathfrak{H}$ (cf. theorem 6) we conclude $\overline{\mathfrak{D}}_{A^*} = \mathfrak{H}$. The converse assertion is proved below under b).

b) If $\overline{\mathfrak{D}}_{A^*} = \mathfrak{H}$, then the operator A^{**} is a closed extension of A (cf. § 17 theorem 4, § 20 theorem 3). If B is any other closed linear operator containing A, then as in a) we conclude $B^* \subset A^*$, and, as a consequence, $B = B^{**} \supset A^{**}$ (cf. theorem 6). $\square$

If the linear operator A in $\mathfrak{H}$ admits a closure $\bar{A} = A^{**}$, then from the fact that $\mathfrak{G}(\bar{A})$ is a subspace containing $\mathfrak{G}(A)$ it follows that

$$\mathfrak{G}(A) \subset \overline{\mathfrak{G}(A)} \subset \mathfrak{G}(\bar{A})$$

(closure of $\mathfrak{G}(A)$ taken in $\mathfrak{H} \times \mathfrak{H}$). Since $\mathfrak{G}(\bar{A})$ is the graph of a linear operator, so is the subspace $\overline{\mathfrak{G}(A)}$ of $\mathfrak{H} \times \mathfrak{H}$ (cf. remark 1). The corresponding linear operator B in $\mathfrak{H}$ has to be closed since $\mathfrak{G}(B) = \overline{\mathfrak{G}(A)}$ is a subspace (cf. theorem 3). From the minimality of $\bar{A}$ (cf. corollary 6.1 b)) we conclude that $B \supset \bar{A}$ and $\mathfrak{G}(B) = \overline{\mathfrak{G}(A)} \supset \mathfrak{G}(\bar{A})$. As a consequence we get

$$\mathfrak{G}(\bar{A}) = \overline{\mathfrak{G}(A)}.$$

THEOREM 7 (CLOSED GRAPH THEOREM). *A closed linear operator defined every-where on $\mathfrak{H}$ is bounded.*

PROOF. By § 17 theorem 8 the operator A^* is bounded. Since A^* is also closed, its domain $\mathfrak{D}_{A^*}$ is a subspace of $\mathfrak{H}$ (cf. § 20 theorem 3, § 20 theorem 1). Since $\mathfrak{D}_{A^*}$ is also everywhere dense in $\mathfrak{H}$ (cf. theorem 6) we conclude $\mathfrak{D}_{A^*}=\mathfrak{H}$. Applying § 17 theorem 8 to the operator A^* in place of A we obtain that the operator $A^{**}=A$ is bounded. $\square$

COROLLARY 7.1. *Let A be a closed one-to-one linear operator in $\mathfrak{H}$ such that $A\mathfrak{D}_A=\mathfrak{H}$. Then A^{-1} is bounded.*

PROOF. The linear operator A^{-1} is closed (cf. § 20 theorem 2) and defined everywhere on $\mathfrak{H}$ by hypothesis. The assertion then follows from theorem 7. $\square$

The last corollary below uses some concepts and results of § 22 and § 23. It should be compared with (and could be used to prove) § 23 exercise f), § 24 theorem 5, and § 27 theorem 5.

COROLLARY 7.2. *Let A be a closed linear operator in $\mathfrak{H}$ and suppose $\lambda \in \mathbf{C}$ is not an eigenvalue of A. Then λ is a regular value for A iff $(A-\lambda I)\,\mathfrak{D}_A=\mathfrak{H}$.*

PROOF. The operator $A-\lambda I$ is closed (cf. the proof of § 23 theorem 6) and one-to-one (cf. § 23 theorem 1). By corollary 7.1 the equation $(A-\lambda I)\,\mathfrak{D}_A= =\mathfrak{H}$ implies that $(A-\lambda I)^{-1}$ is a bounded linear operator on $\mathfrak{H}$ and therefore λ is regular. The converse assertion is a trivial consequence of § 23 definition 1. $\square$

RIEMANN-STIELTJES AND LEBESGUE INTEGRATION

The purpose of this section is to recall some definitions and – without proof but with references – some theorems about Riemann-Stieltjes and Lebesgue integration which are referred to in the preceding text.

For technical reasons let us agree to use alternatively the notation f or $f(x)$ for the function which at the point $\xi \in \mathbf{C}$ assumes the value $f(\xi)$. In particular, by x we denote the function which at ξ assumes the value $x(\xi) = \xi$ (cf. § 3 remark 1). An analogous notation is used for functions in more than one variable.

B1. Weierstrass' approximation theorem

Let f be a continuous complex-valued function on the closed disk of radius ϱ about the origin in the complex plane $\mathbf{C}$. For every $\varepsilon > 0$ there exists a complex polynomial

$$p(x) = \sum_{k=0}^{n} \sum_{l=0}^{n} \alpha_{k,l} x^k \bar{x}^l$$

($\alpha_{k,l} \in \mathbf{C}$ for $0 \leqslant k \leqslant n$, $0 \leqslant l \leqslant n$; $\bar{x} =$ complex conjugate of x) such that

$$|f(\xi) - p(\xi)| \leqslant \varepsilon \quad \text{for all } \xi \in \mathbf{C} \text{ such that } |\xi| \leqslant \rho$$

(WEIERSTRASS' APPROXIMATION THEOREM).

If the complex variable ξ is restricted to the real axis or to the complex unit circle then the above theorem gives rise to the following more specialized versions of WEIERSTRASS' approximation theorem:

If f is a continuous complex-valued function on the interval $[\alpha, \beta] \subset \mathbf{R}$ and if $\varepsilon > 0$ is given, then there exists a complex polynomial $p(x) = \sum_{n=1}^{\infty} \alpha_k x^k$ ($\alpha_k \in \mathbf{C}$ for $0 \leqslant k \leqslant n$) such that

$$|f(\xi) - p(\xi)| \leqslant \varepsilon \quad \text{for all } \xi \in [\alpha, \beta].$$

If f is a continuous complex-valued function on the interval $[\alpha, \beta] \subset \mathbf{R}$ such that $f(\alpha) = f(\beta)$ and if $\varepsilon > 0$ is given, then there exists a trigonometric polynomial

$$p\left(e^{2\pi i\,(x-\alpha)/(\beta-\alpha)}\right) = \sum_{k=-n}^{n} \alpha_k e^{2\pi i k\,(x-\alpha)/(\beta-\alpha)} \quad (\alpha_k \in \mathbf{C} \text{ for } -n \leqslant k \leqslant n)$$

such that

$$\left| f(\xi) - p\left(e^{2\pi i\,(\xi-\alpha)/(\beta-\alpha)}\right)\right| \leqslant \varepsilon \quad \text{for all } \xi \in [\alpha, \beta].$$

Reference: [8] IV § 5 theorem 2, theorem 4;
[6] (7.30), (7.31), (7.34), (7.35).

B2. Riemann-Stieltjes integration

In what follows, suppose w is a given non-decreasing real-valued function on $\mathbf{R}$ (i.e., $w(\xi) \leqslant w(\eta)$ for $\xi < \eta$).

DEFINITION 1. A complex valued function f on the interval $[\alpha, \beta] \subset \mathbf{R}$ is called *Riemann-Stieltjes integrable* with respect to the function w over $[\alpha, \beta]$ if there exists a number $\omega \in \mathbf{C}$ with the following property:

For every $\varepsilon > 0$ there exists a $\delta > 0$ such that

$$\left| \omega - \sum_{k=1}^{n} f(\xi_k') \left[w(\xi_k) - w(\xi_{k-1}) \right] \right| \leqslant \varepsilon$$

whenever

(1)
$$\begin{cases} \alpha = \xi_0 < \xi_1 < \cdots < \xi_n = \beta, \\ \xi_k - \xi_{k-1} \leqslant \delta & \text{for } 1 \leqslant k \leqslant n, \\ \xi_k' \in [\xi_{k-1}, \xi_k] & \text{for } 1 \leqslant k \leqslant n. \end{cases}$$

The number ω is called the *Riemann-Stieltjes integral* of f with respect to w over $[\alpha, \beta]$ and denoted by

$$\omega = \int_{\alpha}^{\beta} f(\xi)\, dw(\xi).$$

In particular, if $w(x) = x$, then f is called *Riemann integrable* over $[\alpha, \beta]$ and

$$\omega = \int_{\alpha}^{\beta} f(\xi)\, d\xi$$

is called the *Riemann integral* of f over $[\alpha, \beta]$.

Every continuous complex-valued function f on $[\alpha, \beta]$ is Riemann-Stieltjes integrable with respect to w over $[\alpha, \beta]$. Complex linear combinations of two functions f, g which are Riemann-Stieltjes integrable over $[\alpha, \beta]$ are again Riemann-Stieltjes integrable, and the Riemann-Stieltjes integral depends linearly on the two functions to be integrated, i.e.

$$\int_\alpha^\beta [\lambda f(\xi) + \mu g(\xi)]\,dw(\xi) = \lambda \int_\alpha^\beta f(\xi)\,dw(\xi) + \mu \int_\alpha^\beta g(\xi)\,dw(\xi).$$

Moreover, for $\alpha < \gamma < \beta$ we have

$$\int_\alpha^\beta f(\xi)\,dw(\xi) = \int_\alpha^\gamma f(\xi)\,dw(\xi) + \int_\gamma^\beta f(\xi)\,dw(\xi).$$

DEFINITION 2. A complex-valued function f on $\mathbf{R}$ is called *improperly Riemann-Stieltjes integrable* over $\mathbf{R}$ with respect to w if f is Riemann-Stieltjes integrable over every interval $[\alpha, \beta] \subset \mathbf{R}$ and if there exists the limit (then called the *improper Riemann-Stieltjes integral* of f over $\mathbf{R}$)

$$\int_{-\infty}^{+\infty} f(\xi)\,dw(\xi) = \lim_{\substack{\alpha \to -\infty \\ \beta \to +\infty}} \int_\alpha^\beta f(\xi)\,dw(\xi).$$

The concept of the Riemann-Stieltjes integral can be carried over to the case of functions defined on a rectangle $X = [\alpha_1, \beta_1] \times [\alpha_2, \beta_2] \subset \mathbf{R}^2$ (or, more generally, on a *parallelotope* $X = [\alpha_1, \beta_1] \times \cdots \times [\alpha_n, \beta_n] \subset \mathbf{R}^n$). To this end w has to be replaced by a suitable real-valued function on X which, for every fixed value of one variable, is a non-decreasing function in the other variable, while the subdivision (1) has to be replaced by an analogous rectangular subdivision of X. The Riemann integral over X is obtained by choosing $w(x, y) = (x - \alpha_1)(y - \alpha_2)$ (i.e., $w(\xi, \eta) = (\xi - \alpha_1)(\eta - \alpha_2)$ for all $(\xi, \eta) \in X$).

Reference: [8] VIII § 6, § 9;
   [6] § 8, in particular (8.1)–(8.13);
   [9] § 49, § 59.

B3. Lebesgue measurable sets

Let $\mathbf{R}^n$ be the n-dimensional Euclidean space, in particular $\mathbf{R}^1 = \mathbf{R}$. Then there is a family $\mathscr{M}^n$ of subsets of $\mathbf{R}^n$, called the *σ-algebra of Lebesgue measurable sets*, with the following properties (which also motivate the term "σ-algebra"):

a) Every open set and every closed set is Lebesgue measurable. In particular, the empty set $\emptyset$ and the entire space $\mathbf{R}^n$, considered as a set, are Lebesgue measurable.

b) If $X \in \mathscr{M}^n$ and $Y \in \mathscr{M}^n$, then the set

$$X \setminus Y = \{(\xi_1, \ldots, \xi_n) \in X : (\xi_1, \ldots, \xi_n) \notin Y\}$$

is Lebesgue measurable. In particular, the *complement* $Y^c = \mathbf{R}^n \backslash Y$ of Y in $\mathbf{R}^n$ is Lebesgue measurable.

c) If $\{X_k\}_{k=1}^{\infty} \subset \mathcal{M}^n$, then $\bigcup_{k=1}^{\infty} X_k \in \mathcal{M}^n$ and $\bigcap_{k=1}^{\infty} X_k \in \mathcal{M}^n$.

d) If the subset $Y \subset \mathbf{R}^n$ may be covered, given any $\varepsilon > 0$, by countably many *n*-dimensional parallelotopes $X_k (1 \leqslant k < \infty)$ of total *n*-dimensional content $\sum_{k=1}^{\infty} \lambda^n(X_k) < \varepsilon$ (cf. section B4 a)), then Y is Lebesgue measurable.

Not every subset of $\mathbf{R}^n$ is Lebesgue measurable. The σ-algebra $\mathcal{M}^n$ is not yet uniquely defined by a), b), c), d) but is, in some measure theoretic sense, defined to be the "smallest" family of subsets of $\mathbf{R}^n$ having the properties a), b), c), d).

If $X \in \mathcal{M}^n$ and $Y \in \mathcal{M}^m$, then the "*Cartesian product*"

$$X \times Y = \{(\xi_1, ..., \xi_n, \eta_1, ..., \eta_m) \in \mathbf{R}^{n+m} : (\xi_1, ..., \xi_n) \in X, (\eta_1, ..., \eta_m) \in Y\}$$

belongs to $\mathcal{M}^{n+m}$.

Reference: [8] III § 4, § 9, XI § 3;
            [4] § 4, § 5, § 15, § 33;
            [6] § 1, in particular (1.11)–(1.13); § 10, in particular
            (10.5–10.11); (21.2), (21.3).

B4. Lebesgue measure

The *n-dimensional Lebesgue measure* λ^n is defined for every Lebesgue measurable set X in such a way that the following requirements are satisfied (properly speaking λ^n is an extended real-valued function on $\mathcal{M}^n$):

a) If X is a product of intervals $[\alpha_j, \beta_j]$ $(1 \leqslant j \leqslant n)$, i.e. a parallelotope of the form

(2)
$$X = [\alpha_1, \beta_1] \times \cdots \times [\alpha_n, \beta_n]$$
$$= \{(\xi_1, ..., \xi_n) : \alpha_j \leqslant \xi_j \leqslant \beta_j, \alpha_j \in \mathbf{R}, \beta_j \in \mathbf{R} \quad \text{for } 1 \leqslant j \leqslant n\}$$

(in particular for $n=1$ if $X=[\alpha_1, \beta_1]$) then

$$\lambda^n(X) = \prod_{j=1}^{n} (\beta_j - \alpha_j).$$

b) $0 \leqslant \lambda^n(X) \leqslant \infty$ for all $X \in \mathcal{M}^n$.

c) If $\{X_k\}_{k=1}^{\infty}$ is a sequence of disjoint Lebesgue measurable sets (i.e. $X_k \cap X_l = \emptyset$ for $k \neq l$), then

$$\lambda^n \left(\bigcup_{k=1}^{\infty} X_k \right) = \sum_{k=1}^{\infty} \lambda^n(X_k)$$

(by definition we have $\infty + \alpha = \infty$ for $\alpha \in \mathbf{R}$).

As a consequence of b) and c) the union of countably many Lebesgue measurable sets of measure zero is again a set of Lebesgue measure zero. It is already used as part of the definition of Lebesgue measurable sets (which we have not given here explicitly; cf. section B3 d)) that every subset of a Lebesgue measurable set of Lebesgue measure zero is again Lebesgue measurable.

A statement made about *"almost all"* (a.a.) points of a set $X \in \mathcal{M}^n$ is meant to hold for all points in X except possibly for a subset of X of Lebesgue measure zero. By the same token *"almost everywhere* (a.e.) *in X"* means "for a.a. points in X". The terms *"almost no"* and *"almost nowhere"* are to be interpreted analogously.

Reference: [8] III § 1, § 2, § 4, § 9, XI § 3;
             [4] § 7, § 8, § 15.

B5. Lebesgue measurable functions

In the following definition we use a short cut through part of the theory by defining *"measurability"* only for functions in a class which is particularly important for Lebesgue integration.

DEFINITION 3. Let X be a fixed measurable subset of $\mathbf{R}^n$ (e.g. $X = [\alpha, \beta]$ for $n = 1$). A function f which is *real-valued* and defined a.e. on X is called *Lebesgue measurable* on X if for every $\gamma \in \mathbf{R}$ the set

$$\{(\xi_1, \ldots, \xi_n) \in X : f(\xi_1, \ldots, \xi_n) \leqslant \gamma\}$$

is Lebesgue measurable. A function f which is *complex-valued* and defined a.e. on X is called *Lebesgue measurable* on X if its real and imaginary parts $\operatorname{Re} f$ and $\operatorname{Im} f$ ($f = \operatorname{Re} f + i \operatorname{Im} f$) are Lebesgue measurable on X.

A Lebesgue measurable function f on X (without the attribute "real") will further be understood to be complex-valued and defined a.e. on X. Note that this tacitly implies that also the set of points in X for which $|f|$ assumes the "value" $+\infty$ has Lebesgue measure zero. The set of all Lebesgue measurable functions on X will be denoted by $\mathfrak{F}(X)$.

Every continuous complex-valued function on X is Lebesgue measurable. The same is true for every Riemann integrable function on X if X is a product of intervals (2). Conversely, if X is a product of intervals (2), then every Lebesgue measurable function on X may be "approximated" by continuous functions in the following sense: for every $f \in \mathfrak{F}(X)$ and every $\varepsilon > 0$ there exists a continuous complex-valued function g on X coinciding with f on X except on a subset $Y \in \mathcal{M}^n$ of measure $\lambda^n(Y) < \varepsilon$ (LUZIN'S THEOREM).

For $f \in \mathfrak{F}(X)$ and $g \in \mathfrak{F}(X)$ also the functions $|f|, \alpha f, f+g, f \cdot g \ (=fg)$ defined by

$$\left.\begin{array}{l} |f|(\xi) = |f(\xi)| \\ (\alpha f)(\xi) = \alpha f(\xi) \end{array}\right\} \quad \text{whenever } f(\xi) \in \mathbf{C}$$

$$\left.\begin{array}{l} (f+g)(\xi) = f(\xi) + g(\xi) \\ (f \cdot g)(\xi) = f(\xi) \cdot g(\xi) \end{array}\right\} \quad \text{whenever } f(\xi) \in \mathbf{C} \text{ and } g(\xi) \in \mathbf{C}$$

are Lebesgue measurable. Consequently $\mathfrak{F}(X)$ is a complex linear space (even a complex algebra). If $f \in \mathfrak{F}(X)$ is different from zero a.e. on X, then the function $1/f$ defined by

$$\left(\frac{1}{f}\right)(\xi) = \frac{1}{f(\xi)} \quad \text{whenever } 0 \neq f(\xi) \in \mathbf{C}$$

is Lebesgue measurable. If $f \in \mathfrak{F}(X)$ and $g \in \mathfrak{F}(X)$ are real, then the functions $\min(f, g)$ and $\max(f, g)$ defined by

$$\left.\begin{array}{l} \min(f, g)(\xi) = \min(f(\xi), g(\xi)) \\ \max(f, g)(\xi) = \max(f(\xi), g(\xi)) \end{array}\right\} \quad \text{whenever } f(\xi) \in \mathbf{R} \text{ and } g(\xi) \in \mathbf{R}$$

are Lebesgue measurable.

Let $\{f_k\}_{k=1}^{\infty}$ be a sequence of real Lebesgue measurable functions on X with the property that the function $\liminf_{k \to \infty} f_k$ or the function $\limsup_{k \to \infty} f_k$ defined by

$$\left.\begin{array}{l} (\liminf_{k \to \infty} f_k)(\xi) = \liminf_{k \to \infty} f_k(\xi) \\ (\limsup_{k \to \infty} f_k)(\xi) = \limsup_{k \to \infty} f_k(\xi) \end{array}\right\} \quad \text{whenever all } f_k(\xi) \in \mathbf{R}$$

is finite a.e. on X. Then $\liminf_{k \to \infty} f_k \in \mathfrak{F}(X)$ or, respectively, $\limsup_{k \to \infty} f_k \in \mathfrak{F}(X)$. As a consequence, if the sequence $\{f_k\}_{k=1}^{\infty} \subset \mathfrak{F}(X)$ has the property that $\lim_{k \to \infty} f_k(\xi)$ exists and is finite a.e. on X, then $\lim_{k \to \infty} f_k \in \mathfrak{F}(X)$.

Suppose X is a product of (possibly infinite) intervals of which we single out one, say the last one:

$$-\infty \leqslant \alpha_j < \beta_j \leqslant +\infty \quad \text{for } 1 \leqslant j \leqslant n,$$
$$X = \{(\xi_1, \ldots, \xi_n): \alpha_{j\,(\leqq)} \xi_{j\,(\leqq)} \beta_j \text{ for } 1 \leqslant j \leqslant n\}$$
$$= X^{(n)} \times X_n,$$
$$X^{(n)} = \{(\xi_1, \ldots, \xi_{n-1}): \alpha_{j\,(\leqq)} \xi_{j\,(\leqq)} \beta_j \text{ for } 1 \leqslant j \leqslant n-1\},$$
$$X_n = \{\xi_n: \alpha_{n\,(\leqq)} \xi_{n\,(\leqq)} \beta_n\}.$$

Furthermore, let $f \in \mathfrak{F}(X)$ be given and for fixed $(\eta_1, \ldots, \eta_{n-1}) \in X^{(n)}$ and fixed $\eta_n \in X_n$ define the functions $f_{\eta_1, \ldots, \eta_{n-1}}(x_n) = f(\eta_1, \ldots, \eta_{n-1}, x_n)$ and $f^{\eta_n}(x_1, \ldots, x_{n-1}) = f(x_1, \ldots, x_{n-1}, \eta_n)$ by

$$f_{\eta_1, \ldots, \eta_{n-1}}(\xi_n) = f(\eta_1, \ldots, \eta_{n-1}, \xi_n),$$
$$f^{\eta_n}(\xi_1, \ldots, \xi_{n-1}) = f(\xi_1, \ldots, \xi_{n-1}, \eta_n)$$

whenever the right hand side is defined. Then $f^{\eta_n}(x_1, \ldots, x_{n-1}) \in \mathfrak{F}(X^{(n)})$ for almost all $\eta_n \in X_n$ and $f_{\eta_1, \ldots, \eta_{n-1}}(x_n) \in \mathfrak{F}(X_n)$ for almost all $(\eta_1, \ldots, \eta_{n-1}) \in X^{(n)}$.

Conversely, if $g \in \mathfrak{F}(X^{(n)})$ and $h \in \mathfrak{F}(X_n)$ are given, then the functions g_1 and h_1 defined on X by

$$g_1(\xi_1, \ldots, \xi_{n-1}, \xi_n) = g(\xi_n, \ldots, \xi_{n-1}),$$
$$h_1(\xi_1, \ldots, \xi_{n-1}, \xi_n) = h(\xi_n)$$

whenever the right hand side is defined, belong to $\mathfrak{F}(X)$.

Reference: [8] IV § 1, § 2, § 4, § 6; XII § 1;
      [4] § 18, § 19, § 20, § 34, § 35;
      [6] § 11, in particular (11.1)–(11.24), (11.36); (12.51e),
      (21.4), (21.5).

B6. Lebesgue integrable functions

We shall call a function $f \in \mathfrak{F}(X)$ *simple* if it assumes only a finite number of complex values. A special example of a simple function is the *characteristic function* c_Y of a Lebesgue measurable subset $Y \subset X$ defined by

$$c_Y(\xi_1, \ldots, \xi_n) = \begin{cases} 1 & \text{for } (\xi_1, \ldots, \xi_n) \in Y, \\ 0 & \text{for } (\xi_1, \ldots, \xi_n) \in X \backslash Y. \end{cases}$$

DEFINITION 4. If $f \in \mathfrak{F}(X)$ is a non-negative simple function assuming the values $\alpha_1, \ldots, \alpha_m$ on the sets $Y_1, \ldots, Y_m$ respectively (i.e. $f = \sum_{k=1}^{m} \alpha_k c_{Y_k}$), then

$$\int_X f \, d\lambda^n = \int_X f(\xi_1, \ldots, \xi_n) \, d(\xi_1, \ldots, \xi_n) = \sum_{k=1}^{m} \alpha_k \lambda^n(Y_k).$$

If $f \in \mathfrak{F}(X)$ is non-negative, then

$$\int_X f \, d\lambda^n = \int_X f(\xi_1, \ldots, \xi_n) \, d(\xi_1, \ldots, \xi_n)$$

$$= \sup \left\{ \int_X g \, d\lambda^n : 0 \leqslant g \leqslant f, g \in \mathfrak{F}(X), g \text{ is simple} \right\}.$$

In particular, for $Y \subset X$, $Y \in \mathcal{M}^n$ we get

$$\int_X c_Y d\lambda^n = \lambda^n(Y).$$

If $f \in \mathfrak{F}(X)$ is non-negative, then $\int_X f d\lambda^n = 0$ iff $f(\xi) = 0$ a.e. on X. For a non-negative function $f \in \mathfrak{F}(X)$ it may also very well happen that $\int_X f d\lambda^n = \infty$.

DEFINITION 5. A *non-negative* function $f \in \mathfrak{F}(X)$ is called *(Lebesgue) integrable* (on X) if $\int_X f d\lambda^n < \infty$.

A *real valued* function $f \in \mathfrak{F}(X)$ is called *integrable* if the functions $f^+ = \max(f, 0) \in \mathfrak{F}(X)$ and $f^- = \max(-f, 0) \in \mathfrak{F}(X)$ are integrable, and the *Lebesgue integral* of f is defined by

$$\int_X f d\lambda^n = \int_X f(\xi_1, ..., \xi_n) \, d(\xi_1, ..., \xi_n) = \int_X f^+ d\lambda^n - \int_X f^- d\lambda^n.$$

A *(complex valued)* function $f \in \mathfrak{F}(X)$ is called *integrable* if its real and imaginary parts $\operatorname{Re} f \in \mathfrak{F}(X)$ and $\operatorname{Im} f \in \mathfrak{F}(X)$ $(f = \operatorname{Re} f + i \operatorname{Im} f)$ are integrable and the *Lebesgue integral* of f is defined by

$$\int_X f d\lambda^n = \int_X f(\xi_1, ..., \xi_n) \, d(\xi_1, ..., \xi_n) = \int_X \operatorname{Re} f \, d\lambda^n + i \int_X \operatorname{Im} f \, d\lambda^n.$$

A function $f \in \mathfrak{F}(X)$ is integrable iff $\int_X |f| \, d\lambda^n < \infty$. If two functions coincide a.e. on X, then if one is integrable, so is the other and the Lebesgue integrals of both functions coincide. For the purposes of Lebesgue integration it is therefore legitimate to identify any two functions in $\mathfrak{F}(X)$ which coincide a.e. on X. With this identification performed, the set of all integrable functions on X is denoted by $\mathfrak{L}_1(X)$:

$$\mathfrak{L}_1(X) = \{ f \in \mathfrak{F}(X) : \int_X |f| \, d\lambda^n < \infty \}.$$

For $f \in \mathfrak{L}_1(X)$ and $Y \subset X$, $Y \in \mathcal{M}^n$, also $f c_Y$ is integrable and by definition

$$\int_Y f d\lambda^n = \int_X f c_Y d\lambda^n.$$

If X is a product of intervals (2) and if f is Riemann integrable over X (in particular if f is continuous on X), then f is (Lebesgue) integrable on X and the Riemann integral of f over X coincides with the Lebesgue integral. In particular, for $n=1$ this motivates the notation

$$\int\limits_{[\alpha,\,\beta]} f \mathrm{d}\lambda = \int\limits_{\alpha}^{\beta} f(\xi)\,\mathrm{d}\xi\,.$$

Still for $n=1$, if f is Riemann integrable over every finite interval $[\alpha, \beta]$ and if the improper Riemann integral

$$\int\limits_{-\infty}^{+\infty} |f(\xi)|\,\mathrm{d}\xi = \lim\limits_{\substack{\alpha\to-\infty \\ \beta\to+\infty}} \int\limits_{\alpha}^{\beta} |f(\xi)|\,\mathrm{d}\xi$$

is finite, then f is (Lebesgue) integrable on $\mathbf{R}$ and the Lebesgue integral of f coincides with the improper Riemann integral of f

$$\int\limits_{\mathbf{R}} f \mathrm{d}\lambda = \int\limits_{-\infty}^{+\infty} f(\xi)\,\mathrm{d}\xi = \lim\limits_{\substack{\alpha\to-\infty \\ \beta\to+\infty}} \int\limits_{\alpha}^{\beta} f(\xi)\,\mathrm{d}\xi\,.$$

Reference: [6] § 12, in particular (12.1), (12.2), (12.4), (12.13)–(12.18),
                (12.26), (12.28), (12.31), (12.32), (12.51f);
              [4] § 23, § 25;
              [8] V § 1, § 2, § 4, VI § 1, § 2, § 4, XII § 2.

B7. Properties of the Lebesgue integral

Under pointwise addition and scalar multiplication $\mathfrak{L}_1(X)$ is a complex linear space, and the Lebesgue integral depends linearly on the functions to be integrated (in other words the Lebesgue integral is a linear functional on $\mathfrak{L}_1(X)$):

$$\int\limits_{X} (\alpha f + \beta g)\,\mathrm{d}\lambda^n = \alpha \int\limits_{X} f \mathrm{d}\lambda^n + \beta \int\limits_{X} g \mathrm{d}\lambda^n$$

$$\text{for all } f \in \mathfrak{L}_1(X) \text{ and all } g \in \mathfrak{L}_1(X)\,.$$

Similarly, if $f \in \mathfrak{L}_1(X)$, then

$$\int\limits_{Y\cup Z} f \mathrm{d}\lambda^n = \int\limits_{Y} f \mathrm{d}\lambda^n + \int\limits_{Z} f \mathrm{d}\lambda^n \quad \text{if } \begin{cases} Y \subset X,\ Y \in \mathscr{M}^n, \\ Z \subset X,\ Z \in \mathscr{M}^n, \end{cases} \quad Y \cap Z = \emptyset\,.$$

If $f \in \mathfrak{L}_1(X)$ and $g \in \mathfrak{L}_1(X)$ are real-valued and $f \leqslant g$, then

$$\int\limits_{X} f \mathrm{d}\lambda^n \leqslant \int\limits_{X} g \mathrm{d}\lambda^n\,.$$

If $f \in \mathfrak{F}(X)$, if $g \in \mathfrak{L}_1(X)$ is non-negative, and if $|f| \leqslant g$, then $f \in \mathfrak{L}_1(X)$. Furthermore

$$\left| \int_X f d\lambda^n \right| \leqslant \int_X |f|\, d\lambda^n \quad \text{for all } f \in \mathfrak{L}_1(X).$$

If $\{f_k\}_{k=1}^{\infty} \subset \mathfrak{F}(X)$ is a monotone sequence of non-negative functions such that $\lim_{k \to \infty} f_k$ is finite a.e. on X (and if $f_1 \in \mathfrak{L}_1(X)$ in case of a decreasing sequence), then

$$\int_X \lim_{k \to \infty} f_k d\lambda^n = \lim_{k \to \infty} \int_X f_k d\lambda^n$$

(LEBESGUE's THEOREM ON MONOTONE CONVERGENCE). If the sequence $\{f_k\}_{k=1}^{\infty} \subset \mathfrak{L}_1(X)$ has the property that $\lim_{k \to \infty} f_k$ exists and is finite a.e. on X and if $|f_k| \leqslant h$ for some non-negative function $h \subset \mathfrak{L}_1(X)$ and all $k \geqslant 1$, then $\lim_{k \to \infty} f_k \in \mathfrak{L}_1(X)$ and

$$\int_X \lim_{k \to \infty} f_k d\lambda^n = \lim_{k \to \infty} \int_X f_k d\lambda^n$$

(LEBESGUE's THEOREM ON DOMINATED CONVERGENCE).

These theorems admit equivalent formulations in terms of converging series if we put $g_1 = f_1$, $g_k = f_k - f_{k-1}$ for $k \geqslant 2$, $f_n = \sum_{k=1}^{n} g_k$. If $\{g_k\}_{k=1}^{\infty} \subset \mathfrak{F}(X)$ is a sequence of non-negative functions and if $\sum_{k=1}^{\infty} g_k$ converges a.e. on X, then

$$\int_X \left(\sum_{k=1}^{\infty} g_k \right) d\lambda^n = \sum_{k=1}^{\infty} \int_X g_k d\lambda^n$$

(LEBESGUE's THEOREM ON MONOTONE CONVERGENCE). If the sequence $\{g_k\}_{k=1}^{\infty} \subset \mathfrak{L}_1(X)$ has the property that $\sum_{k=1}^{\infty} g_k$ converges a.e. on X and if $|\sum_{k=1}^{n} g_k| \leqslant h$ for some non-negative function $h \in \mathfrak{L}_1(X)$ and all $n \geqslant 1$, then $\sum_{k=1}^{\infty} g_k \in \mathfrak{L}_1(X)$ and

$$\int_X \left(\sum_{k=1}^{\infty} g_k \right) d\lambda^n = \sum_{k=1}^{\infty} \int_X g_k\, d\lambda^n$$

(LEBESGUE's THEOREM ON DOMINATED CONVERGENCE). Conversely, if the sequence $\{g_k\}_{k=1}^{\infty} \subset \mathfrak{L}_1(X)$ has the property that

$$\sum_{k=1}^{\infty} \int_X |g_k|\, d\lambda^n < \infty,$$

then $\sum_{k=1}^{\infty} g_k$ converges a.e. on X to an integrable function and

$$\int_X \left(\sum_{k=1}^{\infty} g_k \right) d\lambda^n = \sum_{k=1}^{\infty} \int_X g_k \, d\lambda^n$$

(B. LEVI'S THEOREM).

     Reference: [6] § 12, in particular (12.20)–(12.33);

                  [4] § 23–§ 27;

                  [8] V § 2, § 3, VI § 2, § 3 theorem 1, § 4.

B8. Fubini's theorem

We now formulate FUBINI'S THEOREM on integration in product spaces in a specialized version adapted to its application in the preceding text. On the one hand this theorem asserts that under certain conditions a Lebesgue integral over a product set may be decomposed into an iterated integral. On the other hand it sums up the conditions under which in such an iterated integral the order of integration may be reversed.

Suppose X is a product of (possibly unbounded) intervals of which we single out one, say the last one:

$$-\infty \leqslant \alpha_j < \beta_j \leqslant +\infty \quad \text{for } 1 \leqslant j \leqslant n,$$
$$X = \{(\xi_1, \ldots, \xi_n): \alpha_j \, (\leqslant) \, \xi_j \, (\leqslant) \, \beta_j \quad \text{for } 1 \leqslant j \leqslant n\}$$
$$= X^{(n)} \times X_n,$$
$$X^{(n)} = \{(\xi_1, \ldots, \xi_{n-1}): \alpha_j \, (\leqslant) \, \xi_j \, (\leqslant) \, \beta_j \quad \text{for } 1 \leqslant j \leqslant n-1\},$$
$$X_n = \{\xi_n: \alpha_n \, (\leqslant) \, \xi_n \, (\leqslant) \, \beta_n\}.$$

Furthermore, let $f \in \mathfrak{F}(X)$ be given and suppose at least one of the integrals

$$\int_X |f(\xi_1, \ldots, \xi_n)| \, d(\xi_1, \ldots, \xi_n),$$

$$\int_{X_n} \int_{X^{(n)}} |f(\xi_1, \ldots, \xi_{n-1}, \xi_n)| \, d(\xi_1, \ldots, \xi_{n-1}) \, d\xi_n,$$

$$\int_{X^{(n)}} \int_{X_n} |f(\xi_1, \ldots, \xi_{n-1}, \xi_n)| \, d\xi_n d(\xi_1, \ldots, \xi_{n-1})$$

is finite. Then all three integrals are finite, $f \in \mathfrak{L}_1(X)$, and

     a) for almost every fixed $\xi_n \in X_n$ we have $f(x_1, \ldots, x_{n-1}, \xi_n) \in \mathfrak{L}_1(X^{(n)})$;

b) for almost every fixed $(\xi_1, \ldots, \xi_{n-1}) \in X^{(n)}$ we have

$$f(\xi_1, \ldots, \xi_{n-1}, x_n) \in \mathfrak{L}_1(X_n);$$

c) $\displaystyle\int_{X^{(n)}} f(\xi_1, \ldots, \xi_{n-1}, x_n) \, d(\xi_1, \ldots, \xi_{n-1}) \in \mathfrak{L}_1(X_n);$

d) $\displaystyle\int_{X_n} f(x_1, \ldots, x_{n-1}, \xi_n) \, d\xi_n \in \mathfrak{L}_1(X^{(n)});$

e) $\displaystyle\int_{X} f(\xi_1, \ldots, \xi_n) \, d(\xi_1, \ldots, \xi_n) =$

$$= \int_{X_n} \int_{X^{(n)}} f(\xi_1, \ldots, \xi_{n-1}, \xi_n) \, d(\xi_1, \ldots, \xi_{n-1}) \, d\xi_n$$

$$= \int_{X^{(n)}} \int_{X_n} f(\xi_1, \ldots, \xi_{n-1}, \xi_n) \, d\xi_n d(\xi_1, \ldots, \xi_{n-1}).$$

Reference: [6] (21.13);
            [4] § 36;
            [8] XII § 3, § 4.

B9. Absolutely continuous functions

Using Lebesgue integrals, the connection between integration and differentiation as inverse operations of each other can be extended beyond the class of functions for which this connection is usually established in elementary analysis.

DEFINITION 6. A complex-valued function f on a closed interval $[\alpha, \beta] \subset \mathbf{R}$ is called *absolutely continuous* on $[\alpha, \beta] \, (=X)$ if there exists a function $g \in \mathfrak{L}_1(\alpha, \beta) \, (= \mathfrak{L}_1(X))$ such that

(3) $$f(\xi) = f(\alpha) + \int_{\alpha}^{\xi} g(\eta) \, d\eta \quad \text{for all } \xi \in [\alpha, \beta].$$

Suppose f is absolutely continuous on $[\alpha, \beta]$ and given by (3). Then f is continuous on $[\alpha, \beta]$ and differentiable a.e. on $[\alpha, \beta]$. Its derivative f' is defined a.e. on $[\alpha, \beta]$ and coincides a.e. in $[\alpha, \beta]$ with g. As a consequence, we have $f' \in \mathfrak{L}_1(\alpha, \beta)$ and

$$f(\xi) = f(\alpha) + \int_{\alpha}^{\xi} f'(\eta) \, d\eta \quad \text{for all } \xi \in [\alpha, \beta].$$

If f and g are absolutely continuous on $[\alpha, \beta]$, then $fg' \in \mathfrak{L}_1(\alpha, \beta)$, $f'g \in \mathfrak{L}_1(\alpha, \beta)$, and

$$\int\limits_\alpha^\beta f(\xi) g'(\xi) \, d\xi + \int\limits_\alpha^\beta f'(\xi) g(\xi) \, d\xi = f(\xi) g(\xi) \Big|_\alpha^\beta = f(\beta) g(\beta) - f(\alpha) g(\alpha).$$

Reference: [6] § 18, in particular (18.10), (18.13), (18.17), (18.20);
          [8] IX § 1, § 2, § 4 theorem 1 – theorem 3.

B10. Differentiation under the integral sign

The following theorem describes a situation in which the order of integration and differentiation may be reversed.

THEOREM. *Let* $X = [\alpha, \beta] \times \mathbf{R} \subset \mathbf{R}^2$ *(*$\alpha \in \mathbf{R}$*,* $\beta \in \mathbf{R}$*,* $\alpha < \beta$*) and let* $f = f(x, y)$ *be a Lebesgue measurable function on X with the following properties:*
a) *for every fixed* $\xi \in [\alpha, \beta]$ *we have* $f(\xi, y) \in \mathfrak{L}_1(\mathbf{R})$*;*
b) *for almost every fixed* $\eta \in \mathbf{R}$ *the function $f(x, \eta)$ is absolutely continuous on* $[\alpha, \beta]$*;*

c)
$$\frac{\partial f(x, y)}{\partial x} \in \mathfrak{L}_1(X).$$

Then

$$\frac{d}{d\xi} \int\limits_{-\infty}^{+\infty} f(\xi, \eta) \, d\eta = \int\limits_{-\infty}^{+\infty} \frac{\partial f(\xi, \eta)}{\partial \xi} \, d\eta \quad \textit{for almost all } \xi \in [\alpha, \beta].$$

PROOF. By property c) and FUBINI's theorem (assertion d) in section B8) we have

$$\int\limits_{-\infty}^{+\infty} \frac{\partial f(x, \eta)}{\partial x} \, d\eta \in \mathfrak{L}_1(\alpha, \beta)$$

and the function

$$g(x) = \int\limits_\alpha^x \int\limits_{-\infty}^{+\infty} \frac{\partial f(\xi, \eta)}{d\xi} \, d\eta \, d\xi$$

is absolutely continuous on $[\alpha, \beta]$ (cf. section B9). Moreover, for every $\zeta \in [\alpha, \beta]$ we get

$$g(\zeta) = \int\limits_{\alpha}^{\zeta} \int\limits_{-\infty}^{+\infty} \frac{\partial f(\xi, \eta)}{\partial \xi} \, d\eta \, d\xi = \int\limits_{-\infty}^{+\infty} \int\limits_{\alpha}^{\zeta} \frac{\partial f(\xi, \eta)}{\partial \xi} \, d\xi \, d\eta$$

$$= \int\limits_{-\infty}^{+\infty} [f(\zeta, \eta) - f(\alpha, \eta)] \, d\eta = \int\limits_{-\infty}^{+\infty} f(\zeta, \eta) \, d\eta - \int\limits_{-\infty}^{+\infty} f(\alpha, \eta) \, d\eta.$$

Differentiating on both sides with respect to ζ we obtain

$$\int\limits_{-\infty}^{+\infty} \frac{\partial f(\zeta, \eta)}{\partial \zeta} \, d\eta = \frac{d}{d\zeta} \int\limits_{-\infty}^{+\infty} f(\zeta, \eta) \, d\eta \quad \text{for almost all } \zeta \in [\alpha, \beta]. \quad \square$$

BIBLIOGRAPHY

A. References cited in the text

[1] ACHIESER, N. I. and I. M. GLASMANN, Theorie der linearen Operatoren im Hilbert-Raum (Akademie-Verlag, Berlin 1960).

[2] HALMOS, P. R., Introduction to Hilbert space and the theory of spectral multiplicity (Chelsea Publ. Comp., New York 1951).

[3] HALMOS, P. R., Lectures on ergodic theory (Publ. Math. Soc. Japan, Tokyo 1956).

[4] HALMOS, P. R., Measure theory (Van Nostrand, Princeton 1956).

[5] HELSON, H., Lectures on invariant subspaces (Academic Press, New York 1964).

[6] HEWITT, E. and K. STROMBERG, Real and abstract analysis (Springer-Verlag, Berlin 1965).

[7] NAIMARK, M. A., Normed rings (P. Noordhoff, Groningen 1959).

[8] NATANSON, I. P., Theory of functions of a real variable I, II (Frederick Ungar, New York 1955, 1960).

[9] RIESZ, F. and B. SZ.-NAGY, Functional analysis (Frederick Ungar, New York 1955).

[10] SZ.-NAGY, B., Introduction to real functions and orthogonal expansions (Oxford University Press, New York 1965).

[11] TITCHMARSH, E. C., Introduction to the theory of Fourier integrals (Clarendon Press, Oxford 1950).

[12] TRICOMI, F. G., Vorlesungen über Orthogonalreihen (Springer-Verlag, Berlin 1955).

B. Other books dealing entirely or partially with Hilbert space, spectral theory, and integral equations

BACHMAN, G. and L. NARICI, Functional analysis (Academic Press, New York 1966).

BANACH, S., Théorie des opérations linéaires (Chelsea Publ. Comp., New York 1955, reprint).

BERBERIAN, S. K., Introduction to Hilbert space (Oxford University Press, New York 1961).

BERBERIAN, S. K., Notes on spectral theory (Van Nostrand, Princeton 1966).

BÔCHER, M., An introduction to the study of integral equations (Hafner Publ. Comp., New York 1960, reprint).

BOURBAKI, N., Théories spectrales, Éléments de mathématique XXXII (Hermann, Paris 1967).

BRANGES, L. DE, Hilbert spaces of entire functions (Prentice-Hall, London 1968).

COOKE, R. G., Infinite matrices and sequence spaces (Dover Publ. Inc., New York 1965, reprint).

COOKE, R. G., Linear operators (MacMillan and Co., London 1953).

COURANT, R. and D. HILBERT, Methods of mathematical physics I (Interscience Publishers, New York 1953).

DIEUDONNÉ, J., Foundations of modern analysis (Academic Press, New York 1960).

XMIER, J., Les algèbres d'opérateurs dans l'espace Hilbertien (Algèbres de Von Neumann), authier-Villars, Paris 1957).

DOLEZAL, V., Introduction to functional analysis for scientists and engineers (Czechoslovak Ac. of Sciences, Prague 1967).

DUNFORD, N. and J. T. SCHWARTZ, Linear operators I, II (Interscience Publishers, New York 1958, 1963).

EDWARDS, D. A., Introduction to functional analysis (Lecture notes; Lehigh University 1964).

EDWARDS, R. E., Functional analysis (Holt, Rinehart and Winston, New York 1965).

FRIEDMAN, B., Principles and techniques of applied mathematics (John Wiley & Sons, New York 1956).

FRIEDRICHS, K. O., Spectral theory of operators in Hilbert space (Lecture notes; Institute of Mathematical Sciences, New York University 1960).

FRIEDRICHS, K. O., Perturbation of spectra in Hilbert space (Amer. Math. Soc., Providence 1965).

GOFFMAN, C. and G. PEDRICK, First course in functional analysis (Prentice-Hall, Englewood Cliffs 1965).

GOLDBERG, S., Unbounded linear operators (McGraw-Hill, New York 1966).

HAMBURGER, H. L. and M. E. GRIMSHAW, Linear transformations in n-dimensional vector space (University Press, Cambridge 1956).

HALMOS, P. R., Finite dimensional vector spaces (Van Nostrand, Princeton 1958).

HALMOS, P. R., A Hilbert space problem book (Van Nostrand, Princeton 1967).

HAMEL, G., Integralgleichungen (Springer-Verlag, Berlin 1949).

HELLINGER, E. and O. TOEPLITZ, Integralgleichungen und Gleichungen mit unendlichvielen Unbekannten (Chelsea Publ. Comp., New York 1953, reprint).

HELLWIG, G., Differentialoperatoren der mathematischen Physik (Springer-Verlag, Berlin 1964).

HILBERT, D., Grundzüge einer allgemeinen Theorie der linearen Integralgleichungen (Chelsea Publ. Comp., New York 1953, reprint).

HILLE, E. and R. S. PHILLIPS, Functional analysis and semi-groups (Amer. Math. Soc. Colloquium Publications, Providence 1957).

JÖRGENS, K., Spectral theory of second-order ordinary differential operators (Lecture notes; Aarhus Universitet 1962/63).

JULIA, G., Introduction mathématique aux théories quantiques I, II (Gauthier-Villars, Paris 1949, 1955).

KANTOROWITSCH, L. W. and G. P. AKILOW, Funktionalanalysis in normierten Räumen (Akademie-Verlag, Berlin 1964).

KATO, O., Perturbation theory for linear operators (Springer-Verlag, Berlin 1966).

KELLEY, J. L., I. NAMIOKA and co-authors, Linear topological spaces (Van Nostrand, Princeton 1963).

KOLMOGOROV, A. N. and S. V. FOMIN, Elements of the theory of functions and functional analysis (Graylock Press, Rochester 1957).

KOLMOGOROV, A. N. and S. V. FOMIN, Measure, Lebesgue integrals, and Hilbert space (Academic Press, New York 1961).

LICHNEROWICZ, A., Lineare Algebra und lineare Analysis (VEB Deutscher Verlag der Wissenschaften, Berlin 1956).

LJUSTERNIK, L. A. and W. I. SOBOLEW, Elemente der Funktionalanalysis (Akademie-Verlag, Berlin 1955.)

LORCH, E. R., Spectral theory (Oxford University Press, New York 1962).

LOVITT, W. V., Linear integral equations (Dover Publ. Inc., New York 1950, reprint).

MAURIN, K., Methods of Hilbert spaces (Polish Scientific Publ., Warsaw 1967).

MESCHKOWSKY, H., Hilbertsche Räume mit Kernfunktion (Springer-Verlag, Berlin 1962).

MIKHLIN, S. G., Integral equations (Pergamon Press, London 1957).

MIKHLIN, S. G., Linear integral equations (Hindustan Publ. Corp., Dehli 1960).

MURRAY, F. J., An introduction to linear transformations in Hilbert space (Ann. of Math. Studies, Princeton 1941).

NAGUMO, M., Introduction to the theory of Banach space I, II (Lecture notes, Editôra Meridional „EMMA", Porto Alegre 1961/1965).

NAKANO, H., Modern spectral theory (Maruzen Co., Tokyo 1950).

NAKANO, H., Spectral theory in the Hilbert space (Maruzen Co., Tokyo 1953).

NEUMANN, J. VON, Functional operators, Volume I: Measures and integrals; Volume II: The geometry of orthogonal spaces (Princeton University Press, Princeton 1950).

NEUMANN, J. VON, Mathematical foundations of quantum mechanics (Princeton University Press, Princeton 1955).

NEUMARK, M. A., Lineare Differentialoperatoren (Akademie-Verlag, Berlin 1960).

NIKODÝM, O. M., The mathematical apparatus for quantum-theories (Springer-Verlag, Berlin 1966).

NIRENBERG, L., Functional analysis (Lecture notes; Courant Institute, New York University 1961).

PETROVSKY, I. G., Lectures on the theory of integral equations (Graylock Press, Rochester 1957).

POGORZELSKI, W., Integral equations and their applications (Pergamon Press, Oxford 1966).

PUTNAM, C. R., Commutation properties of Hilbert space operators (Springer-Verlag, Berlin 1967).

RELLICH, F., Perturbation theory of eigenvalue problems (Lecture notes; Institute of Mathematical Sciences, New York University 1953).

RUDIN, W., Real and complex analysis (McGraw-Hill, New York 1966).

SCHATTEN, R., Norm ideals of completely continuous operators, Ergebnisse der Mathematik und ihrer Grenzgebiete (Springer-Verlag, Berlin 1960).

SCHMEIDLER, W., Lineare Operatoren im Hilbertschen Raum (B. G. Teubner Verlagsgesellschaft, Stuttgart 1954).

SCHMEIDLER, W., Integralgleichungen mit Anwendungen in Physik and Technik (Akademische Verlagsgesellschaft Geest & Portig, Leipzig 1955).

SEGAL, I. E. and R. A. KUNZE, Integrals and operators (McGraw-Hill, New York 1968).

SHILOV, G. YE., Mathematical analysis (Pergamon Press, Oxford 1965).

SIMMONS, G. F., Introduction to topology and modern analysis (McGraw-Hill, New York 1963).

SMIRNOV, W. I., Lehrgang der höheren Mathematik, Teil V (VEB Deutscher Verlag der Wissenschaften, Berlin 1962).

SMITHIES, F., Integral equations (University Press, Cambridge 1958).

SOBOLEV, S. L., Applications of functional analysis in mathematical physics (Amer. Math. Soc. Translations of Mathematical Monographs, Providence 1963).

STONE, M. H., Linear transformations in Hilbert space (Amer. Math. Soc. Colloquium Publications, New York 1958).

SZ.-NAGY, B., Spectraldarstellung linearer Transformationen des Hilbertschen Raumes, Ergebnisse der Mathematik und ihrer Grenzgebiete (Springer-Verlag, Berlin 1942).

SZ.-NAGY, B., Prolongements des transformations de l'espace de Hilbert qui sortent de cet espace (Akadémiai Kiadó, Budapest 1955).

SZ.-NAGY, B. and C. FOIAŞ, Analyse harmonique des opérateurs de l'espace de Hilbert (Akadémiai, Kiadó, Budapest 1967).

TAYLOR, A. E., Introduction to functional analysis (John Wiley & Sons, New York 1958).

TRICOMI, F. G., Integral equations (Interscience Publishers, New York 1957).

WILANSKY, A., Functional analysis (Blaisdell Publ. Comp., New York 1964).

WINTNER, A., Spektraltheorie der unendlichen Matrizen (Verlag von S. Hirzel, Leipzig 1929).

WULICH, B. S., Einführung in die Funktionalanalysis I, II (B. G. Teubner Verlagsgesellschaft, Leipzig 1961).

YOSIDA, K., Lectures on differential and integral equations (Interscience Publishers, New York 1960).

YOSIDA, K., Functional analysis (Springer-Verlag, Berlin 1966).

ZAANEN, A. C., Linear analysis (North-Holland Publ. Comp., Amsterdam 1953).

$\bar{A}$	extension of the bounded linear mapping A, 74
	closure of the operator A, 314
$\varphi(A), \psi(A)$	functions of $A(A \in \mathscr{B})$, 228, 232, 245, 249
e^{iA}	special unitary function of $A(A = A^*)$, 234, 307
$\sqrt{A}$	square root of $A(A = A^*, A \geqslant 0)$, 240
$|A|$	absolute value of $A(A \in \mathscr{B})$, 241
A^+, A^-	positive and negative parts of $A(A = A^*)$, 251
$\|A\|$	norm of A, 71
$A = B$	A equal to B, 79
$A \subset B$	A contained in B, 119
$A \leqslant B$	A less than or equal to B $(A = A^*, B = B^*)$, 219
$A + B$	sum of A and B, 79
λA	scalar multiple of A, 79
AB	product of A and B, 79
$A\mathfrak{D}_A$	range of A, 103
A	special compact operator in ℓ_2, 120, 173, 177, 181
B	special unbounded operator in ℓ_2, 73, 121, 140
$\bar{B}$	closed extension of this operator B, 140, 151, 174, 302
D	differentiation operator, 124, 129, 130
D^*	adjoint differentiation operator, 126, 129
D_θ	selfadjoint extension of D, 133
E	multiplication operator, 131, 133
F	Fourier-Plancheral operator, 112
$H_n = H_n(x)$	Hermite polynomial, 60, 61
I	identity operator on $\mathfrak{H}$, 78
I_A	identity operator on $\mathfrak{D}_A$, 119
$I^\times$	identity operator on $\mathfrak{H} \times \mathfrak{H}$, 313
$L_n = L_n(x)$	Laguerre polynomial, 62, 63, 65
O	zero operator, 78
P, Q, R	projection operators, 102, 105, 134, 205
$P_{\mathfrak{M}}$	projection upon the subspace $\mathfrak{M}$, 44, 72
$P^\perp$	projection orthogonal to P, 104
$P \perp Q$	P orthogonal to Q, 106
$P(\lambda), Q(\lambda)$	projections from spectral families, 251, 273, 281, 288, 296
$P(\lambda + 0)$	$= (s) \lim_{n \to 0} P(\lambda + \theta_n)$, 252
$P(\lambda - 0)$	$= (s) \lim_{n \to 0} P(\lambda - \theta_n)$, 252
$P(x)$	projection-valued function, 254

$P(\xi, \eta)$	projections from the spectral family	284
$R(\xi, \eta)$	of a bounded normal operator,	281
U	isomorphism, 67, 69	
	unitary operator $(U^* = U^{-1})$, 98, 154	
$\varphi(U), \psi(U)$	functions of U $(U^* = U^{-1})$, 267, 269, 271, 272	
$V^\times$	special unitary operator on $\mathfrak{H} \times \mathfrak{H}$, 313	
X, Y	Lebesgue measurable sets in $\mathbf{R}^n$, 318	
X	square in $\mathbf{R}^2$, 34	
	parallelotope in $\mathbf{R}^n$, 318	
Y^c	complement of Y in $\mathbf{R}^n$, 319	
$X \backslash Y$	complement of Y in X, 103, 318	
$X \cup Y$	union of X and Y, 15	
$X \cap Y$	intersection of X and Y, 15	
$X \times Y$	Cartesian product of X and Y, 318, 319	
$Y \subset X$	Y contained in X, 15	
$\bigcup_{k=1}^{\infty} X_k$	union of $\{X_k\}_{k=1}^{\infty}$, 319	
$\bigcap_{k=1}^{\infty} X_k$	intersection of $\{X_k\}_{k=1}^{\infty}$, 319	
$\mathbf{R}$	field of real numbers, 1	
$\mathbf{C}$	field of complex numbers, 6	
$\mathbf{F}$	either of the fields $\mathbf{R}$ or $\mathbf{C}$, 6	
$\mathbf{R}^n$	n-dimensional Euclidean space, 1	
$\mathbf{C}^n$	n-dimensional unitary space, 8	
a, b, c	vectors, 2, 6, 7, 36	
$a = a(x, y)$	Hilbert-Schmidt kernel in $\mathfrak{L}_2(X)$, 85, 86	
$a^{(k)} = a^{(k)}(x, y)$	iterated Hilbert-Schmidt kernel, 210	
$c_{[\alpha, \beta]} = c_{[\alpha, \beta]}(x)$	characteristic function of $[\alpha, \beta]$, 104, 322	
$c_{]\alpha, \beta[} = c_{]\alpha, \beta[}(x)$	characteristic function of $]\alpha, \beta[$, 104, 322	
$c_\tau = c_\tau(x)$	$= c_{[0, \tau]}$, 113	
e	$= \sum_{k=0}^{\infty} 1/k!$	
e^x	exponential function $\sum_{k=0}^{\infty} x^k/k!$, 12, 57, 61	
e_k, e_ν	basis vectors, 12, 25, 26	
f, g, h	vectors, 7, 36	
$f = f(x)$		
$g = g(x)$	functions on subsets of $\mathbf{R}$ or $\mathbf{C}$, 9, 15, 27, 316	
$h = h(x)$		
$f(\xi), g(\xi), h(\xi)$	function values at ξ, 9, 15, 316	
$f^\eta(x)$	$= f(x, \eta)$, 64	

$f_\xi(y)$ $=f(\xi, y)$, 64

$g_\tau = g_\tau(x)$ $=(e^{-i\tau x}-1)/(-ix)$, 104, 306

$h_n = h_n(x)$ Hermite function, 60, 61, 108, 109

i $=\sqrt{-1}$

j, k, l, m, n integers, 7

$l_n = l_n(x)$ Laguerre function, 62, 63, 65

o zero element, zero vector, 6

$p = p(x), q = q(x)$ polynomials in x, 109, 316

$p_n = p_n(x)$ Legendre polynomial, 56, 57

x function with values $x(\xi)=\xi$, 15, 316

y, z functions with values $y(\eta)=\eta$, $z(\zeta)=\zeta$, 108, 210

Δ discrete spectrum, 176

Φ linear functional on $\mathfrak{H}$, 76

Φ_g $\Phi_g f = \Phi_g(f) = \langle f, Ag\rangle$, 92

Σ index set, 11

$\Sigma(A)$ spectrum of A, 158

$\left.\begin{array}{l}\alpha, \beta, \gamma \ldots \\ \lambda, \mu, \nu \ldots \\ \xi, \eta, \zeta \ldots\end{array}\right\}$ real or complex numbers, 2, 7, 36

$\bar{\alpha}$ complex conjugate of α, 6

α_1 $=\min \Sigma(A)$ $(A=A^*)$, 199, 225

α_2 $=\max \Sigma(A)$ $(A=A^*)$, 199, 225

α norm of Hilbert-Schmidt kernel, 85

$\delta_{k, l}$ Kronecker symbol ($\delta_{k,l}=0$ for $k \neq l$, $\delta_{kk}=1$), 12

κ symbol for fixed natural number or ∞, 49, 202

λ^n n-dimensional Lebesgue measure, 319

$\varphi = \varphi(,), \psi = \psi(,)$ bilinear forms on $\mathfrak{H}$, 88

$\hat{\varphi} = \hat{\varphi}(), \hat{\psi} = \hat{\psi}()$ associated quadratic forms, 89

$\varphi(x), \psi(x)$ functions in $\mathscr{P}, \mathscr{P}', \mathscr{S}, \mathscr{S}', \mathscr{R}, \mathscr{R}'$, 228

$(\varphi + \psi)(x)$ $=\varphi(x)+\psi(x)$, 228

$(\lambda\varphi)(x)$ $=\lambda\varphi(x)$, 228

$(\varphi\psi)(x)$ $=\varphi(x)\psi(x)$, 228

$\bar{\varphi}(x)$ $=\overline{\varphi(x)}$, 228

$\varphi_\lambda(x), \psi_\lambda(x)$ special step functions on $\mathbf{R}$ and $[0, 2\pi]$, 244, 272, 273

$\varphi_{\lambda, n}(x)$ special continuous functions on $\mathbf{R}$, 244

$\|\varphi(x)\|_A$ $=\sup\{|\varphi(\xi)|:\xi\in\Sigma(A)\}$, 231, 268

$\varphi(\Sigma(A))$ $=\{\varphi(\xi):\xi\in\Sigma(A)\}$, 229, 235, 267, 270

$\varphi(A)$	function of A ($A \in \mathcal{B}$, $A = A^*$), 228, 232, 245, 249
$\varphi(U)$	function of U ($U^* = U^{-1}$), 267, 269–272
a.a.	almost all, 320
a.e.	almost everywhere, 27, 320
lim	limit, 18
(s) lim	strong limit, 223
(w) lim	weak limit, 224
$\{f: \ldots\}$	definition of a set by characterization of its elements, XI
$\aleph_0$	aleph zero (cardinal number of countably infinite sets), 54
$\langle , \rangle$	inner product, 3, 8, 23, 28
$\| \ \|$	norm, 3, 9, 14, 71, 89
$\emptyset$	empty set, 15, 318
$\square$	end of proof ($=$ q.e.d.), XI
$[\alpha, \beta]$	closed interval, 9
$]\alpha, \beta[$	open interval, 27
$]\alpha, \beta]$	left open, right closed interval, 33
$\backslash$	set-theoretic subtraction, 103, 318

SUBJECT INDEX

absolute value of a bounded linear
   operator, 241, 242;
absolutely continuous function, 124, 327;
accumulation point, 16;
adjoint equation, 213;
– matrix, 96;
– operator, 95, 118;
algebra, 80;
–, Banach, 81;
–, commutative, 81;
–, normed, 81;
almost all (a.a), 320;
– everywhere (a.e), 27, 320;
approximate eigenvalue, 159;
approximation theorem, WEIERSTRASS', 316;
associative law, 6, 80;
automorphism, 69, 100;

ball of radius ε, open, 15;
Banach algebra, 81;
– – of bounded linear operators on $\mathfrak{H}$, 81;
Banach space, 20;
Banach-Steinhaus theorem, 122;
basis, 47;
–, polynomial, 55;
–, standard, 48;
BESSEL's inequality, 13;
bilinear form, 88;
– functional, 88;
bounded bilinear form, 89;
– linear mapping, 71;
– quadratic form, 89;
– set, 179;
BUNJAKOWSKY's inequality, 9;

Cartesian product, 310, 319;
CAUCHY's inequality, 3, 9;
– –, generalized, 221;
Cauchy sequence, 5;
Cayley transform, 289, 291;
characteristic function, 104, 250, 322;

– root, 151;
– value, 150;
closed graph theorem, 143, 315;
closed linear operator, 140, 312;
– set, 16;
– sphere, 17;
– unit sphere, 17, 177, 180, 181, 185;
closure of an operator, 314;
– of a set, 17;
commutative algebra, 81;
– group, 6;
commutative law, 6, 81;
commuting operators, 148;
compact linear operator, 181, 187, 208;
– set, 179;
– –, relatively, 178;
complement in $\mathfrak{L}$, 15;
– in $\mathbf{R}^n$, 319;
– in $\mathfrak{H}$, orthogonal, 40;
complete normed linear space, 5, 20;
completely continuous linear operator, 181;
completion, 21;
complex Banach space, 20;
– Hilbert space, 23;
– linear space, 7;
– rational numbers, 17;
contained in B, operator A, 119;
continuous functions, absolutely, 124, 327;
– – on $[\alpha, \beta]$, 9, 20, 33;
– – of a bounded selfadjoint operator, 230;
– – of a unitary operator, 269;
continuous linear mapping, 73;
– linear operator, completely, 181;
– spectrum, 173;
convergent sequence or series, 4, 18;
– – in $\mathscr{B}$, strongly, 223;
– – in $\mathscr{B}$, uniformly, 223;
– – in $\mathscr{B}$, weakly, 224;
– – in $\mathfrak{H}$, strongly, 184;
– – in $\mathfrak{H}$, weakly, 184;
cube, Hilbert, 177;

Engineering

DE RE METALLICA, Georgius Agricola. The famous Hoover translation of greatest treatise on technological chemistry, engineering, geology, mining of early modern times (1556). All 289 original woodcuts. 638pp. 6¾ x 11. 0-486-60006-8

FUNDAMENTALS OF ASTRODYNAMICS, Roger Bate et al. Modern approach developed by U.S. Air Force Academy. Designed as a first course. Problems, exercises. Numerous illustrations. 455pp. 5⅜ x 8½. 0-486-60061-0

DYNAMICS OF FLUIDS IN POROUS MEDIA, Jacob Bear. For advanced students of ground water hydrology, soil mechanics and physics, drainage and irrigation engineering and more. 335 illustrations. Exercises, with answers. 784pp. 6⅛ x 9¼.
0-486-65675-6

THEORY OF VISCOELASTICITY (Second Edition), Richard M. Christensen. Complete consistent description of the linear theory of the viscoelastic behavior of materials. Problem-solving techniques discussed. 1982 edition. 29 figures. xiv+364pp. 6⅛ x 9¼. 0-486-42880-X

MECHANICS, J. P. Den Hartog. A classic introductory text or refresher. Hundreds of applications and design problems illuminate fundamentals of trusses, loaded beams and cables, etc. 334 answered problems. 462pp. 5⅜ x 8½. 0-486-60754-2

MECHANICAL VIBRATIONS, J. P. Den Hartog. Classic textbook offers lucid explanations and illustrative models, applying theories of vibrations to a variety of practical industrial engineering problems. Numerous figures. 233 problems, solutions. Appendix. Index. Preface. 436pp. 5⅜ x 8½. 0-486-64785-4

STRENGTH OF MATERIALS, J. P. Den Hartog. Full, clear treatment of basic material (tension, torsion, bending, etc.) plus advanced material on engineering methods, applications. 350 answered problems. 323pp. 5⅜ x 8½. 0-486-60755-0

A HISTORY OF MECHANICS, René Dugas. Monumental study of mechanical principles from antiquity to quantum mechanics. Contributions of ancient Greeks, Galileo, Leonardo, Kepler, Lagrange, many others. 671pp. 5⅜ x 8½. 0-486-65632-2

STABILITY THEORY AND ITS APPLICATIONS TO STRUCTURAL MECHANICS, Clive L. Dym. Self-contained text focuses on Koiter postbuckling analyses, with mathematical notions of stability of motion. Basing minimum energy principles for static stability upon dynamic concepts of stability of motion, it develops asymptotic buckling and postbuckling analyses from potential energy considerations, with applications to columns, plates, and arches. 1974 ed. 208pp. 5⅜ x 8½.
0-486-42541-X

METAL FATIGUE, N. E. Frost, K. J. Marsh, and L. P. Pook. Definitive, clearly written, and well-illustrated volume addresses all aspects of the subject, from the historical development of understanding metal fatigue to vital concepts of the cyclic stress that causes a crack to grow. Includes 7 appendixes. 544pp. 5⅜ x 8½. 0-486-40927-9

Mathematics

FUNCTIONAL ANALYSIS (Second Corrected Edition), George Bachman and Lawrence Narici. Excellent treatment of subject geared toward students with background in linear algebra, advanced calculus, physics and engineering. Text covers introduction to inner-product spaces, normed, metric spaces, and topological spaces; complete orthonormal sets, the Hahn-Banach Theorem and its consequences, and many other related subjects. 1966 ed. 544pp. 6⅛ x 9¼. 0-486-40251-7

ASYMPTOTIC EXPANSIONS OF INTEGRALS, Norman Bleistein & Richard A. Handelsman. Best introduction to important field with applications in a variety of scientific disciplines. New preface. Problems. Diagrams. Tables. Bibliography. Index. 448pp. 5⅜ x 8½. 0-486-65082-0

VECTOR AND TENSOR ANALYSIS WITH APPLICATIONS, A. I. Borisenko and I. E. Tarapov. Concise introduction. Worked-out problems, solutions, exercises. 257pp. 5⅝ x 8¼. 0-486-63833-2

AN INTRODUCTION TO ORDINARY DIFFERENTIAL EQUATIONS, Earl A. Coddington. A thorough and systematic first course in elementary differential equations for undergraduates in mathematics and science, with many exercises and problems (with answers). Index. 304pp. 5⅜ x 8½. 0-486-65942-9

FOURIER SERIES AND ORTHOGONAL FUNCTIONS, Harry F. Davis. An incisive text combining theory and practical example to introduce Fourier series, orthogonal functions and applications of the Fourier method to boundary-value problems. 570 exercises. Answers and notes. 416pp. 5⅜ x 8½. 0-486-65973-9

COMPUTABILITY AND UNSOLVABILITY, Martin Davis. Classic graduate-level introduction to theory of computability, usually referred to as theory of recurrent functions. New preface and appendix. 288pp. 5⅜ x 8½. 0-486-61471-9

ASYMPTOTIC METHODS IN ANALYSIS, N. G. de Bruijn. An inexpensive, comprehensive guide to asymptotic methods—the pioneering work that teaches by explaining worked examples in detail. Index. 224pp. 5⅜ x 8½ 0-486-64221-6

APPLIED COMPLEX VARIABLES, John W. Dettman. Step-by-step coverage of fundamentals of analytic function theory—plus lucid exposition of five important applications: Potential Theory; Ordinary Differential Equations; Fourier Transforms; Laplace Transforms; Asymptotic Expansions. 66 figures. Exercises at chapter ends. 512pp. 5⅜ x 8½. 0-486-64670-X

INTRODUCTION TO LINEAR ALGEBRA AND DIFFERENTIAL EQUA-TIONS, John W. Dettman. Excellent text covers complex numbers, determinants, orthonormal bases, Laplace transforms, much more. Exercises with solutions. Undergraduate level. 416pp. 5⅜ x 8½. 0-486-65191-6

RIEMANN'S ZETA FUNCTION, H. M. Edwards. Superb, high-level study of landmark 1859 publication entitled "On the Number of Primes Less Than a Given Magnitude" traces developments in mathematical theory that it inspired. xiv+315pp. 5⅜ x 8½. 0-486-41740-9

CALCULUS OF VARIATIONS WITH APPLICATIONS, George M. Ewing. Applications-oriented introduction to variational theory develops insight and promotes understanding of specialized books, research papers. Suitable for advanced undergraduate/graduate students as primary, supplementary text. 352pp. 5⅜ x 8½.
0-486-64856-7

COMPLEX VARIABLES, Francis J. Flanigan. Unusual approach, delaying complex algebra till harmonic functions have been analyzed from real variable viewpoint. Includes problems with answers. 364pp. 5⅜ x 8½. 0-486-61388-7

AN INTRODUCTION TO THE CALCULUS OF VARIATIONS, Charles Fox. Graduate-level text covers variations of an integral, isoperimetrical problems, least action, special relativity, approximations, more. References. 279pp. 5⅜ x 8½.
0-486-65499-0

COUNTEREXAMPLES IN ANALYSIS, Bernard R. Gelbaum and John M. H. Olmsted. These counterexamples deal mostly with the part of analysis known as "real variables." The first half covers the real number system, and the second half encompasses higher dimensions. 1962 edition. xxiv+198pp. 5⅜ x 8½. 0-486-42875-3

CATASTROPHE THEORY FOR SCIENTISTS AND ENGINEERS, Robert Gilmore. Advanced-level treatment describes mathematics of theory grounded in the work of Poincaré, R. Thom, other mathematicians. Also important applications to problems in mathematics, physics, chemistry and engineering. 1981 edition. References. 28 tables. 397 black-and-white illustrations. xvii + 666pp. 6⅛ x 9¼.
0-486-67539-4

INTRODUCTION TO DIFFERENCE EQUATIONS, Samuel Goldberg. Exceptionally clear exposition of important discipline with applications to sociology, psychology, economics. Many illustrative examples; over 250 problems. 260pp. 5⅜ x 8½.
0-486-65084-7

NUMERICAL METHODS FOR SCIENTISTS AND ENGINEERS, Richard Hamming. Classic text stresses frequency approach in coverage of algorithms, polynomial approximation, Fourier approximation, exponential approximation, other topics. Revised and enlarged 2nd edition. 721pp. 5⅜ x 8½. 0-486-65241-6

INTRODUCTION TO NUMERICAL ANALYSIS (2nd Edition), F. B. Hildebrand. Classic, fundamental treatment covers computation, approximation, interpolation, numerical differentiation and integration, other topics. 150 new problems. 669pp. 5⅜ x 8½. 0-486-65363-3

THREE PEARLS OF NUMBER THEORY, A. Y. Khinchin. Three compelling puzzles require proof of a basic law governing the world of numbers. Challenges concern van der Waerden's theorem, the Landau-Schnirelmann hypothesis and Mann's theorem, and a solution to Waring's problem. Solutions included. 64pp. 5⅜ x 8½.
0-486-40026-3

THE PHILOSOPHY OF MATHEMATICS: AN INTRODUCTORY ESSAY, Stephan Körner. Surveys the views of Plato, Aristotle, Leibniz & Kant concerning propositions and theories of applied and pure mathematics. Introduction. Two appendices. Index. 198pp. 5⅜ x 8½. 0-486-25048-2

INTRODUCTORY REAL ANALYSIS, A.N. Kolmogorov, S. V. Fomin. Translated by Richard A. Silverman. Self-contained, evenly paced introduction to real and functional analysis. Some 350 problems. 403pp. 5⅜ x 8½.   0-486-61226-0

APPLIED ANALYSIS, Cornelius Lanczos. Classic work on analysis and design of finite processes for approximating solution of analytical problems. Algebraic equations, matrices, harmonic analysis, quadrature methods, much more. 559pp. 5⅜ x 8½.
0-486-65656-X

AN INTRODUCTION TO ALGEBRAIC STRUCTURES, Joseph Landin. Superb self-contained text covers "abstract algebra": sets and numbers, theory of groups, theory of rings, much more. Numerous well-chosen examples, exercises. 247pp. 5⅜ x 8½.   0-486-65940-2

QUALITATIVE THEORY OF DIFFERENTIAL EQUATIONS, V. V. Nemytskii and V.V. Stepanov. Classic graduate-level text by two prominent Soviet mathematicians covers classical differential equations as well as topological dynamics and ergodic theory. Bibliographies. 523pp. 5⅜ x 8½.   0-486-65954-2

THEORY OF MATRICES, Sam Perlis. Outstanding text covering rank, nonsingularity and inverses in connection with the development of canonical matrices under the relation of equivalence, and without the intervention of determinants. Includes exercises. 237pp. 5⅜ x 8½.   0-486-66810-X

INTRODUCTION TO ANALYSIS, Maxwell Rosenlicht. Unusually clear, accessible coverage of set theory, real number system, metric spaces, continuous functions, Riemann integration, multiple integrals, more. Wide range of problems. Undergraduate level. Bibliography. 254pp. 5⅜ x 8½.   0-486-65038-3

MODERN NONLINEAR EQUATIONS, Thomas L. Saaty. Emphasizes practical solution of problems; covers seven types of equations. ". . . a welcome contribution to the existing literature...."–*Math Reviews*. 490pp. 5⅜ x 8½.   0-486-64232-1

MATRICES AND LINEAR ALGEBRA, Hans Schneider and George Phillip Barker. Basic textbook covers theory of matrices and its applications to systems of linear equations and related topics such as determinants, eigenvalues and differential equations. Numerous exercises. 432pp. 5⅜ x 8½.   0-486-66014-1

LINEAR ALGEBRA, Georgi E. Shilov. Determinants, linear spaces, matrix algebras, similar topics. For advanced undergraduates, graduates. Silverman translation. 387pp. 5⅜ x 8½.   0-486-63518-X

ELEMENTS OF REAL ANALYSIS, David A. Sprecher. Classic text covers fundamental concepts, real number system, point sets, functions of a real variable, Fourier series, much more. Over 500 exercises. 352pp. 5⅜ x 8½.   0-486-65385-4

SET THEORY AND LOGIC, Robert R. Stoll. Lucid introduction to unified theory of mathematical concepts. Set theory and logic seen as tools for conceptual understanding of real number system. 496pp. 5⅜ x 8¼.   0-486-63829-4

Math–Decision Theory, Statistics, Probability

ELEMENTARY DECISION THEORY, Herman Chernoff and Lincoln E. Moses. Clear introduction to statistics and statistical theory covers data processing, probability and random variables, testing hypotheses, much more. Exercises. 364pp. 5⅜ x 8½. 0-486-65218-1

STATISTICS MANUAL, Edwin L. Crow et al. Comprehensive, practical collection of classical and modern methods prepared by U.S. Naval Ordnance Test Station. Stress on use. Basics of statistics assumed. 288pp. 5⅜ x 8½. 0-486-60599-X

SOME THEORY OF SAMPLING, William Edwards Deming. Analysis of the problems, theory and design of sampling techniques for social scientists, industrial managers and others who find statistics important at work. 61 tables. 90 figures. xvii +602pp. 5⅜ x 8½. 0-486-64684-X

LINEAR PROGRAMMING AND ECONOMIC ANALYSIS, Robert Dorfman, Paul A. Samuelson and Robert M. Solow. First comprehensive treatment of linear programming in standard economic analysis. Game theory, modern welfare economics, Leontief input-output, more. 525pp. 5⅜ x 8½. 0-486-65491-5

PROBABILITY: AN INTRODUCTION, Samuel Goldberg. Excellent basic text covers set theory, probability theory for finite sample spaces, binomial theorem, much more. 360 problems. Bibliographies. 322pp. 5⅜ x 8½. 0-486-65252-1

GAMES AND DECISIONS: INTRODUCTION AND CRITICAL SURVEY, R. Duncan Luce and Howard Raiffa. Superb nontechnical introduction to game theory, primarily applied to social sciences. Utility theory, zero-sum games, n-person games, decision-making, much more. Bibliography. 509pp. 5⅜ x 8½. 0-486-65943-7

INTRODUCTION TO THE THEORY OF GAMES, J. C. C. McKinsey. This comprehensive overview of the mathematical theory of games illustrates applications to situations involving conflicts of interest, including economic, social, political, and military contexts. Appropriate for advanced undergraduate and graduate courses; advanced calculus a prerequisite. 1952 ed. x+372pp. 5⅜ x 8½. 0-486-42811-7

FIFTY CHALLENGING PROBLEMS IN PROBABILITY WITH SOLUTIONS, Frederick Mosteller. Remarkable puzzlers, graded in difficulty, illustrate elementary and advanced aspects of probability. Detailed solutions. 88pp. 5⅜ x 8½. 65355-2

PROBABILITY THEORY: A CONCISE COURSE, Y. A. Rozanov. Highly readable, self-contained introduction covers combination of events, dependent events, Bernoulli trials, etc. 148pp. 5⅜ x 8¼. 0-486-63544-9

STATISTICAL METHOD FROM THE VIEWPOINT OF QUALITY CONTROL, Walter A. Shewhart. Important text explains regulation of variables, uses of statistical control to achieve quality control in industry, agriculture, other areas. 192pp. 5⅜ x 8½. 0-486-65232-7

TENSOR CALCULUS, J.L. Synge and A. Schild. Widely used introductory text covers spaces and tensors, basic operations in Riemannian space, non-Riemannian spaces, etc. 324pp. 5⅜ x 8¼. 0-486-63612-7

ORDINARY DIFFERENTIAL EQUATIONS, Morris Tenenbaum and Harry Pollard. Exhaustive survey of ordinary differential equations for undergraduates in mathematics, engineering, science. Thorough analysis of theorems. Diagrams. Bibliography. Index. 818pp. 5⅜ x 8½. 0-486-64940-7

INTEGRAL EQUATIONS, F. G. Tricomi. Authoritative, well-written treatment of extremely useful mathematical tool with wide applications. Volterra Equations, Fredholm Equations, much more. Advanced undergraduate to graduate level. Exercises. Bibliography. 238pp. 5⅜ x 8½. 0-486-64828-1

FOURIER SERIES, Georgi P. Tolstov. Translated by Richard A. Silverman. A valuable addition to the literature on the subject, moving clearly from subject to subject and theorem to theorem. 107 problems, answers. 336pp. 5⅜ x 8½. 0-486-63317-9

INTRODUCTION TO MATHEMATICAL THINKING, Friedrich Waismann. Examinations of arithmetic, geometry, and theory of integers; rational and natural numbers; complete induction; limit and point of accumulation; remarkable curves; complex and hypercomplex numbers, more. 1959 ed. 27 figures. xii+260pp. 5⅜ x 8½. 0-486-63317-9

POPULAR LECTURES ON MATHEMATICAL LOGIC, Hao Wang. Noted logician's lucid treatment of historical developments, set theory, model theory, recursion theory and constructivism, proof theory, more. 3 appendixes. Bibliography. 1981 edition. ix + 283pp. 5⅜ x 8½. 0-486-67632-3

CALCULUS OF VARIATIONS, Robert Weinstock. Basic introduction covering isoperimetric problems, theory of elasticity, quantum mechanics, electrostatics, etc. Exercises throughout. 326pp. 5⅜ x 8½. 0-486-63069-2

THE CONTINUUM: A CRITICAL EXAMINATION OF THE FOUNDATION OF ANALYSIS, Hermann Weyl. Classic of 20th-century foundational research deals with the conceptual problem posed by the continuum. 156pp. 5⅜ x 8½. 0-486-67982-9

CHALLENGING MATHEMATICAL PROBLEMS WITH ELEMENTARY SOLUTIONS, A. M. Yaglom and I. M. Yaglom. Over 170 challenging problems on probability theory, combinatorial analysis, points and lines, topology, convex polygons, many other topics. Solutions. Total of 445pp. 5⅜ x 8½. Two-vol. set. Vol. I: 0-486-65536-9 Vol. II: 0-486-65537-7